IN THE FOOTSTEPS OF

Joseph Dalton Hooker

A Sikkim adventure

IN THE FOOTSTEPS OF
Joseph Dalton Hooker
A Sikkim adventure

Seamus O'Brien

Kew Publishing
Royal Botanic Gardens, Kew

First published in 2018 by the Royal Botanic Gardens, Kew, Richmond, Surrey, TW9 3AB, UK www.kew.org

ISBN 978-1-84246-656-8
e-ISBN 978-1-84246-677-3

Distributed on behalf of the Royal Botanic Gardens, Kew in North America
by the University of Chicago Press, 1427 East 60th St, Chicago, IL 60637, USA.

British Library Cataloguing in Publication Data
A catalogue record for this book is available from the British Library.

Production Management: Georgie Hills
Design and page layout: Christine Beard
Copy-editing: Alison Rix
Proofreading: Matthew Seal

Printed in Italy by Printer Trento srl

Endpapers: A view of Tibet and the Cholamu Lake. W. H. Fitch's watercolour of Hooker's original sketch.

Frontispiece: A portrait of Joseph Dalton Hooker by George Richmond (1809–1896), dated 1855.

Title page: Prayer wheels on the wall of the Buddhist monastery at Tashiding.

Dedication page: Detail of the superb frontispiece of *Illustrations of Himalayan Plants* (1855).

For information or to purchase all Kew titles please visit shop.kew.org/kewbooksonline or email publishing@kew.org

Kew's mission is to be the global resource in plant and fungal knowledge, and the world's leading botanic garden.

Kew receives about one third of its running costs from Government through the Department for Environment, Food and Rural Affairs (Defra). All other funding needed to support Kew's vital work comes from members, foundations, donors and commercial activities, including book sales.

For my Mother

Contents

Dzongri campsite

Patrons

The author would like to acknowledge the following individuals and organisations, whose generosity made the publication of this book possible:

The Alpine Garden Society (Dublin Group)

The Alpine Garden Society (Ulster Group)

Mr Peter Cuthbert

Sir David Davies

The Hon. Mrs Helen Dillon

Mrs Tracy Hamilton

The International Dendrology Society

The Irish Garden Plant Society

The Irish Society of Botanical Artists

Mrs Carmel Naughton

The Stanley Smith (UK) Horticultural Trust

Mr Robert Wilson-Wright

LEFT Dark trees of the Sikkim fir, *Abies densa*, seen here silhouetted against the snow-capped peaks of the Chola range of mountains in east Sikkim.

Foreword

Who better to tread in the footsteps of Joseph Dalton Hooker than Seamus O'Brien of the National Botanic Gardens, Kilmacurragh, where so many of the Victorian botanist's introductions can still be admired? In the warm, damp climate of coastal south-east Ireland, plants that started their lives in India really thrive. Robin Lane Fox described Kilmacurragh as 'hallowed ground' where 'the greatest finds of the greatest plant hunters are still in the long grass around us. It is a holy site, a place to hunt the hunters.' Hunting perhaps the greatest hunter of them all, this book is a scholarly homage to Hooker, as well as a fascinating account of four expeditions which Seamus O'Brien led to Sikkim in the Himalaya. He travelled with groups of like-minded friends to see the plants that Hooker found in the wild, which was clearly a thrilling experience for everyone who went. Part biography, part travelogue, with plenty of plant portraits, maps and pictures, the book brings the Himalaya to life for the armchair traveller. Lovers of history, adventure and science, as well as botanists and gardeners will all find something to enjoy in its pages.

Joseph Dalton Hooker (1817–1911) was a Victorian polymath, a close friend of Charles Darwin. He had a medical training, but the branches of science which continued to engage him in later life were zoology, paleontology (in the form of paleobotany) and plant taxonomy. Botany is perhaps the branch for which Hooker is best known. Like his father, William Jackson Hooker, he became Director of the Royal Botanical Gardens, Kew and his many published floras relating to New Zealand, Tasmania and India, for example, are still consulted today.

Hooker could sketch and write and was also an accomplished cartographer, as well as an intrepid traveler. He thought nothing of hacking his way through fields of snow, or being dragged through freezing water courses. The Himalayan expeditions he embarked upon were conducted on the backs of elephants. His 21st century followers were able to take advantage of travel on modern roads, but they also trekked for miles to be rewarded with sights like a forest of *Magnolia campbellii*, clothed with an understorey of *Daphne bholua*. In the epilogue to this book, Seamus O'Brien looks back on his great adventures to Sikkim, with some sense of foreboding at the destruction which may come with climate change and an expanding population.

We owe so much to the great plant hunters and this book is a marvellous tribute to Joseph Dalton Hooker who introduced so many rhododendrons, as well as humbler plants, to the gardens and botanical collections of Britain and Ireland. The appendix to the book has a helpful list with descriptions of some of Hooker's rhododendrons, but perhaps the best way to appreciate these Himalayan emperors of the plant world is to visit the gardens at Kilmacurragh in County Wicklow, Ireland. Or of course travel, as the author did, to see them in their natural habitat.

MARY KEEN
Garden designer and author
July 2018

LEFT *Rhododendron campanulatum* seen here beneath jagged peaks in the Yumthang valley in north Sikkim. Joseph Hooker introduced superior lavender-blue and rose-purple forms that were immediately seized upon by hybridisers, including Dr David Moore, the Curator of the Royal (now National) Botanic Gardens, Glasnevin, who created the superb *Rhododendron* 'Thomas Acton'.

Preface

The bicentenary of the birth of Sir Joseph Dalton Hooker, the greatest exploring botanist of the 19th century, fell on 30 June 2017. Hooker was born into fortunate circumstances: his family came from a relatively wealthy, respectable background and his father, Sir William Jackson Hooker, was well placed in society and was on familiar terms with many of the leading botanical figures of the day including eminent men like Sir Joseph Banks and Robert Brown.

His father's friendship with influential members of the British and Irish aristocracy smoothed the way for the young Joseph's brilliant career, and among his many journeys of exploration, his expeditions to the Antarctic and to the Himalaya remain his greatest feats. Hooker was lucky enough to have been born into a great golden era of exploration, when frantic measures were put in place to catalogue the zoology and botany of Britain's burgeoning empire, and large areas of the globe remained unmapped, providing immeasurable possibilities of adventure to spirited explorers.

Even in his youth, closer to home, Hooker was a pioneering scientific explorer. In 1836, the same year his father was knighted, he travelled to Connemara, in the West of Ireland, to collect specimens of the newly discovered *Erica mackayana*, a Lusitanian species found simultaneously in northern Spain and Ireland in 1835. Thus Hooker's specimen, collected when he was just 19, was one of the earliest known collections of the species, and this pioneering rush to catalogue all things new and unknown stayed with him throughout his long and extremely productive life.

Among his many other attributes, Hooker was also a keen artist, a useful talent in an era before photography had become widespread. His original field sketches are held in the vast archives at the Royal

Julia, William and Martin Hooker, the great-great-great grandchildren of the Sikkim explorer, pictured here, in August 2016, beneath Joseph Hooker's 1850 introduction of *Rhododendron grande* at the National Botanic Gardens, Kilmacurragh.

Botanic Gardens, Kew; these watercolours of the remote interior valleys of Sikkim are places that as modern-day explorers, we would visit over 160 years later.

Hooker was a polymath, a scientist of breath-taking ability. Cartography was another of his many skills, and in Sikkim he created the first comprehensive map of the kingdom, highlighting mountain passes that would be of enormous strategic value in the decades to come. He recorded everything from the geology of this small pocket of the Himalaya, to its zoology, and even went as far as discovering a species of freshwater snail that bears his name when he made his famous unauthorised crossing of the border into Tibet, then a place absolutely forbidden to Western travellers.

Hooker was one of those early explorers fortunate enough to reach a previously untouched flora. Almost everything they found was unknown to Western science and he would spend the next 50 years describing his plant collections from Sikkim. This same collection amounted to thousands of species, all pressed and dried and sent back to Kew for naming and further scientific study.

Taxonomy, the naming of these plants, was not his only concern. Hooker was not oblivious to the fact that many of these species had enormous horticultural potential. The hot tropical valleys of Sikkim abounded in spectacularly exotic plants, particularly orchids which were highly suited to the great glasshouses of Europe, while the higher valleys offered a plethora of new, highly garden-worthy plants that would prove absolutely hardy in Britain and Ireland.

Of these, it was to be Hooker's *Rhododendron* collections that wowed the Victorian public. His sumptuous publication, *Rhododendrons of the Sikkim Himalaya*, portrayed these species at their brilliant best, and, from his Sikkim seed collections, gardens across Europe were filled with new, often spectacular tree-like species, thus kick-starting a craze that became known as rhododendromania, a passion for the genus that is still widespread even today.

In Ireland, the most famous remaining collection of Hooker rhododendrons is at Kilmacurragh in east County Wicklow, and in 2006 I found myself living on the estate. Formerly the seat of the Acton family, as early as 1820, the Actons were growing *Rhododendron arboreum*, the first record of this Himalayan species in Ireland. By the 1850s, David Moore, Curator of what was then the Royal Botanic Gardens, Glasnevin, had begun to advise on garden matters at Kilmacurragh, supplying seedlings of Joseph Hooker's Sikkim rhododendrons.

As a result of being surrounded by these veteran rhododendron species, I yearned to see them in their wild haunts, where a young Joseph Hooker blazed a trail during the late 1840s. This led to not one, but four expeditions to this botanically rich pocket of the eastern Himalaya, to a region aptly described by Hooker as 'a perfect microcosm of the Himalaya'.

In the past, several individuals have attempted to re-trace Hooker's route, the earliest, of course, being Sir Henry Elwes, though no one until now has completely covered his historic route. The tale of our travels, and those of Hooker's, are therefore intertwined in the following narrative.

Unlike Hooker, our mission was not to collect, but to study and compare places he visited and to record how they had fared and appeared over 160 years later. In some ways Sikkim has changed little over the course of time. Certainly the Bhutia villages of north Sikkim have not altered substantially, and *Himalayan Journals*, Hooker's account of the region, is still the visiting naturalist's best guide to this remote area.

To a group of visiting Irish botanists and horticulturists Sikkim held many attractions, a spectacular landscape, varied ethnic groups and gentian-blue November skies. The region's flora was of course the greatest draw. Just as it fascinated Joseph Hooker in Victorian times, we were very much left in awe of the biodiversity of this tiny Himalayan state where tropical plants lie in deep valleys beneath the shadow of mighty Kangchenjunga.

I still remember trekking above Yuksam, on the track leading ultimately towards the alpine meadows of Dzongri. Below us plunged the Rathong River with tropical figs and bananas luxuriating on its banks, yet just a few miles north of our position lay the greatest glaciers in the eastern Himalaya. Sikkim is a land of botanical paradoxes, and to anyone botanically minded, it is one of the most fascinating places on the planet. The following chapters recount the story of Hooker's momentous visit to Sikkim and East Nepal, and the tale of how we followed 160 years later. Happy reading!

SEAMUS O'BRIEN
July 2018

Acknowledgements

A great many people and organisations assisted in the preparation of this book, none more so than staff from the library and archives at the Royal Botanic Gardens, Kew. There, I am especially grateful to Lynn Parker, Curator of Illustrations and Artefacts, for allowing me to reproduce so many of Hooker's original field sketches of India, Sikkim and Nepal. I also received valuable assistance at Kew from Acting Archivist Lorna Cahill and Librarian Craig Brough. Also at Kew, my former tutors, Dr Colin Clubbe and Dr Pat Griggs, offered constant encouragement as did Head of Arboretum, Tony Kirkham.

To the team at Kew Publishing for all their hard work during the production stage, especially Head of Publishing, Gina Fullerlove, Christine Beard for her work on the book's design, Sales and Marketing Manager Lydia White and Production Manager Georgie Hills, thank you all for a superb job done.

At the Royal Botanic Garden Edinburgh, David Long, an authority on the flora of Bhutan and Sikkim, identified a number of critical and obscure species. Roy Lancaster, always a stalwart friend, offered much encouragement and I am most grateful to him and his wife Sue for their help and advice over the past few years. Another really good friend, Lady Mary Keen, made time to write a foreword. Thanks must also go to Ken Cox, at Glendoick in Perthshire, and to Dr Hartwig Schepker, Scientific Director of the Rhododendron-Park, Bremen, for comments on critical *Rhododendron* species and naturally occurring hybrids.

At the National Botanic Gardens, Glasnevin, Dublin I am most grateful to Director Dr Matthew Jebb, Curator Paul Maher, librarian Alexandra Caccamo and her assistant, Colette Edwards, for their help in the archives and for facilitating my study of a considerable number of rare books. In the National Herbarium at Glasnevin I wish to acknowledge the help of Dr Noeleen Smyth who guided me through the Indian fern collections, particularly those of Harry Corbyn Levinge. I also wish to express my thanks to Margie Phillips from the College of Amenity Horticulture at the National Botanic Gardens, Glasnevin for her company on not one, but two expeditions to Sikkim. Lynn Stringer, the Wicklow-based botanical artist, has painted several of Hooker's rhododendrons at Kilmacurragh and very kindly allowed several of her works to be reproduced in Appendix 1.

In Sikkim I am indebted to the Pradhan family, particularly our knowledgeable and patient guide, Sailesh, and his parents, the great plantsman Keshab Pradhan, and his wife Shanti, who were so hospitable during our several visits to Gangtok. On Sailesh Pradhan's staff, our group is most grateful to the many drivers, porters, ponymen and cooks, and especially to Thupden Tsering and Neelam Basnett, who proved to be superb botanical guides.

Numerous people in Ireland offered assistance and I am most grateful to friend, neighbour and fellow author Patricia Butler and her son Dr Charles Butler, for their help and advice. Mary Bradshaw

May 2015. Our group enjoying an alfresco lunch in the Yumthang valley. Left to right: The author, Averil Milligan, Daphne Levinge Shackleton, Orlaith Murphy, Derek Halpin, Bruce Johnson. Thupden Tsering stands behind near a dense thicket of flowering *Rhododendron hodgsonii*.

kindly read the manuscript and made suggestions. The Hooker family also provided much encouragement while visiting Kilmacurragh to see the living collections of their illustrious ancestor.

Peter Hooker, from County Tipperary, offered much support and enthusiasm towards this publication about his great-great grandfather, as did his wife Eleanor and his brother Philip. It is also pleasing to see the younger generation of Hookers making their pilgrimage to the gardens here at Kilmacurragh, particularly Julia, Martin and William Hooker, who are pictured in the Preface of this book beneath their great-great-great grandfather's *Rhododendron grande*.

Several of my fellow travellers in Sikkim gave much help and inspiration during the compilation of the manuscript, none more so than Kristin Jameson, Helen Dillon and Robert Wilson-Wright. I owe much to their words of encouragement and friendship.

Daphne Levinge Shackleton, a scion of two great Irish gardening families, joined our party in May 2015 to trace the route of her forebear, Harry Corbyn Levinge. It made for an interesting adventure and I was pleased to have such a knowledgeable plantswoman on our travelling party. A special word of thanks also goes to David Koning, Billy Alexander and Alan Ryder, the trio with whom I ultimately trekked to the Goecha La, the highest point of our travels, where we reached the great glaciers of Kangchenjunga and stood within a few miles of the mountain's colossal and icy eastern flank; a memory that will stay with me for all time. Another magical moment in Sikkim, was, when trekking in the Yumthang valley with Lesley Fennell and Bruno Nicolai, we discovered *Rhododendron × thupdenii*, having just walked through an entire valley painted by the blossoms of *Primula denticulata*. It was a memorable day, shared with two knowledgeable companions and I am grateful to them for their help in compiling an on-the-spot description of this previously unknown natural hybrid.

Several friends from County Down, Northern Ireland also participated in the various expeditions to Sikkim; thanks are due to Tracy Hamilton from Ringdufferin House, Averil Milligan from Rowallane Gardens, and to Alan Ryder, Paul Stewart and Neil Porteous from Mount Stewart House and Gardens, on the scenic Ards Peninsula.

I'd also like to express my appreciation to Wyn Hughes, Esther Shickling, Bruce Johnson, Adam Whitbourn, Eileen Murphy, Gail Gilliland, Debbie Bailey, Terry Smith, Thomas Pakenham, Orlaith Murphy, Octavia Tulloch and Derek Halpin. Without you, this story could never have been told.

In the United States, Philip Hooker very kindly supplied a copy of George Richmond's 1855 portrait of a young Joseph Hooker. Kathe Scullion offered constant words of support from Pennsylvania and I am also extremely grateful to Dan Hinkley for initially putting me in touch with Sailesh Pradhan.

And finally, here at the National Botanic Gardens, Kilmacurragh in County Wicklow, I'd like to thank my colleagues, Claire Mullarney,

Gráinne Larkin and Philip Quested, for joining me during our travels in West Bengal and Sikkim. It was a wonderful adventure inspired by Hooker's veteran rhododendrons here at our base at Kilmacurragh.

Photographic acknowledgements

I am most grateful to the following institutions and individuals for permission to use their images in this book:

Sir Henry Elwes, Colesbourne Park, Gloucestershire, UK: 35.

Peter Hooker: 176 (top and bottom), 264 (bottom right), 284.

Philip Hooker: ii.

Gráinne Larkin: xiv, 249 (bottom left).

Liam McCaughey: 88, 181 (bottom left), 190 (both images).

Claire Mullarney: 12.

National Botanic Gardens, Glasnevin, Dublin 9, Ireland: 6, 14 (top left), 22 (bottom right), 24 top right), 26, 28, 29, 31, 32, 34, 41 (bottom right), 44 (bottom), 47, 51 (bottom right), 52 (bottom right), 73 (middle right), 74 (top left), 85 (bottom left), 86 (bottom), 88 (top left), 90 (top), 91 (bottom), 95 (top left), 120 (top left), 141 (bottom), 156, 157, 161 (top left), 162, 166 (bottom), 172 (right), 177, 179, 181 (top right), 189 (bottom), 192 (right), 213 (top left), 241 (bottom), 259 (bottom left and right), 261 (bottom right), 262, 281 (bottom left), 289 (bottom right), 290, 292 (right), 293).

National Botanic Gardens, Kilmacurragh, Co. Wicklow, Ireland: 42, 44 (top left), 50, 260 (top left),

Charles Nelson, 42 (bottom).

Terry Smith: jacket (front flap), 129 (top and bottom), 135 (bottom), 143 (top and bottom left), 147 (bottom and middle), 148 (top), 151 (top), 152.

Pitt-Rivers Museum, Oxford: 275 (bottom).

Royal Botanic Garden Edinburgh, 20A Inverleith Row, Edinburgh, Scotland: 64.

Royal Botanic Gardens, Kew, Richmond, Surrey, UK: 3 (bottom), 5, 6 (bottom), 9, 13, 18, 19, 20 (top), 38, 41 (top left), 43 (top), 56 (bottom right), 72 (bottom right), 87, 88 (top right), 89 (bottom left), 94 (bottom), 96, 98 (bottom), 99 (top), 100, 101 (top), 123, 154, 155 (top), 158 (bottom left), 159 (top right), 165 (bottom), 175 (bottom right), 184, 187 (bottom), 212 (right), 214, 216, 223, 226, 229, 265 (top), 267 (bottom), 273 (bottom), 295, 296.

Thupden Tsering: 171.

St Paul's School, Darjeeling, West Bengal, India: 26, 341.

Palace Collection/Keshab Pradhan, Gangtok, Sikkim, India: 22 (top), 266 (bottom).

Neil Porteous: 111.

Adam Whitbourn: 282 (bottom left).

Maps

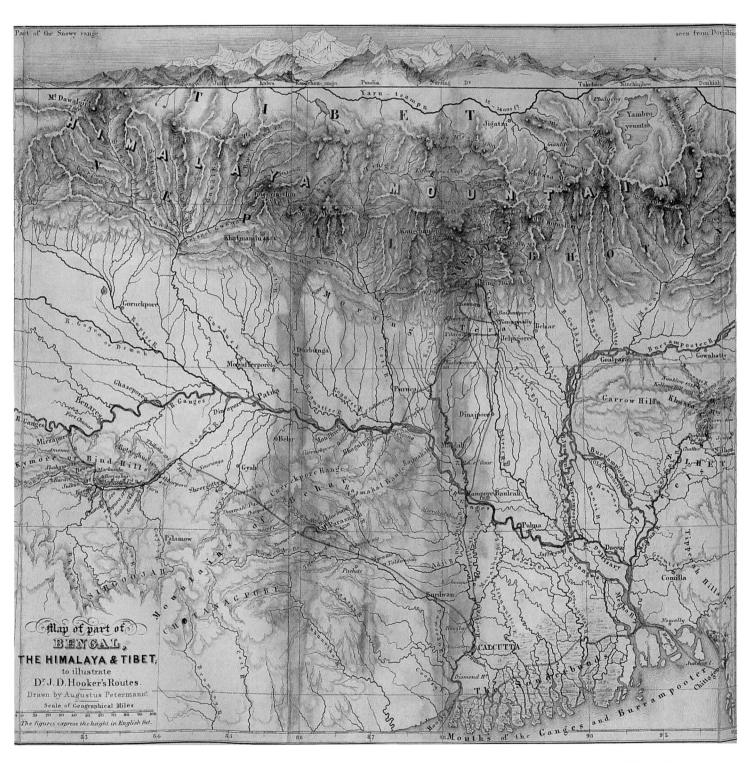

ABOVE Joseph Hooker's map of India and Sikkim, published in *Himalayan Journals* (1854), outlining his travelling routes (in red) from Calcutta across the plains of India, to Sikkim and Assam.

RIGHT A map of Sikkim and Darjeeling District highlighting the places visited by Joseph Hooker in the mid-19th century and by the author and his colleagues 160 years later.

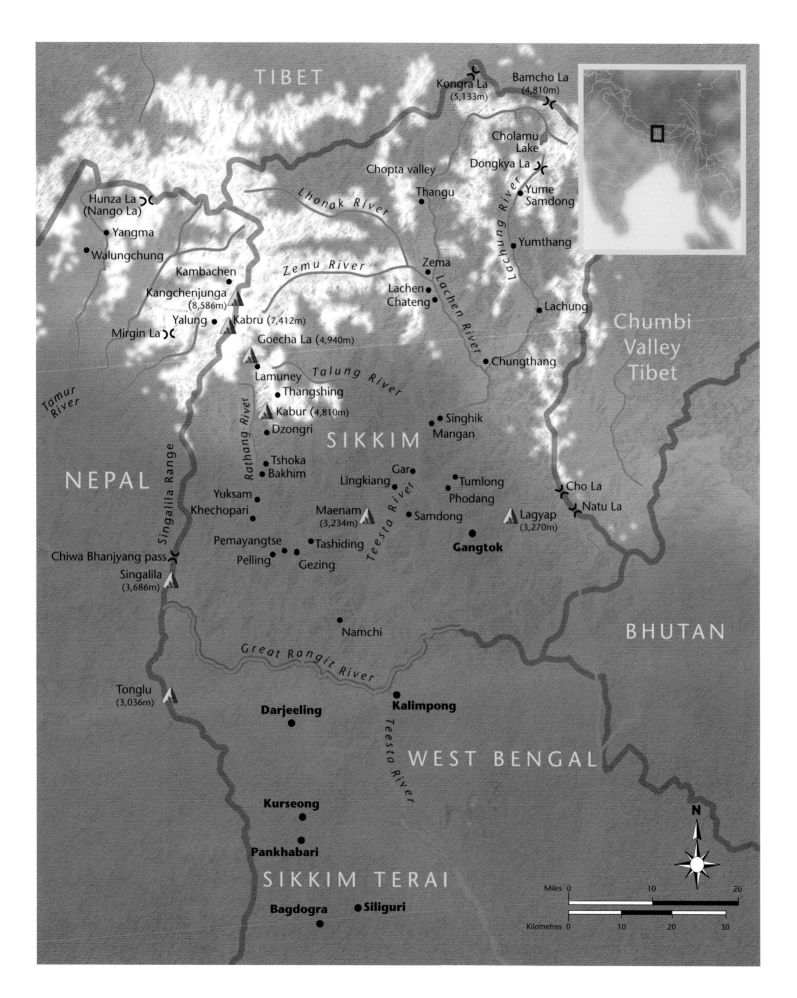

TIBET

Kongra La
(5,133m)

Bamcho La
(4,810m)

Cholamu
Lake

Chopta valley

Dongkya La

Thangu

Yume
Samdong

Hunza La
(Nango La)

Lhonak River

Lachung River

Yangma

Yumthang

Walungchung

Kambachen

Zemu River

Zema

Lachen
Chateng

Kangchenjunga
(8,586m)

Lachen River

Yalung

Kabru (7,412m)

Lachung

Mirgin La

Goecha La (4,940m)

Chumbi
Valley
Tibet

Tamur
River

Lamuney

Talung River

Thangshing

Chungthang

Kabur (4,810m)

Singhik

Rathang River

Dzongri

SIKKIM

Mangan

NEPAL

Tshoka
Bakhim

Gar

Singalila Range

Yuksam

Lingkiang

Tumlong

Cho La

Khechopari

Maenam
(3,234m)

Teesta River

Phodang

Lagyap
(3,270m)

Natu La

Pemayangtse

Tashiding

Samdong

Chiwa Bhanjyang pass

Pelling

Gezing

Gangtok

Singalila
(3,686m)

Namchi

BHUTAN

Great Rangit River

Tonglu
(3,036m)

Teesta River

Darjeeling

Kalimpong

WEST BENGAL

Kurseong

Pankhabari

N

SIKKIM TERAI

Bagdogra

Siliguri

Miles 0   10   20

Kilometres 0   10   20   30

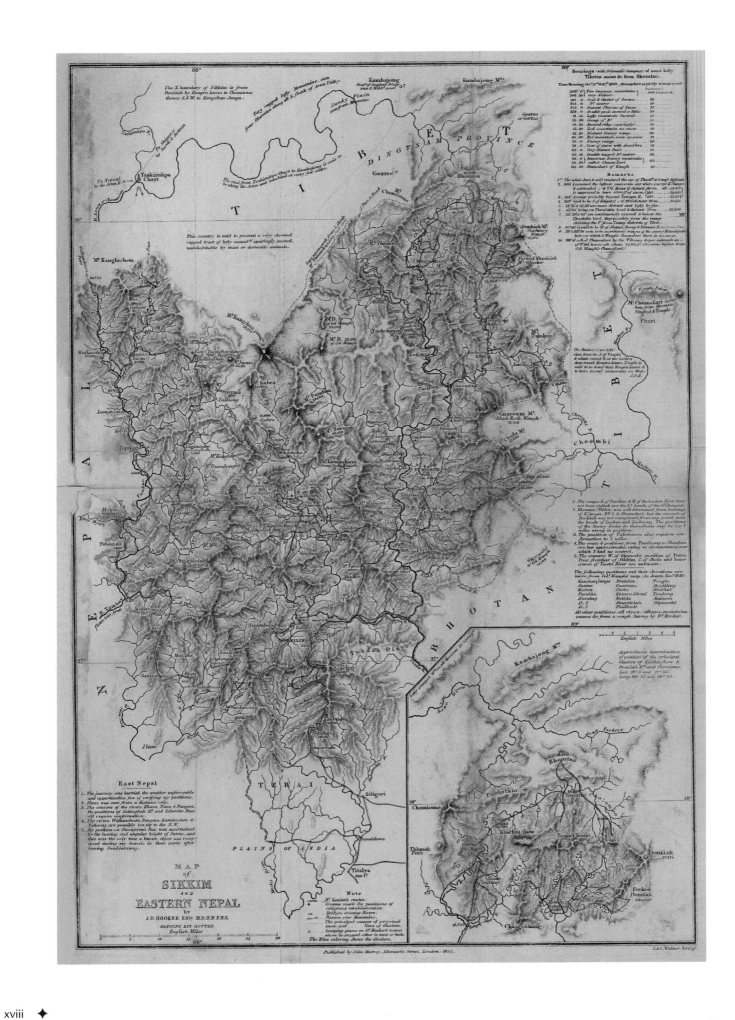

MAP
of
SIKKIM
and
EASTERN NEPAL
by
J.D. HOOKER ESQ. M.D. R.N. F.R.S.
SHEWING HIS ROUTES.
English Miles

Published by John Murray, Albemarle Street, London, 1854.

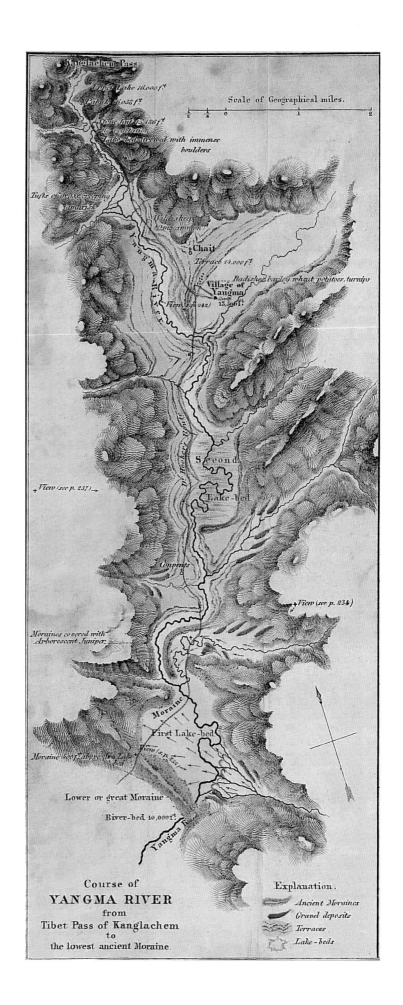

LEFT Hooker's historic map of Sikkim, charting his progress from the plains of India and the Sikkim Terai, north to the hill station at Darjeeling, and, beyond there, the northern valleys, the frontier region and the Tibetan passes.

RIGHT The Yangma River valley in north-east Nepal with details of glacial landmarks and the village of Yangma Gola.

The view from the Goecha La at 16,207 ft.
Regarded as one of the most challenging treks
in the Himalaya, we succeeded in ascending the
pass in November 2014. This glaciated landscape
lies close to the Tibetan frontier.

CHAPTER 1

A PASSAGE TO INDIA

Joseph Hooker (1817–1911) was already a seasoned explorer by the time he set sail for India in November 1847. Between 1839 and 1843, he had travelled on a British Naval expedition to Antarctica, the primary purpose of which was to determine the exact location of the South magnetic pole. Sir James Clark Ross (1800–1862) commanded the expedition comprising the former bomb ships HMS *Erebus* and HMS *Terror,* and the 22 year-old Hooker, the youngest of a crew of 128, sailed as assistant surgeon and botanist.

Departing England in September 1839, the two sailing ships visited Madeira, Tenerife and the Cape Verde archipelago, before continuing to St Helena in the South Atlantic Ocean and then on to the Cape of Good Hope. The young Hooker botanised *en route* collecting at each point.

Having established a familiarity with the polar floras of the Southern Hemisphere, Joseph Hooker then sought to visit the tropics. He had two choices: India or the Andes, and he chose the former, based on the fact that Hugh Falconer (1808–1865), Superintendent of the Royal Botanic Garden, Calcutta, had promised him every assistance in planning and carrying out his expedition in India. Falconer was keen to point out that to the north lay the great Himalayan mountain range of which very little was known, especially of its central and eastern parts, while Tibet, which lay further north still, was little explored and all but a mystery to Europeans.

Hooker's chosen area for exploration of the eastern Himalaya later brought him great fame as a botanical explorer. Seen here is one of his discoveries, *Myricaria rosea,* in the Prek Chu valley in north Sikkim, with the looming mass of Mount Pandim in the distance.

The exact area of the Himalaya Hooker was to explore was ultimately decided by Hugh Falconer and Lord Auckland, who had served as Governor-General of India between 1836 and 1842. Both men recommended Sikkim, a tiny kingdom sandwiched between Nepal and Bhutan, and bordering, to the south, the tropical plains of Bengal in British-controlled India. Sikkim was almost entirely unexplored and a blank spot on maps of the eastern Himalaya.

On his return from India in March 1851, Hooker gained further support from the most notable scientific figures in Britain and Ireland; eminent botanists including Dr John Lindley (1799–1865) from the Horticultural Society, Robert Brown (1773–1858) President of the Linnean Society and the third Earl of Rosse, William Parsons (1800–1867), President of the Royal Society, all helped to procure additional government funding to arrange, name and distribute his collections. Rosse was a keen scientist, and at the family seat, Birr

Hugh Falconer, Superintendent of the British East India Company's Botanic Garden in Calcutta, had much influence on Hooker's choice of Sikkim as a base for exploration

Joseph Hooker's grandfather, Dawson Turner (1775–1858). Seeing pictures of Cook's voyages in his grandfather's library inspired Hooker from childhood to become an explorer.

Hooker's proposed journey to India gained support from the leading men of science of the day, and not just those in Britain. The Prussian naturalist and explorer, Alexander von Humboldt (1769–1859), whose extensive travels in Latin America during the early 19th century laid the foundations for the field of biogeography, was extremely keen. Biogeography, the study of the distribution of species and ecosystems in geographic space and geological time, was a subject that Hooker was also keen to pursue.

The seventh Earl of Carlisle, George Howard (1802–1864), using his position as Chief Commissioner of Woods and Forests, secured Hooker's collections for the Royal Gardens at Kew, while the Earl of Auckland, John Eden (1784–1849), then First Lord of the Admiralty, was instrumental in gaining a Treasury grant of £400 per year for Hooker's expedition to India. The Admiralty had also supported the plan for an expedition to Borneo following on from India, though this never came to fruition and Hooker was instead allocated a further £300 for a third year in India.

Castle in Co. Offaly, Ireland, he built the famous 'Leviathan', the world's largest telescope during the 19th century.

The possibility of visiting Tibet was raised at the earliest stage of the expedition's planning, and Hooker hoped to study Jomolhari (Chumulari), the great mountain peak on the Tibet–Bhutan border that rose to 7,326 m. His earliest recollections of reading, as a young boy, were of Captain James Cook's voyages and of Samuel Turner's (c. 1749–1802) embassy to Tibet in 1783. Thus fired by early descriptions of the mysterious and forbidden plateau, Hooker was determined from the outset to cross the border.

On 11 November 1847 Hooker sailed from Portsmouth for India, initially embarking for Alexandria, his passage provided by the Admiralty aboard the first-class paddle steam frigate HMS *Sidon*. The *Sidon* was due to carry the new Governor-General of India, the first Marquis of Dalhousie, James Andrew Broun Ramsay (1812–1860), to his post in Calcutta. The ship's route brought the group to Lisbon, where Hooker had time to explore the Mediterranean flora near Sintra, then sailing past the Rock of Gibraltar, and from there stopping at a coaling-station in Malta before finally reaching Egypt.

The Marquess of Dalhousie, James Broun Ramsay. Joseph Hooker travelled with him and Lady Dalhousie to India where Dalhousie assumed the position of Governor-General of India.

Egypt and the Great Pyramids

While travelling through Egypt, Hooker struck up a great friendship with Lord Dalhousie, and was invited by the latter to travel out to India as part of his suite. At Alexandria he left the *Sidon* for a brief exploration of the city sights, among them Pompey's Pillar and the slave market.[1]

From Alexandria, the party continued by steamer along the Mahmoudieh Canal, and then sailed up the Nile, stopping at Giza where Hooker sketched the Great Pyramids and the famous Sphinx. Curiously, the journey along the Egyptian canal reminded him of the stretch of the Grand Canal through the Bog of Allen in central Ireland, which he had sailed when botanising in the west of Ireland in August 1836.

In a letter to his father, Sir William Hooker, written on board the East India Company's steam frigate *Moozuffer*, Hooker related that his first view of Cairo was at sunset when 'the sinking sun darted forth golden beams along the desert, lighting up the pyramids on the right bank of the river which appear in strong relief and gilding the white hill that overtops Cairo, mosques and larger buildings'.[2]

Hooker made arrangements to visit the Pyramids, hiring two small boats, guides and interpreters, bringing provisions and several donkeys, the main means of conveyance. Crossing the Nile, he beheld the Pyramids rising above palm trees. He sketched the two

The Great Pyramids and the Sphinx, a rapid sketch made by Hooker. The young traveller wasted no opportunities; from their summits he collected lichens in an otherwise almost barren, desert landscape.

William Jackson Hooker, an engraving dated 1813, by Mrs Dawson Turner, from an original by J. S. Cotman.

Hills. Looking up the Nile, the ribband of verdure appears to dwindle to nothing, as the river retreats into the desert, its course *buoyed out*, so to speak, where it traverses the sandy plain, by two other groups of pyramids on its banks ... I took a few sketches of these scenes ... and after collecting all the lichens I could find on the stones near the summit, I descended, and made arrangements for visiting the interior.[3]

Hooker's many interests included paleobotany, and the previous year he had accepted a post with the Geological Survey preparing a catalogue of British fossil plants, visiting coal fields in England and Wales to study fossil plants *in situ*. He had also investigated fossil timber from Tertiary basalt lava on the Macquarie Plains in Tasmania, in this case a large trunk of a coniferous tree, on which he published a paper in 1842. In 1904, this fossilised conifer was formally named *Cupressinoxylon hookeri* in recognition of Hooker's contribution to paleobotany.[4]

One of the highlights of Hooker's brief tour of Egypt was a trip into the desert 16 km to the south-east of Cairo to see a fossil forest. Several officers from the *Sidon* joined him and he also took two mules and attendants to carry back pieces of petrified timber.

Approaching the site, bits of fossil wood became increasingly frequent, until suddenly the desert was strewn with enormous fallen trunks, half buried in the sand. Some of these were up to 43 m long and presented a strange contrast to the sterility of the desert. They in

William Roxburgh, one of the founding fathers of Indian botany. He was one of several prominent botanists to manage the Botanic Gardens in Calcutta.

grand and lesser Pyramids, the Sphinx below them on the slope of the desert dunes, and the mouth of the catacombs on the cliff. He was greatly impressed by the scene:

As to the Sphinx, it is truly stupendous, and looks larger and larger as you approach; no doubt because it is an object directly comparable with that ever-present standard, *one's self* ... Like the pyramids, it is wonderful and suggestive to an educated individual ... The poor face is terribly knocked to pieces, and as it can never [have] had any loveliness to spare, you may guess how flat and unengaging an object it is, buried up to its throat in sand and rubbish ... One likes to relieve a noble piece of art, but it is impossible to pity the Sphinx.

The bases of the pyramids are covered deeply with rubbish; so that the rock on and with which they are built, and which forms a core, eight feet high, in the centre of the largest, is nowhere visible. I had only time to go over one properly, the Pyramid of Cheops, whose dimensions you doubtless know, 456 feet high, and each base 763 feet ...

The view from the summit is magnificent. Beneath, looking westward, lies the emerald plain, through which sweeps the mighty Nile, sparkling in the sun, as it winds through groves, gardens, and cultivated land. Beyond lies the city of Cairo, a dense mass of white houses, and minarets like spear-heads, crowned by the citadel, with its monster castles, domes and pinnacles, and backed by the white cliffs of the Mohattem [*sic*]

Garden Reach, the official residence of the Superintendent of the East India Company's Botanic Garden. It sits on the banks of the Hooghly River and was originally constructed for the Scottish botanist William Roxburgh.

fact formed a record of a once luxuriant vegetation that grew there 35 million years before, during a wetter period in Egyptian history.

Hooker stayed six days in Egypt. Taking an overland route to Suez (the canal had not yet been constructed), he departed with Lord Dalhousie's party on the *Moozuffer* and crossed the Red Sea. During the next two or three days, the wind blew strongly and the boatswain was lost overboard, struck by the paddle-wheel and killed on the spot.

In mid-December the *Moozuffer* anchored at the British coaling station at Aden in Yemen. Aden had been under British rule since 1839, when the British East India Company landed marines there to secure the territory and to stop attacks by pirates against Mumbai (then Bombay).

This was the last port of call before crossing the Arabian Sea to Sri Lanka (then Ceylon). Hooker was slightly disappointed with the East India Company's choice of vessel, used to convey Lord Dalhousie to India:

> The "Moozuffer" which was sent to Suez for us, is in one sense a splendid vessel, more like a yacht than a man-of-war, but neither fitted nor provided with any accommodation suited to the Governor-General of India. The Captain has the only table to supply, &c., and this he has done very well. Anything more sumptuous in the way of fare on board ship I never met with; but there are neither cabins nor bedding for any of his Lordship's suite; and even the Captain gives up his cabin to Lord and Lady Dalhousie. We lie on mattresses on the deck and 'tis all we can do to turn out tidy for meals in the cabin, for breakfast at 9 o'clock, tiffin at noon, dinner at 4, and then we spend the evening any way we can. The motion of her powerful engine is such that we cannot write without difficulty, and we have no private cabins to sit in ...

At Aden I was far too busy botanising; though, alas! Nearly all my collections have been since destroyed by the salt water

getting into our wretched dormitory on board the "Moorzuffer". Not only did my *Hortus Siccus* suffer, but my spare paper also; so that in Ceylon I was unable to preserve a sinthing.[5]

During his brief stay in Sri Lanka, Hooker took the opportunity to visit the Scottish botanist, George Gardner (1810–1849), who managed the Royal Botanic Garden in Peradeniya. From there the *Moozuffer* set sail again to Chennai (then Madras), arriving there on 12 January 1848, where they were met by Lady Dalhousie's father, the Marquis of Tweeddale (then Governor of Madras), who invited the group to Government House.

There was great excitement in the city at the arrival of the new Governor-General and the guns in the battery saluted their arrival:

> We were no sooner in motion than a thousand carriages full of gaily dressed people started with us, together with horsemen and mounted ladies, and running natives, who escorted us the whole way to the Governor-General's house; ourselves being immediately surrounded by the staff-officers and aides-de-camp, splendidly dressed, and mounted on iron-grey Arab horses. The troops occupied a mile and a half on both sides, first the splendid Madras cavalry, then the European, and lastly the native infantry ... It was a gorgeous and stunning sight, but marred in some degree by the clouds of red dust which were carried along the road, and by the immoderate heat of the weather.[6]

It took just three days for the *Moozuffer* to reach the Sunderbans (Sunderbunds), an enormous expanse of mangrove forest at the mouth of the Ganges and the world's largest river delta. These littoral forests were chiefly composed of *Terminalia catappa*, a pantropic tree, whose fruits are dispersed by ocean currents, and also a variety of palms including *Phoenix paludosa* and the toddy palm, *Borassus flabellifer*. The nipa palm, *Nipa fruticans*, was also common in this region and its large fruits were continually tossed up by the paddles of the steamer. These wild populations of *Nipa* interested Hooker, since fossilised nuts of the palm dating to the Eocene had been found in sandbeds in the Thames Estuary.

Above the mangroves the frigate steamed into the Hooghly, a major distributary of the Ganges, and soon large trees, farmland and villages replaced the marshy jungles of the Gangetic delta. Just below the city the river narrowed, and here the banks were lined with palatial colonial-style houses surrounded by gardens. One of the finest of these was the mansion belonging to the Royal Botanic Garden, built by William Roxburgh (1751–1815).

A short distance upriver the city of Calcutta came into view, dominated by the batteries of Fort William, scene of the infamous 'black hole of Calcutta' almost a century previously.

The party reached Calcutta in January 1848, at a time when the city was enjoying a golden age as the finest colonial city in India. During his brief stay there Hooker spent his time between the opulent grandeur of Government-House, and the residence of Sir Lawrence Peel (1799–1884), the judge and Chief Justice of Calcutta. Peel's house lay directly across the river from the botanic gardens, and was surrounded by gardens unrivalled throughout India for their beauty. Hooker went as far as to call them the 'Chatsworth of Bengal'. The best view of the botanic garden was from Peel's house, the situation reminding Hooker of the view of Kew Gardens from the Brentford side of the Thames.

Rhododendron griffithianum. Hooker named this species for William Griffith who had discovered it in Bhutan in 1838.

William Griffith, an explorer of outstanding ability. Joseph Hooker gave due credit to his enormous collections from India and beyond.

At the botanic garden Joseph Hooker met Sir John McClelland (1805–1883), who was serving as interim Superintendent in Hugh Falconer's absence. In 1835 McClelland had been sent on a mission to see if tea could be grown in Assam and to inspect the tea plants said to be indigenous to that region.[7] He had travelled with two of the greatest figures in Indian botany, Nathaniel Wallich (1786–1854) and William Griffith (1810–1845). Griffith had died just three years earlier and McClelland was busy preparing the botanical papers and drawings of his late friend for posthumous publication. During this mission, a mutual hatred had developed between Griffith and Wallich, Griffith deeming his own scientific abilities far superior to those of his Danish counterpart. For the earlier part of the Assam expedition Wallich managed to maintain cordial relations with his two younger colleagues, but he soon became jealous of Griffith's extraordinary ability as a plant collector. Resentment lingered over the next ten years.

The Company's Garden — the Royal Botanic Garden, Calcutta

The Royal Botanic Garden, Calcutta was founded in 1787 by Robert Kyd (1746–1793), an army officer in the British East India Company. Kyd established the garden as both a base for the introduction and trial of new crops, in an attempt to reduce the famines that plagued Bengal, and also as a trial ground for plants of potential economic importance, such as teak and spices, from which the British East

India Company would reap benefits. The garden's greatest triumph was the introduction of tea from China, and its successful transfer from Calcutta to the tea plantations of Assam and Darjeeling.

With the establishment of the botanic gardens, Calcutta soon became the centre of botanical investigation in Asia, orchestrated by the British East India Company. In 1784 the Asiatic Society was founded and its annual journals published the papers of amateur British and Irish enthusiasts in the field of natural history, including botanical exploration.

The gardens soon became an important base for the comprehensive botanical exploration of the subcontinent. William Roxburgh had been appointed to the role of Superintendent of the new botanic garden in 1794. Using an Indian artist trained to paint in European botanical art techniques, he published the beautifully illustrated three-volume *Plants of the Coromandel* (1795–1819), one of the earliest major works on the flora of India, published by the Court of Directors of the East India Company under the approbation of Sir Joseph Banks.

In 1817, the gardens were placed in the charge of Nathaniel Wallich, an Assistant Surgeon in the East India Company's service. Wallich had travelled extensively on botanical expeditions to Nepal, West Hindustan and lower Burma (now outer Myanmar), and one of his more important roles was the assistance he offered to other plant collectors passing through Calcutta on their way to explore the Himalaya.

The Kyd monument seen here in the Calcutta Botanic Garden, with the famous avenue of royal palms, *Roystonea regia* to the rear.

Nathaniel Wallich, another great name in the annals of Indian, particularly Himalayan, botanical exploration.

Wallich was born in Copenhagen, leaving that city when he was 20 to take up employment as a surgeon at the Danish settlement of Serampore in Bengal. Part of his commission entailed the collection of plants and seeds for the botanic gardens in Copenhagen. His career prospects soared, when, in 1808, the British East India Company annexed Serampore and Wallich and his fellow Danes found themselves prisoners of war. Luckily, his talents as a botanist were recognised and William Roxburgh campaigned for his release so that he could be employed at the botanic gardens in Calcutta. Wallich filled the void left by Roxburgh's departure from India, and in 1815 he was appointed Acting Superintendent, a position made permanent in 1817. During his time the gardens opened as a public pleasure grounds for the entertainment of the growing European population of the capital of British India.

Like his predecessor, Wallich was keen to publish his work on the flora of India, and between 1829 and 1832 he produced the lavishly illustrated three-volume *Plantae Asiaticae Rariores*, using a circle of talented Indian artists employed at Calcutta Botanic Garden. Foremost among these artists was Vishnupersaud (Vishnu Prasad), who had worked for the eminent botanists John Forbes Royle (1798–1858) and Francis Buchanan-Hamilton (1762–1829), and whose illustrations also appear in William Roxburgh's earlier flora. Buchanan-Hamilton (later known simply as Hamilton) had served

The Victoria Memorial at Calcutta. At the time of Hooker's visit Calcutta was the capital of British India, a country regarded as the jewel in the crown of the Empire. The East India Company, and later the British Raj, saw the end of Mughal rule.

very briefly as Superintendent of the Calcutta Botanic Garden in 1814, where he succeeded Roxburgh.

Wallich's work was seen as a natural successor to Roxburgh's *Plants of the Coromandel* and was sponsored by the East India Company, giving enthusiasts back in Europe an early hint of the botanical riches of India. He also sent vast quantities of seeds and living plants to botanic gardens and similar institutions around the world. Throughout the 19th century, the then Royal Botanic Gardens, Glasnevin, in Dublin and its sister gardens at Kew and Edinburgh were receiving regular consignments of seeds from Calcutta.

Wallich rapidly expanded the herbarium in the Calcutta Botanic Garden, training and employing local collectors, and preparing a catalogue of more than 20,000 specimens known to this day as the 'Wallich Catalogue'. The specimens were either collected by Wallich himself, or by others such as Roxburgh or Griffith. Both his personal collection (the Wallich Collection) and his duplicate herbarium set, sent to Sir Joseph Banks, are housed at Kew.

During Wallich's time, the East India Company's botanic garden was justifiably celebrated as the most beautiful in the East. The garden had been laid out in the form of an English country park, with generously proportioned drives, grand formal palm-lined walks, and many artificial lakes, which hosted new exotic aquatic species. The dense canopies of wide-spreading tropical trees protected both plants and visitors alike from the scorching rays of the Calcutta sun.

By contrast, at the time of his arrival in Calcutta, Hooker found the gardens in a state of disarray. Between 1842 and 1844 William Griffith had been placed in temporary charge, while Nathaniel Wallich was recuperating from illness at the Cape.

Griffith disagreed with the naturalistic parkland style of the garden, and attempted to impose a more scientific systematic design, similar to that at Glasnevin or Kew. To carry out his scheme, large numbers of mature trees were indiscriminately felled, resulting in an unsightly wilderness without shade. Hooker praised Griffith's botanical abilities and his zeal as a collector, though he admitted that the latter lacked the eye of a landscape gardener, or the education of a horticulturist.

Whatever his failings, Griffith was undeniably one of the most accomplished botanists of his day, certainly the most eminent in the East, and no one at that time had ever pursued botanical researches over such an extensive field.[8] He had enormous energy and travelled widely through Asia collecting for the East India Company and also for his own private herbarium.

Born in Surrey on 4 March 1810, Griffith completed his medical studies at University College, London where he attended classes in botany given by John Lindley and pursued further studies in medical botany at the Chelsea Physic Garden. In 1832 he was appointed as Assistant Surgeon with the British East India Company and while on the subcontinent travelled and collected widely, using a large number of trained native collectors to work further afield.

Griffith managed the botanic gardens at Calcutta between 1842 and 1844, before being transferred to Malacca in Malaysia. It is probably true to say that, even to this day, no individual botanist has ever been able to accomplish so much in so short a time. His botanical travels took him from as far afield as the Caucasus to Malaysia. During these travels he was exposed to all sorts of hardships and privations, which ultimately caused his death from hepatitis in Malacca in February 1845.

His loss, at the age of just 34, in the prime of his life and with so much more to offer, was deeply shocking at the time, in botanical circles in India and Europe. Hooker estimated that his collections from Assam, Bengal, Bhutan, present-day Pakistan, Afghanistan, Myanmar and the Malay Peninsula comprised about 9,000 species, the largest collection ever obtained by individual exertions.

Griffith had also formed a separate collection of ferns and mosses, the duplicates of which he sent to Sir William Hooker and to William Henry Harvey (1811–1866), Keeper of the Herbarium at Trinity College, Dublin. Part of his fern collections from NE India and the Nilgiri mountains of southern India are housed in the National Herbarium at Glasnevin, Dublin.

The Royal Botanic Garden, Calcutta became one of the largest and most important botanic gardens in the Empire, and a major distributor of new exotic plants to public and private gardens across the world, particularly during the years when Wallich was at the helm. For example, between 1836 and 1840, 189,932 plants were distributed free of charge to almost 2,000 different gardens, all at the expense of the East India Company.

Across the Plains of India

Hooker spent less than a month in Calcutta before departing, in late January 1848, to join an expedition to Bijalgarh (Bidjegur) on the banks of the Son River. The trip was led by David Hiram Williams, who was employed by the East India Company's geological survey to search for coal in the eastern parts of India.

For this expedition, the botanic garden at Calcutta had procured Hooker a native plant collector and a porter to carry a large ream of paper for plant pressing. They were assisted by John Hoffman, Hooker's 28 year-old Calcutta-born servant, who had previously spent ten years at sea, and now assisted in pressing the thousands of specimens Hooker gathered across the tropical plains of India.

Hooker received much help and advice in preparing for this trip from Lord Auckland's nephew, Sir James Colvile (1810–1880), Puisne Judge to the Supreme Court of Bengal and President of the Asiatic Society. Colvile supplied Hooker with eight bearers and a palanquin (equipped with tea, sugar, plates, cups and saucers), in which Hooker was carried during the earlier part of his journey from Calcutta to Hooghly (Hoogly), and from there along the Grand Trunk Road to Bardhaman (Burdwan).

Along with five further porters, Hooker's team consisted of 15 men, though frustratingly, 13 of them were changed every ten miles, or so. Hoffman carried out much of the plant pressing as they travelled north. Once on the Grand Trunk Road his mode of transport was often on the back of an elephant, a novel experience which he described to his father as follows:

Parasnath, Hooker's watercolour of the sacred Jain mountain in Lower Bengal. On the summit he found an interesting flora full of temperate elements.

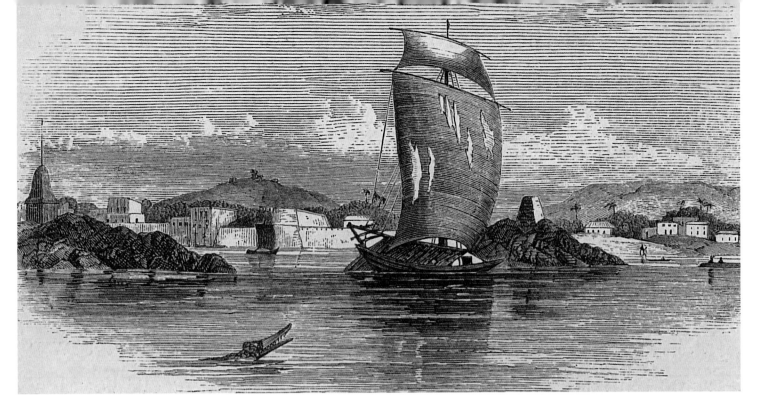

On the Ganges. Hooker's 'floating thatched cottage' at Munger with Kharagpur Hills in the background. A mugger crocodile patrols the waters.

After breakfast Mr. Williams and myself started after camp to Gyna, twelve miles distant; and I mounted an elephant, for the first time since you lifted me upon one at Wombwell's show, [a] good twenty years ago. The docility of these animals is an old story, but it loses so much in the telling, that their gentleness, obedience, and sagacity seemed as strange to me as if I had never heard or read of these attributes. At the word of command my elephant knelt down, and I crawled or rather clomb [sic] up by his hind foot ... and reached a broad pad, or in plain English, a mattress, lashed to his back, holding on by the ropes as he rose, and jogged off at an uncomfortable shuffling pace of four or five miles an hour, and (I took the trouble to count) forty-five paces a minute. The swinging motion under a hot sun, is very oppressive, but to be so high above the dust is an unspeakable comfort. The mahout or driver sits cross-legged on the shoulder, and guides him by poking his great toes under either ear, enforcing obedience with an iron goad, with which he hammers the unhappy beast's head with quite as much force as you use to break a cocoa-nut....

Our elephant was an excellent one, when he did not take obstinate fits, and so docile as to pick up pieces of stone if desired and with a jerk of his trunk throw them over his head for the rider to catch, thus saving the trouble of dismounting. This is geologizing in true Oriental style, and no traveller's tale, I assure you.[9]

By early February, his route north had taken him via Parasnath (Shikharji), the highest mountain in modern-day Jharkhand (then Lower Bengal), and an outlier of the Vindhya Range. Parasnath was (and still is) a major place of pilgrimage for Jain followers, and dozens of temples lay scattered across its summit. Hooker approached the mountain from the Hindu village of Madhuban (Maddaobund), which lay in a clearing in the jungle, and was dominated by the snow-white domes of surrounding temples. Great peepul trees, *Ficus religiosa*, banyans and tamarinds dominated the entrance to the village. In places, Hooker used the elephants not only for transport, but also to collect more inaccessible plant specimens.

Parasnath broke abruptly from the plains, its rugged, granitic peak rising to 1,366 m above sea level, and was clothed with dense tropical jungle from the base to within a few feet of its craggy summit. At its lowest level this jungle was chiefly composed of sal trees, *Shorea robusta*, great vines of *Bauhinia* and dense groves of tree-like bamboos. *Bombax ceiba*, with massively buttressed trunks, were some of the noblest trees of these forests, and with these Hooker found the mallow-relative *Kydia calycina*, a small tree named by Roxburgh for Colonel Kyd.

The summit afforded glorious views of wooded and desert plains and the sandy beds of dry rivers stretching off into the distance. The peak, meanwhile, presented some interesting plants and hinted towards more temperate floral elements with species of *Berberis, Begonia, Clematis, Disporum, Osbeckia* and *Thalictrum*, part of a relic flora, now regarded as having an affinity with the sub-Himalayan region. The summit provided a welcome relief from the baking heat of the plains; by 3 p.m. the temperature was a mere 18°C with a deliciously cool and pleasant breeze. Hooker returned down the mountain in a dooly, carried on the backs of four porters.

From the sacred slopes of Parasnath, Hooker pressed his elephant further along the Grand Trunk Road, passing tiger-infested jungle, and seeing for the first time wild peafowl. He soon reached present-day Bihar and Jharkhand and reached the mighty Son (Soane) River, at that point over 5 km wide. Behind rose the great Kaimur (Kymore) Range which, in places, presented a straight precipitous wall of stratified sandstone reminding Hooker of the Cape of Good Hope.

On the night of 14 February, Hooker witnessed a spectacular display of what he believed was the *aurora borealis*. What he had spotted in fact, was a circumhorizontal arc, an optical phenomenon, caused by an ice halo, formed by the refraction of sun or moonlight in plate-shaped ice-crystals suspended in the atmosphere. This phenomenon he described as being better than the auroras he had seen in the Northern and Southern Hemisphere, and so he stayed up late to record and describe it. When he did eventually turn in, he awoke to find a rat chewing his hair!

The Son River was not without its dangers. Passing the small village of Chanchee his group encountered a mugger crocodile ('mugger' being a corruption of the Hindi word *magar* or water monster), that had just been killed by two men. The beast had swallowed a child that had been playing on the river bank as his mother was washing utensils in the river. The animal was just dead, its body swollen by its prey. The poor mother stood with her hands clasped in agony over the crocodile, unable to withdraw her eyes from its distorted stomach.

This young woman lived in extreme poverty, making a living from manufacturing catechu from surrounding trees of *Acacia catechu*. She owned no property, except a miserable hut, and used two cattle to bring wood, cut by her husband, from the surrounding hills. Her husband was sick, and her only child, her future, now lay in the belly of a crocodile. Though common at the time of Hooker's visit, fewer than 9,000 mugger crocodiles exist today and the species is officially extinct in Bhutan, Myanmar and Bangladesh.

Beyond Chanchee lay the town of Kota, where Hooker crossed the Son River for the last time, and, through a gap in the Kaimur Hills, made his way to the city of Mirzapur (Mirzapore), which lay on the Ganges. His transportation was again on the back of elephants, and also on camels. On 8 March 1848, the group reached the banks of the Ganges and at Mirzapur Hooker hired a sailing boat to take him downriver to Bhagalpur (Bhagulpore). He compared the 40 foot long vessel rather aptly to that of a floating thatched cottage, the deck, on which the structure was erected, floating just above the water's edge; Hooker's lodgings were lined with reed-work formed by the culms of kans grass, *Saccharum spontaneum*.

The crew and captain comprised six naked natives, one of whom steered with the aid of a huge rudder, while the others pulled the four oars, or tracked the boat along the banks of the Ganges. Hooker's cabin was large enough to accommodate his palanquin, fitted as a bed with mosquito curtains, a chair and table. It also held his papers, journals, plant presses and plants, his food supplies, while from the ceiling hung his telescope, vasculum, dark lantern, barometer, thermometers and myriad other instruments. In the hot dry Indian climate, the vasculum was a vital instrument. Constructed from tinned iron and carried horizontally on a strap, it stored fragile flowers in a cool, crush-proof case, until time allowed to press them.

Hooker enjoyed the rather slow sailing, which allowed him ample time to write up his notes and arrange his dried specimens. He visited the magical city of Varanasi (Benares), the 'Athens of India', the architecture and antiquity of which impressed him greatly. Varanasi is one of the oldest continuously inhabited cities in the

Crossing the Son River. In the background is the Kaimur Range.

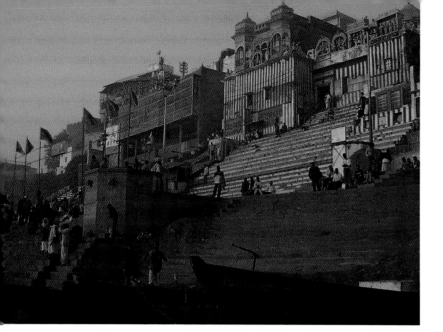

The sacred city of Varanasi, seen here from the banks of the Ganges. Hooker sailed this stretch of the river in March 1848.

Below Ghazipur the muddy waters of the Ganges broadened out to 8 km and were joined to the south, just before Patna, by the mouth of the Son River, on whose vast sandy banks were droves of camels. Below there, to the north, was the mouth of the Ghaghara (Gogra), one of the vast tributaries of the Ganges, originating in the Nepal Himalaya. At the sprawling town of Patna, Hooker inspected yet another of the massive East India Company's opium stores.

By early April, he finally reached the large town of Bhagalpur, where he stayed while his overland transport to Sikkim was arranged. Three days later he journeyed 48 km down the Ganges to the village of Caragola Ghat, where his dawk awaited. At this point the Koshi (Cosi) River, another great tributary originating in the Nepal Himalaya, entered the Ganges, and in the distance lay views of a sweep of the snowy Himalaya, from Gosainkund (Gossain-Than) in Nepal, to Kangchenjunga in Sikkim.

The following day Hooker reached the large town of Purnia (Purnea). The climate was gradually growing more humid and the countryside greener, and he was excited to be within the influence of the Himalaya. The remainder of the journey followed the course of the Mahananda (Mahanuddy) River through the towns of Kishanganj (Kishengunj), Tentulia (Titalya) in present-day Bangladesh, and from there to Siliguri (Siligoree).

Siliguri stood on the verge of the Terai, a low malarial forested belt that skirts the Himalaya from present-day Pakistan to Assam. To an exploring naturalist such as Hooker, the Terai offered much to study. Every feature, botanical, zoological and geological was new to him. The region was often fatal to Europeans and considered unhealthy, though an ethnic group, the Mech, lived there, burning the trees and cultivating the cleared areas. Beyond this jungle area lay the Dak bungalow at Pankhabari (Punkabaree) and above there, the steep and winding road to Darjeeling (Dorjiling).

world, and was (and still is) the spiritual capital of India, the most sacred of Hinduism's seven holy cities, with colourful temples and shrines, sacred bulls and idols. Pilgrims lined the banks of the Ganges in preparation to wash away a lifetime of sins.

From the splendours of Varanasi, he sailed further downriver to the town of Ghazipur (Ghazepore), where in 1820, the British East India Company had established the infamous opium factories that were to lead to the addiction of millions of Chinese men and women, and ultimately, to the Opium Wars with China. The stores there were on an enormous scale, producing a shocking 1,353,000 balls of opium for the Chinese market alone. Hooker questioned the ethics of the East India Company in carrying on such a trade:

> Opium has been a source of enormous revenue to the East India Company, and is still by far its most profitable export. How long it will remain so is now the problem; already the market is considerably fallen, and the Chinese are practising the cultivation of the drug very extensively, and any differences with that Empire are disastrous to the opium-dealers. Under no circumstances can it be expected that China will eventually maintain the ambiguous policy of covertly promoting the import of a pernicious drug, at an enormously high price, the consumption of which she forbids. Her interest plainly is to enforce this prohibition, or to remove it and grow the poppy in the Celestial Empire. I need not trouble you with the *vexata quaestio* of the moral right which the company have to encourage the traffic in this narcotic, in defiance of a nation with which we are at peace, and to whose prejudice it is cultivated.[10]

Ghazipur was surrounded by low fields of roses, red with blossoms in the morning, plucked long before mid-day and employed in the production of rose water and attar of roses. It was also the place where Lord Cornwallis, twice Governor-General of India, had died in 1805, and Hooker made time to visit the elaborate temple-like mausoleum erected in his honour.

An opium store at Patna. Hooker questioned the morality of the East India Company's role in the enormous supply of the narcotic that would ultimately lead to the Opium Wars with China.

CHAPTER 2

A SOJOURN AT THE HILL STATION

The contrast between the flat, plateau-like landscape of the Sikkim Terai and Pankhabari could not have been greater. Leaving the Terai, the ascent to the Dak bungalow at Pankhabari was sudden and steep, with an associated change in soil type and vegetation.

Hooker had finally reached the road to Darjeeling, a narrow winding track that clung to the steep-sided, heavily forested hillsides overlooking the plains of India and the Terai. The latter was cloaked with a dense, stunted forest of *Shorea robusta, Wrightia arborea* and the spectacular *Sterculia villosa*, at that season leafless, but abundantly decorated with large, scarlet star-shaped fruits. Elephants, tigers and occasionally rhinoceros inhabited these foothills, alongside wild boar and leopards, though none were very numerous.

Along the Darjeeling road the bushy timber of the Terai was replaced by giant forest trees, mainly *Duabanga grandiflora*, a tree to 30 m high, with thickly buttressed trunks. With it grew mammoth trees of the Bengal almond, *Terminalia catappa*, and the *Camellia* relative, *Schima wallichii*. The latter, after *Shorea*, was the most important tree of lower forests approaching Darjeeling and ascended the mountain slopes to almost 1,829 m.[1] The toon tree, *Toona ciliata*, was equally abundant and one of the most important of all Indian trees on account of its durable timber. Known as red cedar or Indian mahogany, it was highly prized in cabinet making.

In places bamboos crested the mountain ridges, while torrential streams cut deep gullies down the hillsides and the steep-sided valleys

The Dak bungalow at Pankhabari and the old military road to Darjeeling. Below lies the heavily forested Terai, while the plains of India stretch into the far distance.

Agapetes saligna, as featured in *Illustrations of Himalayan Plants.* It was a common epiphyte in the mountains above Pankhabari.

were choked with riotous vegetation. Trunks of trees were covered with various clubmosses (*Lycopodium* spp.), ferns, *Hoya*, ginger lilies and epiphytic orchids, particularly the spectacular lavender-flowered *Dendrobium aphyllum*.

The bungalow at Pankhabari was surrounded by densely forested hillsides, which rose in places to 1,829 m. The accommodation was spacious and comfortable, with a good view of the road below. Hooker was completely alone, his luggage-bearers lagging far behind, and he was forced to throw away some of his valuable specimens as he had run out of drying paper. In the evenings, the silence was broken only by a deafening chorus of cicadas

To the north of Pankhabari, the great Himalayan range soared skywards, while to the south, the low hills spilled onto the Terai, glinting with numerous streams and rivers. The hot plains of India, seen in the distance, shimmered under a layer of water vapour carried

Agapetes saligna seen here beneath glass at the Royal Botanic Garden Edinburgh.

on breezes from the Indian Ocean more than 644 km away, and further south could be seen the mighty Koshi and Teesta (Tista) Rivers. the main drains for this part of the Himalaya. Hooker marvelled that water vapour could be carried without the loss of a drop of water, to irrigate the luxuriant vegetation of the Himalaya and its foothills.

During his brief time botanising in the forests around Pankhabari Hooker found several new plants, of which the shrubby epiphytic *Agapetes saligna* was the most beautiful. In the mountains above Pankhabari, and in the upper Teesta valley this spectacular shrub abounded on the trunks and thick branches of trees. There, as Hooker explained in *Illustrations of Himalayan Plants*, its lower stem swelled out into a prostrate trunk, as thick as a human arm or leg, and it sent out fibrous roots which attached themselves to the host tree. These swollen trunks, or lignotubers, contain stores of water that sustain the plants during the dry autumn and winter seasons. In spring and early summer, the pendulous branches are wreathed with hundreds of red tubular blossoms. Though it was described from Hooker's Sikkim material, he was keen to point out that it had been found a decade previously by William Griffith in Bhutan. Following his death, Griffith's collections had been placed in storage in the East India Company's 'India House' in London, where they had been more or less forgotten, until rescued and taken to Kew by Hooker on his return from India. It was a generous act on his behalf. He could have chosen to ignore Griffith's enormous herbarium, thereby becoming the sole pioneer in the investigation of the flora of the eastern Himalaya himself, but typical of his absolute honesty, he always gave Griffith the recognition he fully deserved.

This he did by highlighting Griffith's collections in the various Indian floras he published on his return to England, primarily *Flora Indica* (1855), one of his collaborations with Thomas Thomson. Hooker did the lion's share of the research for this work, dedicating it to his father who had taught both him and Thomson botany at Glasgow. Griffith's collections were also liberally quoted in *The Flora of British India*.

Once Hooker's baggage had arrived, plus a pony kindly sent for him by the Darjeeling-based naturalist, Brian Hodgson, he was able to continue his explorations above Pankhabari.

Here he found a new species of olive, *Olea gamblei,* a rare Sikkim endemic. To date its only known natural distribution is at Pankhabari and in hillside forests below Darjeeling. Nearby he also found *Oberonia pachyrachis*, an orchid whose official description was based on these specimens and those of Hugh Falconer, made in the mountains above Garhwal in the western Himalaya.

Hooker recorded the altitudinal sequence of the surrounding vegetation as he ascended higher and higher into the foothills. At 305 m above Pankhabari the trees were gigantic and scaled by enormous vines such as *Bauhinia*, which either sheathed tree trunks or spanned the hot, damp forests with huge cables, joining tree to tree. Their trunks were also densely clothed in epiphytic orchids, wild peppers and various vines, including one of his discoveries, the rampant *Tetrastigma dubium*.

Adding to the jungle-like effect of these tropical forests were the many wild bananas and bamboos that soared to 30 m and enormous screw pines, *Pandanus furcatus*, with massive heads of pineapple-like

Pankhabari as we saw it in November 2012. The forests of the Terai were felled shortly after Hooker's visit to make way for extensive tea plantations. In the background are the same hills that appear in Joseph Hooker's sepia sketch.

foliage on 3 m tall stems. Ferns abounded, and having just travelled the hot, dry Gangetic plains, it must have been exciting to finally enter the damp, tropical and subtropical forests of the outer Himalaya.

Two years later, during his last visit to Pankhabari in April 1850, Hooker found a new species of soldierbush, *Tournefortia hookeri*, a scrambling shrub in the borage family bearing bizarre-looking cymes of lime-green blossoms. There it grew with another of Hooker's namesakes, *Lindenbergia hookeri*, a pretty tropical perennial bearing densely packed racemes of butter-yellow blossoms. Reaching higher still he found sheets of *Lindernia hookeri*, a pretty little carpeting perennial smothered in sky-blue flowers.

At the crest of a ridge at 1,219 m, the vegetation took on a more temperate aspect, and Hooker saw for the first time the yellow Himalayan raspberry, *Rubus ellipticus*, a heavily fruiting species, bearing fine-flavoured orange-yellow fruits. His route continued along the ridge, through forests of the noble *Quercus lamellosa* to the Dak bungalow at Kurseong (Kursiong). The place was named by the Lepcha people, as *kurson-rip* or 'small orchid', because of the abundance of the white-flowered *Coelogyne cristata* on trees in the surrounding valleys.

The tree fern, *Cyathea gigantea*, abounded along the ridge, in places reaching up to 12 m tall and ascending the hills to almost 2,134 m. Beyond Kurseong a steep road zigzagged through a forest of *Castanopsis tribuloides*, *Juglans regia*, *Lithocarpus fenestratus* and various different species of *Lindera*. Enormous vines streamed from tree to tree through the forest, their cable-like lengths weighed down with orchids and ferns, and the dripping, perpetually moist forest canopy was full of pendulous mosses and lichens.

On the forest floor, among emerald green mosses and ferns, Hooker made several notable discoveries including the lovely rose-purple flowered annual gesneriad *Didymocarpus andersonii*, and the shrubby coffee relative, *Pitardella sikkimensis*, at that season bearing dense cymes of large silky tubular blossoms.

The Dak bungalow at Pacheem (present-day Sonada) was the worst Hooker was to stay in. Backed by dark forest and perched on the edge of a deep valley, cold sluggish beetles climbed the interior walls, and these were immediately secured by him for his entomological collections. However dismal his lodgings, though, the surrounding flora was full of interest, and, perched at 2,225 m, plants were taking on a more warm-temperate aspect, with species of *Berberis*, *Dichroa febrifuga*, *Eurya*, *Hydrangea*, *Hypericum*, *Saurauia*, *Skimmia*, *Symplocos*, *Vaccinium*, *Viburnum* and various Araliaceae abounding. On a ridge beyond the bungalow he found the lovely *Dichocarpum adianthifolium*, a pretty little herbaceous woodlander closely related to the buttercups.

Just east of Sonada, Hooker dipped briefly back into the tropics, when he descended into a deep, hot river valley below the town of Mungpoo, there discovering a new dwarf screw pine unlike any other existing species, *Pandanus unguifer*. This species reached little more than a metre tall and first fruited in cultivation at the Royal Botanic Gardens, Kew in July 1873 from plants sent there by Thomas Anderson (1832–1870), Superintendent of the Calcutta Botanic

Coelogyne cristata. This spectacular spring-flowered epiphytic orchid was abundant in the forests surrounding Kurseong.

Darjeeling — the hill station

The little town of Darjeeling lay perched on a narrow ridge that extended from the Senchal range and fell steeply into the valley of the Great Rangit (Rungeet) River below. Forests had been cleared since 1835, to make way for new colonial-style houses that ran along the narrow ridge and along broad flat areas on the western slopes. These ridges were flanked on either side by 1,829 m deep valleys, forested all the way to their bases. Hooker reached Darjeeling on 16 April 1848, to be greeted by showery, cold weather.

Darjeeling had belonged to the Buddhist Chogyals (Rajahs or Kings) of Sikkim until 1780, when it was captured, alongside the Sikkim capital Rabdentse, by invading Gorkhas from Nepal. The British East India Company gained control of the region in 1816 and returned the bulk of the lands to the Sikkim Rajah the following year in exchange for British control over any future border disputes.[2]

The main reason for keeping Sikkim as an independent kingdom, was to retain it as a buffer area between the territorial Nepalese and Bhutanese, since the British East India Company believed that Nepal had plans to overrun the eastern Himalaya as far as present-day Myanmar.

In February 1829, the Governor-General, Lord William Bentinck, sent two officers, Captain George Aylmer Lloyd and J. W. Grant to broker peace following another dispute between Nepal and Sikkim

Garden. It is extremely rare in the wild and was rediscovered at Mungpoo by Indian botanists in 2007, the first time anyone had seen it in the wild since Hooker's visit to the region.

Climbing plants near Sonada included *Ampelocissus latifolia, Hydrangea anomala, Kadsura heteroclita, Smilax lanceifolia* and most beautiful of all, *Holboellia latifolia* with pendulous racemes of lilac blossoms. At this point the forest canopy was composed of several species of oak, including the very handsome *Quercus griffithii* and the closely allied *Lithocarpus pachyphyllus*, alongside *Magnolia doltsopa* and various Lauraceae; among these grew *Acer campbellii, Alnus nepalensis, Betula cylindrostachya*, the Himalayan bird cherry, *Prunus cornuta*, and a previously unknown holly, the rather handsome and distinct *Ilex fragilis*. From Sonada , the old military road to Darjeeling lay due north, along the Balasun River valley (where Hooker found the twining, deciduous vine, *Stephania glandulifera)*, until the saddle of the Senchal (Sinchul) mountain range was crossed at 2,255 m. Beyond lay the hill station of Darjeeling.

The dwarf screw-pine *Pandanus unguifer* was one of Hooker's most remarkable finds. A rare inhabitant of the Terai and lower hills.

Early morning as seen from Darjeeling. Dawn casts a warm glow across the icy peaks of the Kangchenjunga massif.

that had occurred the previous year. During their journey, both men stayed in 'Dorjeling', then inhabited by a hundred Lepchas. They immediately saw the potential for a hill station or sanatorium for recuperating British and Irish soldiers and civil servants employed in British India, who frequently suffered and died from ailments such as dysentery, malaria, cholera, liver disorders and other chronic diseases attributed to the tropical climate of coastal cities. The officers reported to authorities in Calcutta, and their recommendations were promptly followed up.

Given its central position between Nepal, Tibet, Bhutan and British India, the area was also potentially of great strategic military importance. From 1835, the East India Company began to lease a large tract of hillside from the Rajah for an annual fee of three thousand rupees.

Lloyd, 'whose lack of sensitivity in human relations was one of his least attractive traits', informed his superiors that the Rajah had gifted the hill to the British out of friendship. What he did not make clear was, that in line with oriental tradition, the ruler of Sikkim expected a matching gift in return from the Governor-General. The British possession of Darjeeling therefore began with a major grudge on the part of the Sikkim court.

In a very short time, however, a small town developed, shopkeepers and tradesmen poured in with hordes of labourers looking for new opportunities. At the time slavery was common in Nepal, Sikkim and Bhutan, so in addition to genuine migrants, slaves, criminals and political fugitives travelled to Darjeeling in search of shelter.

The hill station at Darjeeling, like those at Shimla, Mussoorie and Dalhousie, for example, was used during the rule of the British East India Company, and later during the British Raj era, to escape the sweltering heat of the plains below. In the following decades

Darjeeling became the summer capital of Bengal, proving the most successful of all the stations, lying as it did just 595 km north of Calcutta, whereas the stations previously mentioned were up to 1,610 km away in the western Himalaya. Hooker predicted it would prosper:

> Its effect on the debilitated frame is marvellous, – really incredible; and if there is truth in children's faces (and where else, if not?) it is undeniably well suited to them too. Its proximity to Calcutta, accessibility at all seasons, the narrowness of the Terai belt, coolness of climate, abundance of vegetables, and great capabilities for further improvement (rapidly as it has progressed) will, no doubt, render it the most crowded of the hill-stations, as it is the only one to which the majority of the Bengal residents can resort, for that short month which may be granted for the re-establishment of health.[3]

On his arrival Hooker stayed for ten weeks with Charles Barnes (who owned large tracts of local land), in a long cottage-like structure, divided into a pair of apartments which he hired to visitors. Barnes was an energetic mountaineer and in the coming weeks they explored many of the surrounding mountain ranges together.

Another resident, Dr Archibald Campbell (1805–1874), the Superintendent of the hill station, organised several local Lepcha boys to be trained as collectors for Hooker. The Lepchas were the aboriginal inhabitants of Sikkim and at the time were the predominant race at Darjeeling where they were usually engaged in outdoor employment. In common with many other races, the Lepchas possessed the legend of a flood, from which their ancestors escaped by climbing to the summit of Mount Tendong in West Sikkim. Their territory once extended as far west as the Arun River in east Nepal, though the Gorkhas had captured much of their land during the 18th century.

Hooker in the Sikkim Himalaya with his Lepcha men, Walter Hood Fitch's re-working of William Taylor's original mezzotint. As his Gorkha soldier looks on, Hooker receives a bough of *Rhododendron arboreum*. The epiphytic *Rhododendron dalhousieae* is in bloom in the trees overhead, and Kinchin, Hooker's Tibetan Mastiff hybrid, keeps close to his master.

Unlike the warlike Gorkhas of Nepal and the Bhutanese, whom Hooker perceived to be quarrelsome, the Lepchas were honest, gentle folk, of diminutive stature, with broad, flat faces and oblique eyes. They had been introduced to Buddhism during the 16th century and about this time they began the practice of plaiting their hair into pigtails. Hooker described them thus:

> Though never really handsome, and very womanish in the cast of countenance, they have invariably a mild, frank, and even engaging expression, which I have in vain sought to analyse, and which is perhaps due more to the absence of anything unpleasing, than to the presence of direct grace or beauty. In like manners, the girls are often very engaging to look upon, though without one good feature: they are all smiles and good nature; and the children are frank, lively, laughing urchins. The old women are thorough hags.[4]

Like the New Zealander, Tasmanian, Fuegian, and natives of other climates, which, though cold, are moist and equable, the Lepcha's dress is very scanty, and when wearing woollen undergarments and hose, he is content with one cotton vesture, which is loosely thrown round the body, leaving one or both arms free; it reaches to the knee, and is gathered round the waist: its fabric is close, the ground colour white, ornamented with longitudinal blue strips, two or three fingers broad, prettily worked with red and white ... A long knife, with a common wooden handle, hangs by the side, stuck in a sheath; he has often also a quiver of poisoned arrows and a bamboo bow across his back. On his right wrist is a curious wooden guard for the bowstring; and a little pouch, containing aconite poison ...[5]

Such are the prominent features of this people, who inhabit the sub-Himalayas, between the Nepalese and Bhotan [*sic*] frontiers, at elevations of 3000 to 6000 feet. In their relations with us, they are conspicuous for their honesty, their power as carriers and mountaineers, and their skill as woodsmen; for they build a waterproof house with a thatch of banana leaves in the lower, or of bamboo in the elevated regions, and equip it with a table and bedsteads for three persons, in an hour, using no implements but their heavy knife. Kindness and good humour soon attach them to your service. A gloomy-tempered or morose master they avoid, an unkind one they flee. If they serve a good hills-man like themselves, they will follow him with alacrity, sleep on a cold, bleak mountain exposed to the pitiless rain, without a murmur, lay down the heavy burden to carry their master over a stream, or give him a helping hand up a rock or precipice ...[6]

Rice was the staple dish of the Lepchas. They also ate elephant meat, and when travelling, lived on whatever the forest could supply, fern tops, roots and flower buds of various members of the ginger family (Zingiberaceae), leaves and fungi, which were invariably chopped up and fried in a little oil.

They drank out of little cups, turned from the knots of maples and other trees. These were often highly polished and mounted in silver, and thus sold for enormous sums. The knots from which they were formed were produced on the roots of several forest trees, by the curious genus of parasites, *Balanophora*.

Some of Hooker's Lepcha boys travelled with him on his excursions into the surrounding glens, or collected by themselves. These boys proved, with constant supervision, to be excellent collectors and extremely good at pressing and drying plants, which left Hooker plenty of time to get on with labelling, drawing and dissecting specimens. Hooker had arrived in Darjeeling in April, just as the pre-monsoon wet season began, so plants and paper had to be dried by a fireside.[7]

Archibald Campbell, a Scot, became the first Superintendent of the hill station in 1839, having been previously based in Kathmandu, Nepal, where he was an assistant to Brian Hodgson. He was entrusted with political relations between the British East India Company and the government of Sikkim and the development of Darjeeling. The military road connecting the town with the plains was constructed the same year. In 1841 Campbell brought tea plants to Darjeeling, growing them in an experimental plot in his garden at Beechwood, thus laying the foundations for the new Indian tea industry.

Darjeeling soon became a boomtown, with the population swollen by visiting officers and civil servants, and the need for labour was met by the mass migration of families from Nepal. Allotments of land were purchased by European settlers and soon covered by houses and barracks, with a bazaar in the centre.

By the time of Hooker's arrival, just 13 years after its handover to the British East India Company, Darjeeling had a population of 4,000, with a considerable trade in musk, salt, borax, soda, gold and Tibetan ponies.

Darjeeling from its earliest days was a melting pot of different Asian ethnic groups. Seen here is a watercolour by Hooker of Limbu and Gurung men, immigrants from Nepal.

The Mech, inhabitants of the low-lying Sikkim Terai. Hooker's copy from Miss Colvile's original watercolour.

The pioneering ethnologist and naturalist Brian Houghton Hodgson befriended the young Hooker, offering lodgings at his house on a ridge above Darjeeling.

Towards the end of June 1848, Hooker moved to stay with the eminent English naturalist and ethnologist Brian Hodgson who lived on a hillside about a mile outside Darjeeling. Hodgson had previously been the British East India Company's British Resident (Ambassador) to the Nepal court at Kathmandu, and, having developed a great love of the Himalaya, rather than retiring to his native Cheshire, chose to settle in Darjeeling in 1845. There he continued his studies of birds, mammals and the many ethnic groups that inhabited the Himalaya. His enormous collection of over 10,000 bird and mammal skins was later donated to the British Museum and a new species of Tibetan antelope, *Pantholhops hodgsonii*, bears his name. Over his lifetime Hodgson discovered 39 species of mammals and 124 species of birds, 79 of which he described himself.

Just three years previously, Hodgson had moved into a spacious house in a forest clearing, on a knoll that commanded the finest prospect of the Himalayan range from any location in Darjeeling. Hooker described the scene as follows:

The view from his windows is one quite unparalleled for the scenery it embraces, commanding confessedly the grandest known landscape of snowy mountains in the Himalaya, and hence the world. Kinchinjunga [sic] (forty-five miles distant) is the prominent object, rising 21,000 feet above the level of the observer out of a sea of intervening wooded hills; whilst, on a line with its snows, the eye descends below the horizon, to a narrow gulf 7000 feet deep in the mountains, where the Great Rungeet, white with foam, threads a tropical forest with a silver line.

To the north-west towards Nepal, the snowy peaks of Kubra and Junno (respectively 24,005 feet and 25,312 feet) rise over the shoulder of Singalelah; whilst eastward the snowy mountains appear to form an unbroken range, trending north-east to the great mass of Donkia (23,176 feet) and thence south-east by the fingered peaks of Tunkola and the silver cone of Chola (17,320 feet) gradually sinking into the Bhotan [sic] mountains at Gipmoochi (14,509 feet).[8]

Hodgson enjoyed Hooker's company, despite their gap in age, and in a letter to his sister in England described his visitor as his 'accomplished and amiable guest'. At this time, he was considering joining Hooker on an expedition to the snows of Kangchenjunga and an unauthorised journey onto the Tibetan Plateau where they would 'slip over one of the passes into Tibet'.[9] Hodgson never did join Hooker in his travels. Constant bouts of ill health may have prevented him from any form of extended, arduous travel.

Hooker was to put Sikkim on the map, so to speak, and in the following decades Darjeeling and its hinterland were visited by dozens of eminent naturalists, botanists and biologists, including the prolific Victorian botanical artist, Marianne North (1830–1890), who arrived there in September 1878 and was left in awe of the mountain views:

... The next day took me over the most glorious road, among forests and mountains, to Darjeeling, the finest hill place in the whole world; and I brought my unusual luck with me, for Kinchinjanga [sic] uncovered himself regularly every day for three hours after sunrise during the first week of my stay, and I did not let the time be wasted, but worked very hard. I had never seen so complete a mountain, with its two supporters, one on each side. It formed the most graceful snow curves, and no painting could give an idea of its size. The best way seemed to me to be to attempt no middle distance, but merely foreground and blue mistiness of mountain over mountain. The foregrounds were most lovely: ferns, rattans, and trees festooned and covered with creepers, also picturesque villages and huts.[10]

The intrepid Victorian artist Marianne North painted several scenes of the mountains and plants at Darjeeling. These are on permanent display in her wonderful gallery at Kew.

New plants from Darjeeling

Brian Hodgson's house lay on a ridge known as Jalapahar, and Hooker made several exciting discoveries there including a wide range of fungi. This included *Strobilomyces polypramis*, the type of a new genus of boletes (mushrooms having a spongy mass of pores under the cap). Hooker's fungi were mostly described by the Rev. Miles Joseph Berkeley (1803–1889), an English cryptogamist and clergyman, and

The view from Brian Hodgson's house
with the snowy peaks of Kangchenjunga
dominating the far distance. The fledgling
hill station is visible on the ridge to the right.

one of the founders of the science of plant pathology. He was well-known as a systematist in mycology with some 600 species of fungi credited to him.

Hooker was to discover dozens of fungi species around Darjeeling, and recorded the host of each. Two species bear his name, the rare *Coprinus hookeri*, discovered on the Jalapahar ridge, and the saprophytic *Panus hookerianus*, abundant throughout the region. Others were named for Hooker's collecting localities, for example, *Peziza darjeleensis*, a saprophytic cup fungus that grew on rotting wood or dung.

Berkeley also described *Lycoperdon sericellum* from Hooker's Darjeeling collections. *Lycoperdon* is a widespread, evil-smelling genus of puffball mushrooms, with about 50 species.

Hooker visited Archibald Campbell and his wife regularly, and in August 1848 he discovered the sweetly scented bracket fungus *Polyporus cremicolor* on decayed timber in the Campbell's garden. The vast majority of plants surrounding Hooker had never been described, and in 1848, for example, he literally had only to step outside on to the verandah of Hodgson's house, to find *Coprinus vellerus*, growing on the mossy edges of the woodwork. It was an exciting era to be an exploring botanist.

The forests that provided a habitat for these new fungi abounded in new tree species, including several species of *Sloanea*, a large group of trees with about 150 species found worldwide. Three species were described from Hooker's Darjeeling collections, *Sloanea sterculiacea*, *S. tomentosa* and *S. dasycarpa*, which Hooker had also collected in the Mai valley in north-east Nepal. Hooker was not their discoverer however; William Griffith had collected all three species some ten years previously in the mountains of Bhutan.

Sloanea was of special interest to one of my travelling companions, Tracy Hamilton, during our 2013 expedition, when we explored the forests that remain around Darjeeling. The genus commemorates the Irish-born physician, collector and naturalist, Sir Hans Sloane (1660–1753), who was born in Killyleagh, Co. Down, just a few miles south of Tracy's home, the Ringdufferin Estate, where she maintains a large garden and arboretum.

The north view of 'Brianstone', Brian Hodgson's house, in 1847. Hooker found a new species of fungi on the verandah.

At higher altitudes Hooker discovered another vigorous climber, *Actinidia strigosa*, an endemic of the eastern Himalaya, scrambling through the trees, and carrying large rounded edible fruits in autumn. Another interesting discovery from Darjeeling was *Leycesteria stipulata*, a handsome shrub with a pendent, cascading habit and leaves densely covered on the under side with a thick grey indumentum. We were later to meet it several times during our own travels through Sikkim.

On the woodland floor were several new aroids, including the strange and very bizarre *Typhonium brevipes*. Hooker found it in fruit only, in 1848, and had to wait some 54 years to see it in bloom, when a tuber sent from Sikkim flowered for the first time in cultivation at Cambridge Botanic Garden. It is a spectacular plant, bearing several crowded, short-stemmed, mustard-yellow, rose-spotted spathes, blotched rose-purple on their interiors.

Begonia gemmipara, an intriguing species that Hooker found in the forests surrounding Darjeeling. He had Walter Hood Fitch re-work paintings by Cathcart's Indian artists and chose to feature it in *Illustrations of Himalayan Plants.*

The Jalapahar ridge on which Brian Hodgson's house sat is also the type locality for *Sorbus rhamnoides*, a whitebeam common at that elevation, and on the ridge below the hill station.

Other new Darjeeling trees included *Machilus gammieana*, a large evergreen to 20 m tall, still in use in the Darjeeling area as a source of timber and charcoal, while the leaves are used for fodder. The specific epithet commemorates James Alexander Gammie (1839–1924), a Kew gardener who moved to Sikkim in 1865 to manage Cinchona plantations. His son, George Alexander Gammie (1864–1935), was later to become Superintendent of the Lloyd Botanic Garden at Darjeeling and Curator of the Royal Botanic Garden, Calcutta.

Another find, the closely related *Persea clarkeana*, was collected by later botanists in Nepal, Assam and Myanmar. This commemorates Charles Baron Clarke (1832–1906), Superintendent of the botanic gardens at Calcutta between 1869 and 1871, who collected plants around Darjeeling and on Tonglu.

Many of the vines scrambling through these trees also proved to be new, including the rampant *Tetrastigma planicaule*, one of Hooker's most curious finds from the subtropical forests below Darjeeling, where it is nowadays endangered. Hooker stated that it formed enormous tropical lianas in the forests of the torrid zone, attracting the attention of every traveller; their remarkable characteristics include the enormous size of the stem, and their shape, of flat, lithe bands, often as much as 46 cm in breadth and not more than 2.5 cm in thickness and with the colour and pliability of gutta-percha. The stems could be seen in the forests at the base of the Himalaya of Sikkim, Nepal and Bhutan, descending from the overhanging limbs of gigantic forest trees. This species was later featured in Kew's *Curtis's Botanical Magazine* from Hooker's 1849 introduction. At Kew it climbed the rafters of the succulent house, while specimens of the trunk were on display in the museum.

The cobra lilies are an abundant group of aroids in Sikkim and the showy, purple-hooded *Arisaema speciosum* var. *mirabile* was one of the most commonly seen of these in the forests close to Darjeeling.

Begonias often favour the same warm, damp, shaded banks as *Arisaema*, and one of the strangest Himalayan species is *Begonia gemmipara*, which Hooker first found near Darjeeling during the autumn of 1848, and in the Lachung valley in August 1849 where it grew beneath a forest of *Picea spinulosa* and *Tsuga dumosa*. Though far from showy, Hooker was fascinated by the strange organs that formed in the axils of the leaves. He stated that it was rare around Darjeeling and that holds true today. It is only occasionally found in Darjeeling District, Sikkim and parts of adjacent Nepal, where it often grows as an epiphyte.

The orchid flora of Sikkim is one of the richest in the Himalaya and it must have been an exciting group for Hooker to study. Most of his new species, including the autumn-flowered terrestrial orchid, *Calanthe biloba,* one of the showiest species in the group, were described by the brilliant English orchidologist John Lindley of the Horticultural Society.

The Senchal Range

Having settled into the little hill station, Hooker began to explore the surrounding peaks. A few miles south-east of Darjeeling lay Senchal (Sinchul), which rose 305 m above the elevation of Brian Hodgson's house, and was easy to climb. The mountain rose well above the level of Darjeeling, thus experiencing colder winters and resulting in a more cold-temperate flora. Dense forest covered the summit, hoary with pendulous lichens and mosses, and masses of *Magnolia* and *Rhododendron* species covered the higher slopes. Ferns abounded, particularly the filmy ferns such as *Hymenophyllum* that grew epiphytically on the tree trunks. The views from the summit were spectacular in all directions. To the north-east, Jomolhari, the famous Bhutanese peak, rose to 7,294 m above the jagged, snow-capped Chola range. In *Himalayan Journals* Hooker elaborates:

> To the north-west again, at upwards of 100 miles distance, a beautiful group of snowy mountains rises above the black Singalelah range, the chief being, perhaps, as high as Kinchinjunga, from which it is fully eighty miles distant to the westward; and between them no mountain of considerable altitude intervenes; the Nepalese Himalaya in that direction sinking remarkably towards the Arun river, which there enters Nepal from Tibet.[11]

Although Hooker didn't realise it at the time, the mountain he could see from the summit of Senchal was what we know today as Mount Everest, which lay 125 km to the north-east of Kangchenjunga (Hooker's Kinchinjunga). It would be another eight years before the height of Everest (then Peak XV) was established, therefore at the time of Hooker's visit to Sikkim, Kangchenjunga was regarded as the tallest mountain in the world.

Senchal was most beautiful in April and May, when the magnolias and rhododendrons were in bloom. Between 2,134–2,438 m the gorgeous *Magnolia doltsopa* was the predominant tree, and in the spring of 1848 it bloomed so profusely that the forests appeared as if sprinkled with snow.

Magnolia doltsopa, once a common tree on Senchal. Seen here in the garden of one of my travelling companions, Esther Shickling, in east Co. Wicklow.

Another species, the spectacular rose-pink-flowered *Magnolia campbellii*, named in 1855 by Hooker for Archibald Campbell, occurred higher up the slopes above 2,438 m, and in early spring bore thousands of great rose-purple to crimson cup-shaped flowers on leafless branches. Regarded by serious magnolia growers as the aristocrat of its group, it was unknown at the time of Hooker's visit to Darjeeling, and one can only imagine his excitement on first meeting this majestic tree. He must also have realised that, due to the altitude at which it grew, it would prove hardy in parts of Britain and Ireland.

During the 1840s *M. campbellii* grew freely on the hills surrounding the hill station, though in the following decades, as tea plantations replaced primeval forest, the great magnolias were felled in shocking numbers. Hooker stated that on the higher ranges, the flanks of hills were turned rose-colour by its abundance. It was first found near Trongsa in central Bhutan by William Griffith, but his specimens were far from perfect and as they were buried in the vaults of the British East India Company's headquarters in London, nothing further was known about it until Hooker's rediscovery near Darjeeling.

Repeated attempts were made to introduce the tree to cultivation, but on every occasion the seed embryos were found to have desiccated. Instead, living plants from Darjeeling were sent in Wardian cases by Thomas Anderson and George King, successive Superintendents of the Royal Botanic Garden, Calcutta, though early introductions (1865) to the east of England proved tender and failed.

It was in Ireland that Campbell's magnolia first succeeded. In 1878 Hooker travelled to Cork to visit Lakelands, the home of William Horatio Crawford (1812–1888), where he saw a young tree, which in March 1884 produced the first blossoms seen in Britain and Ireland. Hooker featured it in *Curtis's Botanical Magazine* (t. 6793) in the following year, thus igniting a passionate interest among gardeners that has never diminished.

Further saplings were later distributed from Calcutta to large gardens in Ireland. Fota, Lord Barrymore's garden in Co. Cork, received a young tree in 1870, and at Kilmacurragh, Co. Wicklow, Thomas Acton planted another of the Calcutta seedlings in his walled garden six years later. In 1895, Frederick William Burbidge, the former Veitchian plant hunter and Curator of the Trinity College Botanic Gardens in Dublin, reported that there were also trees at the Castlewellan Demesne in Co. Down and in England at Kew, but even then, the only trees that had flowered in Britain and Ireland were those at Fota (where it flowered in April 1895) and Lakelands.

The next flowering occurred in 1902, also in Co. Cork, at Belgrove, the famous garden belonging to William Gumbleton (1840–1911), and in the following year his tree carried 147 flowers. From 1904 to 1906 it set no buds, and, in 1907, they were destroyed by a late frost. Gumbleton reckoned that the species would never produce the same

Magnolia campbellii, one of the earliest images of the species from *Illustrations of Himalayan Plants* (1855). This sumptuous work must have raised great excitement among plant lovers in Europe.

A view of Kangchenjunga from the Senchal Range. The outer suburbs of Darjeeling spill across the spur in the middle distance.

Magnolia campbellii, the upper branches of the old tree at Kilmacurragh that was planted in the walled garden in 1876. All of the earliest consignments of living plants were dispatched to Britain and Ireland from the Royal Botanic Garden, Calcutta, having been collected in the mountains surrounding Darjeeling.

glorious display as shown in Cathcart's *Illustrations of Himalayan Plants*, though time has shown that in the milder corners of Britain and Ireland it performs just as well as in the Himalaya, and in the 1970s, his tree was the largest in Britain and Ireland.

Waiting for flowers to appear on *Magnolia campbellii* is a patience game. The old tree at Kilmacurragh, sent from Calcutta and planted in 1876, produced its first blooms in the spring of 1907, a wait of over three decades. Nowadays it is one of the best known trees in Ireland. During the 1940s it was stated to have carried up to a thousand flowers; it probably carries double that in modern times.

In all cases these early flowerings were of the rose-pink flowered clone, though in the wilds of Sikkim the white-flowered form, *Magnolia campbellii* Alba Group, is also relatively common, though it has never really become familiar in cultivation. It was the white-flowered form that was originally found by William Griffith, though in his description of the species, Hooker, based on his knowledge of the trees around Darjeeling stated that the flowers were '*rosei vel rarius albi*'.[12]

It is highly likely that the Royal Botanic Garden, Calcutta obtained the plants they sent to British and Irish gardens from Darjeeling, and taking into account Hooker's comments concerning the rarity of the white-flowered form at Darjeeling, the preponderance of rose-pink flowered trees in European gardens is not surprising. The white-flowered tree does have the advantage of blooming after only 14 years from seed and was first flowered by J. C. Williams in his garden at Caerhays Castle in Cornwall in 1939.[13]

On the branches of Campbell's magnolia, and on those of the evergreen oak, *Lithocarpus pachyphyllus*, and the cinnamon relative, *Cinnamomum glaucescens*, Hooker collected material of a spectacular new epiphytic rhododendron. This slenderly branched species bore clusters of three to six enormous pale lime-yellow, bell-shaped blossoms on the end of branches. It was the first rhododendron species he singled out for attention in *Himalayan Journals*, and he chose to name it *Rhododendron dalhousieae*, for Lady Dalhousie, with whom he had sailed to India. Again, Hooker was not the first to discover this species (that honour must go to William Griffith, who found it in neighbouring Bhutan), but it was he who introduced it to cultivation, where it was to enrapture Victorian enthusiasts.

Higher up the slopes scarlet forms of *Rhododendron arboreum* were occasionally seen, though its beauty was surpassed by *Rhododendron grande*, a superb species forming 12 m tall trees with long, lance-shaped leaves up to 38 cm long, wrinkled above and plastered beneath in a thin silvery-white indumentum. In early spring these trees bore enormous trusses of up to 25 fleshy white blooms with dark purple blotches at their base. Hooker considered it a new species, naming it *Rhododendron argenteum* in his famous publication *Rhododendrons of*

Tab. I.

Rhododendron dalhousieae, a superb epiphytic species that Hooker chose to name for the wife of the Governor-General of India. Frontispiece to Hooker's *Rhododendrons of the Sikkim Himalaya*.

abundant: *Arisaema speciosum,* with broad chocolate-maroon, white-striped spathes and large exotic trifoliate leaves, and the rather dramatic *Arisaema erubescens*, with tall green spathes protected beneath a broad parasol-like formation of umbrella-like foliage.

Hooker referred to them as arums, the genus in which they were originally placed, and had keyed the species out using one of the earliest Himalayan floras as his guide, Nathaniel Wallich's *Tentamen Florae Nepalensis Illustratae* (1824–26). This early work contained good descriptions and plates of newly described Nepalese plants. In a letter to his father, Hooker wrote enthusiastically of the mountain and its many lovely woodland plants:

> The profusion of *Arums* in this region is quite remarkable; the most abundant on Sinchul is triphyllous, and is like Wallich's *A. speciosum* … … they are noble plants, and I hope some of the many roots I have sent down will survive. The enormous clubs of scarlet berries which succeed these cuckoo-flowers are as striking in the woods in September as their blossoms are in April, when the cuckoo cries here as he does with us. *Paris* is another English

Rhododendron grande, a large tree-like species that grows alongside *Magnolia campbellii* in Sikkim Himalaya. The undersides of leaves of Sikkim plants have a brilliant silver hue.

the Sikkim Himalaya, on account of its silver leaf undersides, though it had been described just two years earlier by the Scottish botanist Robert Wight (1796–1872) under its current binomial *Rhododendron grande*. Hooker was sailing to India at the time of Wight's work and thus missed its publication. While he was not the first to find it, he did however introduce it to Western gardens when he sent seeds to Kew where it flowered for the first time in a cool glasshouse in the spring of 1858.

Epiphytic orchids abounded, and, on the thick, gnarled moss-laden trunks of trees he found a diminutive species of *Liparis*, which he named in 1856, *Liparis perpusilla*, the latter epithet appropriately meaning 'exceptionally small'.

The forests on Senchal were formed by a rich mosaic of various tree species, constantly changing in composition as one ascended towards the summit. Hooker revelled in their richness and recorded the sequence of species as altitude was gained. He discovered *Aucuba himalaica* on its damp slopes, and near the base found one of the strangest plants he would encounter in the Darjeeling region: *Helwingia himalaica*, a bushy shrub, bearing tiny clusters of blossoms on the centre of its leaves followed by glossy black berries in autumn. The genus had hitherto been known only from Japan.

In spring immense plants of the spectacular cobra lilies formed enormous colonies in open glades. Two species were particularly

Arisaema speciosum here seen flowering on Senchal's lower slopes in April 2015.

spring genus is now in flower, and very plentiful at this elevation (7–9,000 feet). Falconer tells me it is the *P. polyphylla*; it really is a grand thing, the stem three feet high, a whorl of seven to ten leaves, with three to five sepals, as many petals ... In autumn the fruit is ripe; it attains the size of an apple, bursting by several valves, and exhibiting a profusion of scarlet seeds very like those of a pomegranate. *Disporum* and *Convallaria* are both abundant ... The leaves of an *Ophiopogon* were very abundant, as of various begonias, Didymocarpeae, but none in flower. Mr Edgeworth's genus *Streptolirion,* grows in amazing profusion ...[14]

Some small trees of *Styrax* ... bear a profusion of white flowers, which lie like snow on the ground underneath; there are two species about Darjeeling. But of all things that fall on the ground here, the most remarkable objects are the vivid red outer petals and sepals of the *Magnolia.* This magnificent tree is leafless during the flowering season (April) presenting only a few irregular branches from a trunk sixty to eighty feet in height, covered in a whitish bark. The flowers (resembling a Lotus) are terminal, oddly inserted, and, as well as their peduncles, brittle, and therefore easily damaged by the wind.[15]

The snowy carpet of petals on Senchal's slopes was formed in fact by three species of snowbell tree, the very lovely *Styrax grandiflorus, S. serrulatus* and *S. hookeri,* a small tree to about 9 m, bearing masses of pure white pendulous blossoms in spring. It was probably on Senchal that Hooker found *Sorbus thomsonii,* a small tree to about 10 m, though often smaller. In spring it bears copper-flushed, newly emerged foliage, that matures to a striking silver, typical of the best

of the whitebeams, and the corymbs of scented white flowers are followed by large, pear-like, orange-yellow fruits in autumn.

The cucurbits are a common group of climbers in the eastern Himalaya, and on Senchal and the valleys around Darjeeling Hooker was to collect a previously unknown member of the group, later described, in 1876, by Charles Baron Clarke, as a new genus, *Edgaria darjeelingensis.* The name commemorates J. Ware Edgar, the Deputy Commissioner of Darjeeling during the late 1870s.

Flowering plants and ferns were not Hooker's only interest. Fungi fascinated him and were numerous in Senchal's moist woods, growing on a range of strata, especially dead wood. In the Darjeeling area and on Senchal, piles of logs cut for firewood were often wonderfully phosphorescent, large piles presenting a beautiful appearance in the gloomy forest by night, which Hooker attributed to the presence of bioluminescent mycelial networks.

Around Darjeeling, during the damp, warm summer months (May–October), between altitudes of 1,524–2,438 m, these strange fungi could be easily seen every night by stepping a few yards into the forest. During Hooker's stay there, billets of wood were constantly sent to him by the European residents, to enquire about their luminosity.

A stack of wood collected near Brian Hodgson's cottage apparently presented a striking sight in July and August 1848, and when passing it at night time, Hooker had to calm his pony who was terrified of it. The phenomenon occurred on decaying timber, particularly on oak, birch and several species of *Litsea,* and was most frequent on branches lying close to the ground in wet forest. When camping in the mountains, Hooker often had his Lepcha porters bring phosphorescent timber into his tent, simply for the pleasure of watching its soft, glowing light.

The valley of the Great Rangit

For European residents and visitors to Darjeeling, a popular excursion from the hill station was to a cane bridge over the Great Rangit (Rungeet or Rangeet) River, that lay buried in a deep, tropical valley 1,829 m below the town. In early May 1848 Hooker set off to explore the region with his colleague, Charles Barnes, a major landowner in the Darjeeling area.

This enormously deep valley was a thrilling experience for the young English botanist, enabling him, over a very brief period, to study vegetation types ranging from the temperate flora of Darjeeling to a fully tropical jungle-like flora on the river valley floor.

A traditional Sikkimese cane bridge. One of Joseph Hooker's many delightful sketches from his two-volume classic, *Himalayan Journals.*

Magnolia cathcartii. A showy species that is found across much of the eastern Himalaya and parts of western China. It grows in a number of British and Irish gardens.

Just 305 m below Darjeeling, lay a wooded flat spur called Lebong (Leebong). At that altitude it was already 12°C warmer than Brian Hodgson's house, with considerably more sunshine. Peaches and apples flourished there but never ultimately ripened fruits.

It was at Lebong that the hill station's greatest export succeeded. Archibald Campbell had been experimenting with tea plants since 1841, though his residence lay at 2,134 m and frost, snow and spring hailstorms caused considerable damage to plants. It was on the lower slopes at Lebong therefore, that the first really successful tea gardens were developed in the early 1850s. While descending the slopes during his trip to the Great Rangit in May 1848, Hooker saw the earliest planting in this region and predicted that tea at Darjeeling 'might be cultivated to great profit'.

Joseph Hooker arrived just before the great forests at Lebong were cleared to make way for the tea gardens that today cover the district. It was there that he found the lovely white-flowered *Magnolia cathcartii*, then an abundant tree on the lower, warm hills below the hill station. At the time of Hooker's exploration of the region it was a common tree, flowering in April, with such profusion that it looked as though the branches had been snowed upon. He named it for his friend at Darjeeling, John Ferguson Cathcart (1802–1851), who lived on the Lebong spur, where trees of the magnolia rose to 18 m tall. Hooker claimed it produced fine timber that was favoured by Bengali carpenters for making household goods.

Hooker admired Cathcart's passion for botany and botanical art, the means by which he captured the beauty of hundreds of local plants, using Indian artists whom Hooker helped to train in the more exacting European drawing techniques. Another plant Hooker named for him was a stunning balsam, *Impatiens cathcartii*, a large shrubby species with big showy apricot-yellow blossoms with rose-coloured spurs.

Lebong, later to become the site of the Darjeeling racecourse, was to offer up many new plants including *Xantolis hookeri*, a small tree which Hooker discovered in April 1850, as he departed Darjeeling with Thomas Thomson for Calcutta, and later Assam.

Orchids abounded around Darjeeling. In the upper part of the valley he discovered one of the most beautiful of all the warm-temperate epiphytic species, *Pleione hookeriana*. This lovely plant is highly variable in flower colour, ranging from white to lilac-pink and is the only species to produce stolons. It is by this means that it forms large clumps in the dense moss that clings to the trunks of trees in the damp, monsoon-drenched forests of Sikkim. Hooker's pleione was first flowered in cultivation by Sir Henry Elwes in May 1878 from plants he had sent from Sikkim.

The gorgeous *Pleione hookeriana*. An early summer-flowered species that is still commonly found growing as an epiphyte in the Darjeeling area.

Another epiphyte, *Aeschynanthus hookeri*, was a common undershrub around Lebong, with long pendulous branches that bore clusters of scarlet tubular blossoms in June and July.

It was a beautiful valley. On the steep slopes beneath the experimental tea gardens, the landscape was further enhanced by enormous Buddhist prayer flags, carried on long rows of tall poles. Below the tea gardens, birch and magnolias were replaced by wild bananas, tree ferns and statuesque trees of the graceful fishtail palm, *Caryota urens*.

Below this the flora began to exhibit more subtropical elements, with tall trees of *Schima wallichii* and *Toona ciliata*. In these hot valleys, cobra lilies (*Arisaema*) were replaced by the closely related *Alocasia cucullata* and gigantic bamboos formed extensive thickets. Giant vines raced through the trees including *Tetrastigma hookeri*; although bearing Hooker's name it had been previously found in Assam, in NE India, by William Griffith.

It was in this hot, humid part of the valley that Hooker discovered the pretty, white-flowered orchid, *Calanthe alismifolia*. He later collected it in the Sikkim Terai, and in Sikkim its preferred habitat is hot, humid valleys where it generally grows either in long grass at the base of rocks or in riverine forest. The calanthes are a large group of forest-dwelling orchids in the Himalaya, and it's always a treat to encounter them at flowering time.

Also in the valleys around Darjeeling Hooker found the gorgeous *Cymbidium erythraeum*, a spectacular orchid carrying dense sprays of fragrant, green flowers that are heavily 'tiger striped' red-brown. Popular with orchid growers, this winter-flowered species is now rare around Darjeeling due to forest clearance for tea plantations. The flowers of this species are edible and are a delicacy in Bhutan. The tiny *Nervilia hookeriana* is another of his discoveries from the hot valleys below Darjeeling; it is uncommon there and rarely encountered in the wild.

Perhaps his most spectacular orchid find from this area was the yellow-flowered *Dendrobium hookerianum*, an epiphytic species whose long garlands of golden blooms cascaded from the branches of trees overhead.

In the damp, steaming heat of the lower part of the valley he stumbled across *Amischotolype hookeri*, a tropical perennial in the spiderwort family (Commelinaceae), that was later found to be widely distributed in evergreen tropical and subtropical broad-leaved forest across south-east Asia. Another exotic find from this area was the ginger relative, *Globba racemose* var. *hookeri*. Lovers of exotic foliage will find much to commend this plant, with its lush *Hedychium*-like leaves and striking densely packed racemes of rich yellow, arrow-shaped flowers.

At an elevation of 610 m, and 16 km from Darjeeling, Hooker and Barnes reached a long, low spur that descended to the bed of the Rangit River, at its junction with the Rungmo, whose waters roared down from the summit of Senchal. They were almost on the boundary of the lands leased by the British East India Company and came across a guard-house with two sepoys.

It took the Lepcha men just 20 minutes to construct a table and two bedsteads in Hooker's tent, made from branches, lengths of bamboo and strips of rattan palm, and lined with the soft leaves of bamboo. In this baking hot valley the exotic *Pinus roxburghii* (which

Dendrobium hookerianum, this specimen was illustrated in *Curtis's Botanical Magazine* from material supplied to Kew by F. W. Burbidge.

normally tended to prefer the arid summit of hills and riverside cliffs) was abundant and grew in sal forest alongside the dwarf date palm, *Phoenix acaulis*.

Pinus roxburghii, the chir pine, with flaking red bark reminiscent of the Scots pine, *Pinus sylvestris*, is rather rare in Sikkim, at least in comparison to neighbouring Bhutan, where it is locally abundant.

The Lepchas practised slash and burn cultivation in many of these valleys, though they rarely stayed in one place for more than three years, after which the Rajah increased their rent. They therefore stayed for brief periods before moving on to another place. As Hooker descended through the valley he witnessed the clearing of the jungle by this process. By night it was a spectacular scene; the steep, dry forests blazing all around him, great tongues of fire and enormous boles of trees smouldering like the shale-heaps at a colliery. These great blazes were visible from Darjeeling, but here, within a mile of the scene of destruction, it was an awesome sight:

At this season, firing the jungle is a frequent practice, and the effect by night is exceedingly fine. A forest, so dry and full of bamboo, and spreading over such steep hills, affords grand blazing spectacles. Heavy clouds canopy the mountains above, and, stretching across the valleys, shut out the firmament, the air is dead calm, as usual in such deep valleys, and the fires, invisible by day, are now seen raging all around and … appear in all but dangerous proximity. The voice of insects and birds being hushed, nothing is audible but the harsh roar of the Rungeet and Rungmo, and occasionally, far above it, rises that of the forest-fires. We are literally surrounded with them; some smoulder like the shale-heaps at a colliery … Their triumph is in reaching a great bamboo clump, when the yell of the flames drowns that of the torrents, and as the great stem-joints burst … the noise is that of a salvo from a park of artillery. I have seen houses and ships on fire, but such a jubilee of flame as the burning of the Himalayan forest I never beheld. From Dorjiling, 5,000 feet above this, you may see the blaze and hear the deadening report of the bamboos bursting, all night long; but in the valley, and within a mile of the scene of destruction, the effect is most grand, being heightened by the glare reflected from the masses of mist which hover above.[16]

The temperature of the valley at this time of year was a stifling 35°C. The mountains around Darjeeling left only the narrowest of valleys between them, and at the base of this particular gorge, the raging Rangit spanned a gravel bed some 73 m wide, crossed by a bridge constructed entirely of bamboo and rattan palms that

The confluence of the Great Rangit (seen to the left) and Teesta Rivers. Both rivers snake their way through tropical valleys, the Teesta having originated on the glaciers and icefields that sweep down from the face of Kangchenjunga.

abounded in the river-side jungles. Figs, wild bananas, screw pines and bauhinias battled for space on the river's edge, and beyond the cane bridge Hooker forded the river again by raft and sailed beyond the East India Company's territory and into Sikkim proper, into a region where anyone guiding Europeans was threatened with punishment.

Much of the forest of this region had been burned by the Lepchas, though by riverside flats he encountered 3 m tall plants of *Cycas pectinata* growing among groves of *Pinus roxburghii* and the Indian butter tree, *Diploknema butyracea*. The chaulmoogra tree, *Hydnocarpus wightianus,* was common in the surrounding jungles. In Sikkim its fruits were used to intoxicate fish, and in the 19th century the oil from its seeds was also widely used in Indian medicine to treat leprosy.

Hooker had finally reached the absolute floor of the valley, and the narrowness of the gorge and the steepness of the bounding mountains blocked out any views, except of the opposite mountain face, which was a dense jungle dominated by wild bananas. Spectacular epiphytic orchids, ferns and *Hoya* abounded.

A scene reproduced in *Himalayan Journals*. Hooker's sketch of the chir pine, *Pinus roxburghii,* growing with dwarf plants of *Phoenix acaulis* on the steep slopes of the Great Rangit River valley.

Tropical butterflies were the most spectacular feature of the valley floor. They were everywhere, sailing through the hot still air and settling on the damp sand by the river's edge, where they rested in thousands, in a rocking motion, resembling a crowded fleet of yachts on a calm day.

Hooker camped that night on a ridge above the river, and early the following morning crawled his way out of the jungle by way of the river, when, rounding a bend, the mountains of Bhutan suddenly came into view with the Teesta River below. They emerged from the forest at the confluence of Teesta and Rangit (Rungeet) Rivers, their united waters flowing south.

The Teesta was the colder of the two, being the main drain of the glaciers and icefields of Kangchenjunga. Hooker was struck by the difference in colour of the two rivers, the Teesta being turquoise and muddy, the Great Rangit dark green and crystal clear.

Tonglu — on the frontier with Nepal

Joseph Hooker and Charles Barnes had struck up a friendship and organised several trips away from the hill station during the spring of 1848. Hooker's favourite excursion was to the summit of Tonglu, a mountain in the Singalila (Singalelah) range, 3,070 m high, that lay 19 km south-west of Darjeeling, as the crow flies, though 48 km by path. Only a single European had visited the mountain before Joseph Hooker, in 1847, and that was with the rather reluctant consent of the Rajah.

In May 1848, Barnes took complete charge of a party of 20 men, laden with provisions, instruments, papers for pressing plants, blankets, cooking utensils and a small tent lent to them by a friend at Darjeeling. The men were principally Lepcha porters, each carrying 45–68 kg in large conical baskets strapped to their foreheads and supported by shoulder straps. Hooker was impressed by how far and how steadily the men transported such enormous loads. One of his men carried a wet tent, weighing 91 kg down, a deep descent of 1,524 m, across spurs and rocky streams, before ascending 610 m to their camping ground, all within 12 hours.

Hooker's route took him along a Lepcha track, which plunged through forest along the steep slopes below. At about 1,219 m, stands of the gigantic bamboo, *Dendrocalamus hamiltonii*, the *Pao* of the Lepchas, towered to 18 m overhead, with culms as thick as a human thigh. At Darjeeling its leaves were used for thatching European houses, while its thick, hollow culms were fashioned into large water vessels by the Lepchas. *Schima wallichii* again appeared on the scene, and was much prized by local inhabitants for ploughshares and other instruments requiring a tough, durable wood.

Crossing the Little Rangit River, Hooker and his companions camped at the base of Tonglu overnight, and the following morning, 30 May 1848, began the ascent of the Simonbong spur of Tonglu, named for the tiny village and Buddhist temple of the same name on the spur's summit. Another large bamboo, *Cephalostachyum capitatum*, the *Payong* of the Lepchas, was flowering spectacularly; its 6 m tall culms were transformed into enormous flowering panicles.

Wild strawberries, *Fragaria nubicola*, violets and geraniums marked the beginning of the temperate zone, and potatoes, rice, millet, yams, eggplant, fennel and hemp (for smoking) were cultivated around the little monastery. It was at Simonbong that Hooker discovered *Patrinia monandra*, a vigorous perennial to 1.5 m high bearing densely packed corymbs of pale yellow blossoms. He later introduced it to cultivation via the Royal Gardens at Kew.

The route above Simonbong, on the eastern flank of Tonglu, was one of a number of paths crossing the Singalila range, a high ridge descending like an arm from Kangchenjunga, that formed the border between Sikkim and Nepal, with a series of peaks ranging from 2,134 to 4,572 m. The eastern flank of Tonglu faces directly towards the Indian monsoon and thus benefits from an extraordinary annual rainfall. Sikkim is far wetter than neighbouring Bhutan and Nepal, and the forests on Tonglu reminded Hooker of the damp west coast of Tasmania and of the New Zealand islands that he had explored a decade previously.

The track ran along a series of steep, narrow ridges, though lush humid forests of *Lithocarpus* and large trees of *Magnolia hodgsonii*, which abounded up to 1,829 m. The latter was a spectacular flowering tree with bold obovate leaves almost 60 cm long and had powerfully scented greeny-white flowers with thick, waxy tepals. Hooker first

Magnolia hodgsonii, an evergreen species bearing spectacular foliage and beautifully scented flowers. Hooker was so impressed by it that he named it for his friend at Darjeeling.

found it near Kurseong and on the Lebong spur below Darjeeling, and thought it so spectacular that he named it for Brian Hodgson.

Several laurel family members (Lauraceae) more commonly associated with subtropical regions ascended almost to 2,743 m, particularly the mountain peppers, *Actinodaphne citrata* and *Litsea cubeba* and the 'bara singolee', *Cinnamomum bejolghota*, all with wonderfully aromatic foliage. On those lower, warmer slopes epiphytic shrubs like *Rhododendron dalhousieae* and the spectacular scarlet-flowered *Agapetes serpens* clung to the trunks of forest trees, their fallen blossoms mingling with those of *Magnolia hodgsonii*.

By late afternoon heavy rain drenched the slopes. It was late May and approaching the monsoon season, and the Lepcha men rapidly constructed a hut, thatching it with bamboo and the sail-like leaves of wild bananas. It was at this altitude that Hooker first saw the curious *Wightia speciosissima*, one of the most interesting and showy plants of the warmer forests of Asia. Belonging to the foxglove tree family,

Paulowniaceae, it forms either a tree or a semiepiphytic pseudovine, bearing tubular reddish-pink blossoms in a densely packed thyrse up to 30 cm long. On the slopes of Tonglu it formed large scandent trees up to 15 m high, twisting their trunks around those of host trees, and strangling them over a period of time, gradually leaving only the sheath of the climber.

The approach to the summit the following morning was difficult due to heavy rains. The track was steep and muddy and the weather as gloomy as ever. Above 2,134 m the forests were dominated by three species of oak, of which *Quercus lamellosa*, with its immense lamellated fruits, and leaves more than 40 cm long, was by far the most beautiful. The forest canopy also included the chestnut relative, *Castanopsis tribuloides,* and *Saurauia napaulensis,* with an under-canopy formed by groves of silvery-leaved *Rhododendron arboreum*.

Nearing the summit, at 2,438–2,743 m, the surrounding vegetation finally took on a more temperate element. Gigantic trees of the rose-pink flowered *Magnolia campbellii* grew alongside *Rhododendron grande,* and in late spring, enormous vines of *Holboellia latifolia* bore pendent racemes of lilac-coloured blossoms. It was just one of several showy climbers. Another of Hooker's discoveries from the region was *Clematis tongluensis,* a very showy species allied to the better-known *Clematis montana,* though only reaching 4 m tall, and bearing large white blossoms with contrasting purple-tipped stamens. Hooker also found the scrambling cucumber relative, *Biswarea tonglensis,* a new genus endemic to the eastern Himalaya, though named for the mountain where it was first discovered. It still grows on Tonglu and on Tiger Hill near Darjeeling, though it is endangered throughout its range of distribution.

At 2,743 m, Hooker and his team reached a long boggy flat called 'Little Tonglu,' dominated by enormous trees of *Quercus oxyodon,* *Sorbus hedlundii* (one of his finest Sikkim discoveries), and two showy rhododendrons, the blood-red flowered *R. barbatum* and *R. arboreum* var. *roseum,* both 9–12 m high. Hooker collected several whitebeams during his travels in Darjeeling District, and on Tonglu he took material of *Sorbus griffithii.* Tonglu is the type locality for this lovely tree; William Griffith's collector found it there in spring 1837.

One last steep slope lay between them and the summit and there the woods were dominated by the scarlet-flowered *Rhododendron arboreum* ssp. *cinnamomeum* Campbelliae Group (*syn. R. campbelliae*), which he named for Mrs Archibald Campbell, and the superb blood-red flowered *Rhododendron barbatum*, both of which were laden with flowers. *R. barbatum* is a superb species, certainly one of the finest flowering plants from the Himalaya, where it is rather variable, growing either as a bushy shrub or a tree to 9 m tall. It was discovered by Nathaniel Wallich at Gosainkund, Nepal, in 1821 and described by the Scottish botanist George Don (1798–1856) some 12 years later.

Quercus lamellosa, the most spectacular species of oak around Darjeeling. It is a particularly good host tree for epiphytes. Hooker reckoned it was one of the finest of all the Himalayan oaks and few who know it would disagree. It forms large trees in its native habitat with large impressive leaves, strikingly silver on their undersides. Its fruits are equally fascinating.

Rhododendron barbatum, a superb early-flowered species, seen here in the sheltered confines of the Mount Stewart demesne in Co. Down.

I wrote and told him this morning that I would ask you to confirm the name of a *Rhododendron* on his wife, a little compliment that has touched him to the quick; he is very much attached to his wife, and I never saw a man so heartily appreciate a trifling favor [sic.] Now pray don't forget to attach the name to one of the species sent if the one I have given it to be not new. With regard to all the names, pray alter them as you please or name the plants yourself altogether. I have no ambition that way now, and would indeed rather see your initials at their tails than my own, but, I beseech you, don't forget this MacCallum Morae [for Mrs Campbell].[17]

Next to appear, as the slopes were scaled, was *R. falconeri*. In Hooker's opinion this was, in terms of foliage, the most superb of all the Himalayan species, with trunks fully 9 m high, and bearing leaves up to 45 cm long, deep green above and plastered with a rich rusty-brown indumentum beneath. Above there he encountered an immensely tall Himalayan yew, *Taxus wallichiana*, with an impressively fat trunk measuring 5.5 m in circumference. In Sikkim its red bark was used as a dye, while it stained the foreheads of Brahmins in nearby Nepal.

Another of Hooker's Tonglu discoveries was *Merrilliopanax alpinus*, a small tree in the Araliaceae, which, like its cousin, the rice paper plant, *Tetrapanax papyrifer*, has its stems, petioles and inflorescence densely covered in a yellowish indumentum. It is a wonderful foliage plant that would make an exciting addition to our Western gardens if ever promoted in cultivation.

Rosa sericea, the only species occurring in southern Sikkim, was abundant in open glades, its flowers hanging from the undersides of branches to avoid the pre-monsoon rains. As Hooker pointed out, this is the only rose species to have four petals, rather than the standard five. *R. sericea* was first found in Nepal by Nathaniel Wallich and was introduced to European gardens by Joseph Hooker from north Sikkim.

R. barbatum had been introduced as early as 1829, and Hooker made further seed collections in the autumn of 1849, sending them to his father for distribution. Even out of flower it is a stunning plant with smooth, polished mahogany-red bark and handsome foliage with distinctly 'bearded' petioles (hence *barbatum*, meaning bearded). In Sikkim, *R. barbatum* is not common, and throughout its range in the Himalaya it is under significant threat from habitat degradation, forest loss and because of firewood collection.

R. arboreum ssp. *cinnamomeum* Campbelliae Group commemorates Mrs Archibald Campbell. Hooker was a frequent visitor to the Campbells' home in Darjeeling, where they had a charming home and a young family who were extremely fond of the visiting botanist. The feeling was mutual and on 19 July 1848, Hooker wrote to his father at Kew, who was then preparing Joseph's text for publication in *The Rhododendrons of the Sikkim Himalaya*:

Rhododendron arboreum ssp. cinnamomeum Campbelliae Group, seen here on the summit of Tonglu where Hooker first found it. He chose to name it for the wife of the Superintendent of the hill station.

Even in these cold woods epiphytes clung to the trunks of trees, particularly the Himalayan currant, *Ribes glaciale,* while on the forest floor beneath, *Daphne bholua, Pieris formosa* and several species of *Berberis* vied for attention. One of the best of these, *B. thomsoniana,* was later named for Hooker's school friend and travelling companion in north-east India, Thomas Thomson.

Despite its stunning bright red fruits, Thomson's barberry is extremely rare in gardens, and the only plants I have seen grow at Benmore and Dawyck, the regional gardens belonging to the Royal Botanic Garden Edinburgh.

Hooker named *Rhododendron falconeri* for Hugh Falconer of the Royal Botanic Garden, Calcutta. On Tonglu almost white-flowered clones still exist.

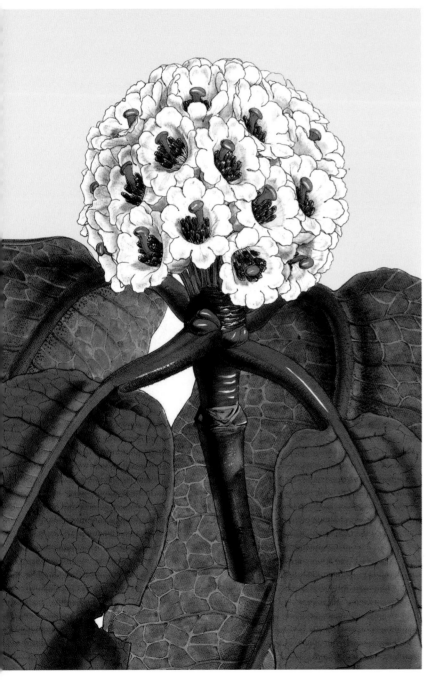

Fitch's illustration of *Rhododendron barbatum* for *Rhododendrons of the Sikkim Himalaya.* Note the bristly 'bearded' petioles, which are a distinctive feature.

Sir Henry Elwes and Sikkim

Thomson's barberry was collected several decades later by Sir Henry John Elwes (1846–1922) on the Singalila range, the long ribbon of mountain spurs dividing Sikkim from Nepal, of which Tonglu forms a part. Elwes was born in May 1846, heir to vast wealth and the family's Colesbourne Estate, near Cheltenham in Gloucestershire. He was a keen big game hunter, and his memoirs leave a fairly grisly account of shooting elephants in the forests of the Sikkim Terai. He was also, though, a superb plantsman, and visited Sikkim on three occasions. He owned a tea garden near Darjeeling and invested heavily in Darjeeling tea, which he claimed was one of his best-ever decisions, making him a fortune.

Elwes visited India for the first time in 1870, having been inspired by Hooker's *Himalayan Journals,* which he found absolutely fascinating. On this trip he was accompanied by the English geologist and naturalist William Thomas Blanford (1832–1905), and they attempted to follow in the footsteps of Joseph Hooker. Elwes was to make the same unauthorised crossing from the Dongkya La pass in north-east Sikkim into Tibet, as Hooker did in 1849 (but more of that later).

Elwes noted the many changes that had occurred since Hooker's earlier visit to the hill station:

> I shall say nothing about the beautiful scenery and surroundings of Darjeeling, which has to me always been the most delightful place in India. Though several books have been written about Sikkim, Hooker's *Himalayan Journals* stands out far and above all others. In the twenty-two years which have elapsed since Hooker was there, a good deal of clearing has been done for tea-planting, but there were few changes compared to those which have come since the railway was made.[18]

Having joined a party of big game hunters on the Terai, Elwes spent May and June 1870 collecting birds, insects and plants around Darjeeling and its hinterlands, hiring Lepcha collectors to extend the scope of his work. Butterflies, one of his great passions, abounded, particularly in the warmer valleys beyond Darjeeling, and Elwes soon realised that Darjeeling and Sikkim were among the richest sites for butterflies and moths in the world.

At the Government cinchona plantation at Mongpo below Darjeeling Elwes met Charles Baron Clarke who then was temporarily managing the plantation. A brilliant botanist, Clarke travelled widely in India, and he and Elwes struck up an immediate friendship. The cinchona gardens had been planted with the South American *Cinchona pubescens*, for the production of quinine, a remedy against malaria, which was rife in the hotter valleys at the time. Gin and tonic was introduced by the army of the British East India Company, to make drinking quinine water more palatable. Prior to the introduction of cinchona to India, the leaves of *Dichroa febrifuga*, a hydrangea relative native to China and the Himalaya were used to treat malaria, hence the specific epithet *febrifuga* or 'fever dispelling'.

In July Elwes was joined by Blanford, of the Geological Survey, and they decided to retrace Hooker's route in Sikkim. Travelling north-east they made a similar unauthorised crossing into Tibet, this time across the Jelep La, where, to the utter shock and surprise of Tibetan guardsmen, Elwes made a bolt across the frontier. They returned to Sikkim to meet the Rajah at Chumanako, where the Rajah's father had seized Hooker and Campbell (more anon) 21 years before.

From there Elwes, Blanford and their party visited the capital at Tumlong, travelling north to Chungthang, and continuing their route up the Lachung valley, stopping to explore the Tangkra Pass, before travelling on to Yumthang and Yume Samdong, beyond where they

Tea plantations below Darjeeling. Henry Elwes invested in the new tea gardens, though sadly their expansion meant the felling of vast forests of *Magnolia campbellii*.

The English naturalist and big game hunter Sir Henry Elwes, photographed about the time of his first expedition to India.

were forbidden to cross the border into Tibet. Reaching the Dongkya Pass, they had planned not to cross without permission, and were instead content to gaze on to the Cholamu Lake. As fate would have it, Elwes did make his crossing into Tibet. He had unintentionally strolled out into a side valley a short distance from the Dongkya Pass, that led to a lesser-known pass, and, seeing the Cholamu Lake beneath him, and no Tibetans in sight, the temptation was irresistible and he crossed over into a landscape that was utterly different from that of Sikkim. They returned to Sikkim by a circuitous route via the Kongra La, and from there down the Lachen valley. They reached Darjeeling that October, following six months of travel in the interior. Birds were another of Elwes's interests, and it was near the Kongra La, a pass leading from north Sikkim into Tibet, that he discovered his horned lark, *Otocorys elwesi*.

Many years later, in a letter to his friend Charles Darwin (1809–1882), Hooker mentions the visit to Kew, in 1871, of 'Mr. Elwes, a guardsman, who had been up my most distant passes in the Himalaya, the first to do it since 1848.' Elwes told Hooker that he found his book a 'miracle of accuracy' and that he could find nothing that Hooker had not taken note of. He also reported that 22 years on, as he travelled through the interior valleys of Sikkim, the Lepchas remembered Hooker with great affection.

Elwes returned to Darjeeling in March 1876 to take charge of a large tea plantation. This of course, allowed for further exploration, during which time he travelled north, climbing Mount Tendong and from there visiting the monasteries at Pemayangtse and Sanga Choeling, and travelling further still to the Chiwa Bhanjyang Pass (Islumbo Pass) where Hooker had re-entered Sikkim from his pioneering tour of east Nepal. From that region he dug tubers of three cobra lilies, namely *Arisaema nepenthoides*, *A. utile* and *A. griffithii*, thus introducing them to cultivation, and Joseph Hooker promptly included them in *Curtis's Botanical Magazine*.[19]

Elwes travelled again to Darjeeling in November 1879, and this time James Alexander Gammie, the Kew-trained gardener, who had lived in Darjeeling for over 30 years, entertained Elwes and provided Lepcha porters for his expeditions.

Arisaema nepenthoides seen here in the Lachen valley in May 2015. Elwes introduced this striking aroid into cultivation.

Buddleja colvilei, one of Hooker's finest discoveries from Tonglu, seen here flowering in David Gilliland's walled garden at Brook Hall in Co. Derry. It is not reliably hardy in the colder parts of Britain and Ireland and needs a sunny sheltered spot.

Returning to Darjeeling, Elwes made an excursion to Tonglu, staying in a newly built stone bungalow, the same building that had provided shelter for Marianne North when she visited the mountain the previous autumn. From there he continued south to climb to the summit of Sandakphu, collecting seeds of interesting plants along the way, and it's likely that the old tree of *Abies spectabilis* in the Deer Park at Kilmacurragh, stated by Augustine Henry to have been raised from seeds sent from the Himalaya in 1879, was from this trip.

Hooker's exploration of Tonglu

One of Hooker's most exciting discoveries from the upper forested slopes on Tonglu was Colvile's Himalayan butterfly tree, *Buddleja colvilei*. There he was to encounter small trees some 9 m tall, which, in July 1849 were smothered with pendulous panicles of superb rose-purple or dark crimson blossoms. Hooker named this lovely species in 1855 for his friend Sir James Colvile, who had given him so much help in Calcutta. It first flowered in Europe in Ireland, in the famous Co. Cork garden of William Gumbleton, from where it was featured in Kew's *Curtis's Botanical Magazine* in 1895 as t. 7449.

Approaching the upper limits of the mountain the forests petered out to be replaced by dense thickets of a new species of holly, *Ilex intricata*. Though rare in gardens, this dwarf, prostrate evergreen shrub is a handsome plant and is still relatively common on Tonglu. Captain Frank Kingdon Ward found it in Myanmar (Burma) in 1931 and introduced it to Western gardens.

On the summit of Tonglu, at 3,070 m, was another flat ridge with depressions and broad pools of water, bordered by the spectacular *Iris clarkei*. This gorgeous iris was discovered by Hooker on Tonglu and in east Sikkim near the Yak La pass. On Tonglu he made field colour

Meconopsis wallichii, a flowering plant at the National Botanic Gardens, Kilmacurragh, where it was raised from seeds obtained from the Royal Botanic Garden Edinburgh. Until recently it was 'lumped' with *Meconopsis paniculata.*

sketches, and it was largely on these that the Kew botanist John Gilbert Baker (1834–1920) based his description of the species. It was later re-collected on Tonglu by Hooker's friend Thomas Thomson in 1857 and again in 1875 by Charles Baron Clarke, for whom the species is named.

Iris clarkei was also collected in the Chumbi valley, Tibet, by Sir George King's Lepcha collectors, and was introduced by him via the Royal Botanic Garden, Calcutta in 1876. Clarke's iris is an extremely showy species. In the wilds of West Bengal and Sikkim it grows to little more than 45 cm, though in cultivation it has reached twice those dimensions. Its flowers vary from blue to red-purple and the markings showed considerable variation. It is at once distinguished from all other Asiatic members of the group by its solid, not hollow, stems.

In the same wet, boggy flats Hooker discovered the delightful little loosestrife, *Lysimachia prolifera*, a pretty little pink (or sometimes white) perennial, also found by later collectors across much of the eastern Himalaya, Myanmar and western China. On steep slopes he also found the gorgeous *Primula irregularis*, a petiolarid species, forming great sheets and smothered in lavender-pink blossoms, nestled in tight nests of densely farinose foliage.

Hooker was also to discover five species of fungi on Tonglu, some of these, including *Microbotryum emodensis*, a parasitic smut, being minute. This particular species grew on the foliage of several species of *Polygonum*. Hooker was interested in all fields of botany and was keen to collect everything from the largest flowering trees to the most obscure fungi, in a bid to classify the flora of this part of the Himalaya.

Two species of Himalayan poppy grew nearby, the yellow-flowered *Meconopsis paniculata*, a monocarpic species bearing panicles of orb-like golden-yellow blossoms on 1.5 m tall stems, and the incomparably beautiful *Meconopsis wallichii*, a similar species to the latter, though bearing purple-blue flowers. Hooker was obviously enchanted by it. He sent seeds to Kew from Sikkim, where it flowered in June 1852, to the astonishment of Victorian visitors. Marianne North climbed Tonglu in the autumn of 1878, where she painted the blue poppies and views of the snowy peaks of Kangchenjunga, which remain on permanent display in her wonderful gallery at Kew. She was spellbound by the surrounding landscape:

From the hills above Jonboo [*sic*] one saw the plains of Bengal like a sea, and mountains on the other three sides. The clouds rolling in and out of the valleys and up into the sky at sunset, quite took one's breath away with their beauty and colours. They were perfect pillars of fire on some evenings, and one thick cloud with a golden edge, just in front of the setting sun, cast wonderful shadows and rays opposite; but the sky was entirely clear overhead on both nights. The road up passed through grand rhododendrons with gigantic leaves with brown and white linings. The hydrangea, too, grew into quite a tree here. All the rhododendron trunks were pinkish, some of them quite satin-like and smooth; no moss or fern could find a hold for their roots in them...

It was too cold for tents. Frost was white on the ground round me when I began, at sunrise, to paint the highest mountain in the world – Deodunga, or Mount Everest as it is now called. It

Marianne North's wonderful depiction of the view of Kangchenjunga from the summit of Tonglu. In the foreground are the dense panicles of blue-flowered *Meconopsis wallichii*, to the right a large flowering tree of *Rhododendron grande.* She stayed overnight in the Dak bungalow.

forms quite a distinct group, detached from Kinchinjanga [*sic*] by a hundred miles at least, and its form is much less graceful and definite ... The trees were scraggy and leafless, and hung with lichen. There were bluish pines (*webbiana*, I believe) with blue cones; blue monk's-hood, and blue poppies – that same preponderance of the colour one finds in other alpine countries ... In the afternoon I tried to drag myself up to the highest point near, only a few hundred feet above the house, but found I could not do it. I sat down once or twice and tried again, but found it too difficult to breathe. It was very cold up there, and there was no furniture in the house, so I piled all the logs close to the fire and within reach of my hand without getting out of my quilt, in which I rolled myself close to it, so that I could poke on the logs and keep the fire alight all night, feeling sure that if it once went out I should be frozen to death. With my hot-water bag under my feet, and the bag which held my clothes for a pillow, that was comfort! [20]

Hooker and his men camped on Tonglu's western side among thickets of rhododendrons. The heavy rains continued, allowing only a transient view into Nepal, which lay just a few miles further west, though most of the time the campsite lay enveloped in a dense mist. Gusts of wind whistled through the ancient, gnarled rhododendron forest, and combined with the damp, made for an uncomfortably cold night. Hooker spent two days here, vainly hoping for the fogs to clear and open out to the views of the distant Himalaya. It never happened, and he would have to wait till the following November, when on a return trip, clear, crisp conditions prevailed and the snowy mountain masses came into splendid view.

The descent of the mountain, in the rains, took just a day, and Hooker and his travelling companions stayed overnight in the little Buddhist monastery at Simonbong, before re-crossing the Little Rangit River and returning to Darjeeling.

CHAPTER 3

IRELAND AND INDIA — THE COLONIAL ERA

To those with an interest in botanical exploration Joseph Dalton Hooker ranks as perhaps the greatest naturalist of the 19th century. My fascination with him began in May 2006, when I moved from my base at the National Botanic Gardens, Glasnevin in Dublin, to manage Kilmacurragh, Glasnevin's country estate in east Co. Wicklow.

For three centuries Kilmacurragh had been the seat of the Acton family, who formed the present estate from the 1640s, and built a pretty Queen Anne house in 1697. The Actons, generation after generation, were passionate plantsmen and women, growing new exotic species, often for the first time in Ireland, and surrounded their house with extensive, well-stocked gardens.

Rhododendron arboreum, the national flower of Nepal, is first recorded as growing in Ireland at Kilmacurragh during the 1820s, and by the close of the 19th century the Himalayan rhododendrons grown there were regarded as one of the best private collections in Europe.[1]

Kilmacurragh's historic Victorian Broad Walk seen here in April with a carpet of fallen petals of *Rhododendron* 'Altaclarense' one of the earliest of the hardy *Rhododendron arboreum* hybrids.

Rhododendron arboreum (the blood-red flowered variant) was discovered in 1796 by Captain (later Major-General) Thomas Hardwicke (1755–1835) of the Bengal Artillery, in the foothills of the north-west Himalaya. The species was described in 1805, and seedlings from Hardwicke's collections raised in the botanic gardens at Calcutta. The first known introduction into British and Irish gardens was in 1803 when Francis Buchanan-Hamilton, an East India Company surgeon and keen botanist, sent seeds home via Calcutta from Nepal. His introduction, from near Kathmandu in the Nepalese foothills, was the scarlet-flowered variant, and the oldest tree at Kilmacurragh, now the tallest plant of *R. arboreum* in Britain and Ireland, is likely to date back to this introduction.

Shortly after moving to Kilmacurragh I set off again, this time to London to complete the Kew International Diploma in Botanic Gardens Management. During that summer, the original plates for Joseph Hooker's *Rhododendrons of the Sikkim Himalaya* were placed on view in the library prior to conservation work, and as I passed by them I was reminded of the many veterans growing at Kilmacurragh

that had been raised from Hooker's Sikkim seed collections.

The Rhododendrons of the Sikkim Himalaya was published between 1849 and 1851, while Joseph Hooker was actively exploring the mountains of NE India. The superb plates depicting his newly named *Rhododendron* species were based on sketches made by Hooker in the field which were re-worked by Kew's excellent, and very prolific, botanical artist, Walter Hood Fitch (1817–1892).

It was hard not to be left in awe by these wonderful works. I can still remember the vibrancy of the colours, particularly the waxy blood-red blooms of *Rhododendron thomsonii* and the strange, yet wonderful glaucous blue seed capsules of *R. fulgens*. Botanical art captures the beauty of a plant in a way modern photography never can, and these early images of Sikkim rhododendrons helped to make them the most sought-after plants of their time.

One of William Hooker's greatest strokes of luck, while Professor of Botany at the University of Glasgow, had been to discover Fitch, then a young apprentice to a firm of local calico designers. Initially, Hooker senior hired Fitch in the evenings, to mount specimens in his

Sisters Kristin Jameson and Tara Lanigan-O'Keeffe in the library at the Royal Botanic Gardens, Kew. Before them is Hooker's field sketch of *Rhododendron maddenii* and Walter Hood Fitch's re-worked image used to illustrate *Rhododendrons of the Sikkim Himalaya*. Kristin travelled on three of our four Sikkim expeditions.

Walter Hood Fitch, the most prolific botanical artist of all time. Both William and Joseph Hooker employed him to provide illustrations for several publications. His images were not just scientifically accurate; they were incredibly beautiful.

During the 19th century, a copy of *Rhododendrons of the Sikkim Himalaya* lay always to hand at Kilmacurragh House should any doubt arise as to the identity of a particular species grown in the gardens. Frederick William Burbidge (1847–1905), the famous professional plant collector employed by the nursery firm Messrs Veitch, was a regular visitor to Kilmacurragh, often in the company of the Irish-born gardener and horticultural journalist William Robinson (1838–1935). Following his travels in SE Asia, Burbidge returned to Europe and later became curator of the Trinity College Botanic Gardens in Ballsbridge in Dublin. He came to know the gardens at Kilmacurragh intimately, stating that the *Rhododendron* collection there 'included one of the most complete series of the Sikkim and Bhotan [*sic*] and Nepalese species that is known'.[3]

Sadly, the house and gardens at Kilmacurragh fell into ruins as a result of the First World War, though the 21st century has been kinder, and despite the decades of decline, an exciting collection survives, including mighty veteran trees of *R. arboreum*, as large as those on Himalayan slopes.

Rhododendron fulgens from *Rhododendrons of the Sikkim Himalaya*. Fitch re-worked Hooker's field sketches and used his herbarium specimens to provide further botanical detail.

herbarium, but Fitch's ability as an artist soon became apparent, and by repaying his apprenticeship fee, Hooker was able to take him into full-time employment.

In 1841, when he was appointed Director of the Royal Gardens at Kew, William Hooker took the young Fitch to London with him. During his long and productive career, Fitch provided illustrations for 35 books and five periodicals, particularly *Curtis's Botanical Magazine*. According to the Kew botanist William Botting Hemsley (1843–1924), Fitch published a staggering 9,960 of his drawings, and some of his finest works are associated with Joseph Hooker. He re-worked Joseph's field sketches for *Rhododendrons of the Sikkim Himalaya* (1849–51), the paintings of Cathcart's Indian botanical artists for *Illustrations of Himalayan Plants* (1855) and provided illustrations for Hooker's *Botany of the Antarctic Voyage* (1844–59).[2]

Fitch also re-worked many of Hooker's field sketches of Sikkim, which appeared in another publication, *Himalayan Journals*. Published in two volumes with detailed maps in 1854, we used this as our guide to re-tracing his route in Darjeeling District and Sikkim. Amazingly, 160 years on, it is still the best guide for visiting naturalists to this part of the Himalaya.

Kilmacurragh House, Co. Wicklow, seat of the Acton family. Before the First World War the gardens displayed one of Ireland's richest plant collections, with a complete range of Himalayan *Rhododendron* species, including those raised from Joseph Dalton Hooker's Sikkim collections.

Dr David Moore, Curator of the botanic gardens at Glasnevin, Dublin. He raised a large consignment of Hooker's *Rhododendron* seedlings and distributed them to several Irish gardens including Fernhill and Kilmacurragh.

As Robin Lane-Fox wrote some years ago in the *Financial Times*, Kilmacurragh is 'hallowed ground' and 'the greatest finds of the greatest [plant] hunters are still in the long grass around us. Kilmacurragh is a holy site, a place to hunt the hunters.'[4] Formed during a golden era of plant exploration, the most famous of all the plant hunters represented in our collections is Joseph Dalton Hooker.

I have the good fortune to live in the centre of the gardens, in an early 18th-century house in the courtyard of Kilmacurragh House. Every spring, when I look out through the upstairs windows of my house, I am confronted by a dazzling, kaleidoscopic array of colour carried above the massive, gnarled peeling trunks of rhododendrons now heading towards their second century in captivity in this old Wicklow garden. If Joseph Hooker had never visited India and the Himalaya, our gardens would undoubtedly be duller places today.

Surrounded by his Sikkim rhododendrons, and those of later collectors, I finally made the decision in the spring of 2012 to visit Sikkim to see 'his rhododendrons' in their native habitat, in the shadows of Kangchenjunga, and over the following years I was to lead four expeditions to Sikkim. During that time we completely retraced the routes blazed by Hooker over 160 years previously.

Glasnevin, David Moore and Hooker's rhododendrons

During the 19th century, Ireland, though at times an unwilling partner, formed an important part of the ever-expanding British Empire. Dublin, one of two viceregal cities within the Empire (the other being Calcutta), was regarded as 'the second city of the Empire' in Georgian times and in 1795, one of its most famous institutions, the Royal Botanic Gardens, was founded at Glasnevin, just north of the city by the Dublin Society (later the Royal Dublin Society).

The gardens soon rose to prominence, becoming a sister garden to the Royal Gardens at Kew and to the Royal Botanic Garden

William Jackson Hooker was Director of the Royal Gardens at Kew during his son's expedition to the Himalaya. Keen to promote Joseph's reputation, he distributed his Sikkim collections across the globe.

Edinburgh, and together the three became the greatest distribution centres of newly discovered plants in the world. It's said however, that whoever chose the sites for these gardens knew little about horticulture. Edinburgh, Glasnevin and Kew are not ideal sites for the cultivation of a wide range of material, and so the gardens began the policy of sending material out to the great landed estates and specialist growers who could offer a more advantageous home to these newly introduced exotics.

David Moore (1808–1879), a Scot, who became Curator in 1838, found that ericaceous plants, particularly rhododendrons, did not thrive in the thin, gravelly alkaline soil, low annual rainfall and relatively cool climate (compared to coastal parts of Ireland) of the Dublin garden, and began to send out plants to large garden owners with better growing conditions and more space.

In 1854 Thomas Acton (1826–1908) inherited the Kilmacurragh Estate, at that time covering almost 2,226 ha (5,500 acres). By then the house and gardens were over 150 years old.[5] Throughout the following decades David Moore began to advise Thomas, and his sister Janet Acton (1824–1906) on garden matters, and thus began the flow of plants from Glasnevin that was to make Kilmacurragh the best-stocked privately owned garden in Ireland.

Richard Turner's curvilinear range of glasshouses at the National Botanic Gardens, Glasnevin, a five-star hotel for plants. David Moore grew many of Hooker's rhododendrons in this glasshouse in large pots while sending duplicate seedlings to Irish gardens with better growing conditions.

The Pond Vista at Kilmacurragh. The gardens
are seen here in their heyday in July 1895.

The earliest recorded dispatch to Kilmacurragh was of seedlings raised from Joseph Hooker's Sikkim collections. Records at Glasnevin show that on 22 April 1850 a consignment of seeds arrived there from Sir William Hooker, of Kew, and this contained '18 papers of seeds of Sikkim Rhododendrons',[6] collected by Joseph the previous autumn.

Later, in that same decade, Glasnevin received further seed consignments from Hooker's childhood friend, fellow botanist and travelling companion in Darjeeling and Assam, Thomas Thomson (1817–1878). Two major seed lots were received from this intrepid explorer. On 7 April 1858, Dr Hooker forwarded from Kew '16 parcels of *Rhododendron* seeds from the eastern Himalaya mnts sent home by Dr Thomson', and on 20 April 1859, a further consignment of '90 papers of seeds from Sikkim Himalaya'[7] reached Glasnevin through the Irish botanist, William Henry Harvey (1811–1866). It may be assumed, that, like Hooker's seedlings, many of Thomson's collections were also later forwarded from Glasnevin to Janet and Thomas Acton at Kilmacurragh.

Glasnevin's 'Plant Book' records the arrival and dispatch of plants during the Victorian period. 22 April 1850 sees the arrival of Joseph Hooker's seed collections of rhododendrons from Sikkim, forwarded from London by his father Sir William Hooker. Three days later Major Madden sends seeds directly from the Himalaya.

Other Irish gardens also received seedlings of Hooker's rhododendrons. Moore was quite friendly with the Darley family, who were creating a new garden at Fernhill near Sandyford in south Co. Dublin. This garden later passed into the hands of the Walker family, and I remember the late Sally Walker (1915–2010) relating the history of *Rhododendron arboreum* var. *roseum* 'Fernhill Silver'. While still based at Glasnevin, I would occasionally receive a note from her letting me know the rhododendrons were in bloom in her garden and I was to visit. 'Fernhill Silver' was (and still is) the finest rhododendron in the collection, and Mrs Walker loved to relate that when Sir Frederick Moore (1857–1949) came to visit, he would take his hat off, as a gesture to this superb rhododendron his father had raised from Hooker's seeds.[8]

Hooker's seedlings grew with great vigour in the damp, mild Irish climate, though it was at Kilmacurragh, where conditions almost matched those of the monsoon-drenched Himalayan slopes of Sikkim, and to Thomas Acton's garden that David Moore referred in 1867, when he reported to Hooker that 'I saw last week eleven kinds of your rhododendrons all growing freely in the open air in County Wicklow. They have been planted out five years and are growing vigorously … The late weather did not affect them in the least, [as] they grow in one of the most favoured spots on the east coast of Ireland.'[9]

Thomas Acton also received seeds from the Himalaya through his brother Colonel Charles Ball-Acton (1830–1897), an officer in the British army in India during the 1860s and 1870s. Charles had grown up at Kilmacurragh, and like his siblings, had inherited a love of plants. When not hunting wild bears in Kashmir, Charles went plant hunting and sent seeds back to Ireland, and it was by this means that a number of Himalayan plants, including the Himalayan barberry, *Berberis angulosa*, entered Western cultivation for the very first time. One of his collections, the west Himalayan fir, *Abies pindrow*, still grows at Kilmacurragh and has formed a fine specimen near the old Oak Avenue.

The Irish in India

India, and the Himalaya in particular, was to become an important source of plants for Kilmacurragh in the following decades. This was at a time when China's 'bamboo curtain' remained solidly shut, and the botanical wealth of that Empire had yet to be revealed. Ireland's role in British India has largely been forgotten, since during the colonial era, Irish citizens were regarded as, and recorded as, being British. For more than two centuries Irish universities supplied men to the Indian civil service. For example, 30% of engineers educated at Trinity College, Dublin in the 1860s settled in India where they played a prominent role in developing the railways and new road networks.

The Honourable British East India Company offered young men the chance to acquire enormous wealth and the opportunities of brilliant careers on the subcontinent, and several of its Irish employees reaped rich rewards. Robert Gregory (1727–1810) joined the Company during the 1740s, becoming a Director in 1769. He returning to Ireland having amassed a vast fortune, to purchase the famous Coole Park Estate of 3,237 ha in his native Co. Galway.

Several Irishmen found themselves at the head of administration in British India. One of these, Francis Rawdon-Hastings, Earl of Moira (1754–1826), grew up in Co. Down and Dublin, and was appointed Governor-General of India in 1812, overseeing victory in the Gurkha War (1814–16), and the return of lands to the Sikkim Rajah. Like many members of the Irish nobility, he took great interest in Dublin's fledgling botanic gardens and in 1816 sent seeds of *Cynodon dactylon* to Glasnevin from his base in Bengal.

Richard Wellesley, first Marquess Wellesley (1760–1842), was born in Co. Meath where his family were part of the Ascendancy. His brother Arthur Wellesley became the first Duke of Wellington.

Richard resolved to annihilate French power in India, consolidating the earlier work of Robert Clive and Warren Hastings, to create an Empire in India; some said that this was to compensate for the loss of the American colonies.

Dubliner, and Trinity College graduate, Richard Bourke, sixth Earl of Mayo (1822–1872) lived for much of his life at Palmerstown House near Naas, Co. Kildare. He became the fourth Viceroy of India in 1869, doing much to consolidate India's frontiers and promote the railways, irrigation, forests and many other public works.

Henry Petty-Fitzmaurice, fifth Marquess of Lansdowne (1845–1927) became Viceroy of India (succeeding Lord Dufferin from the Clandeboye Estate in Co. Down) in 1888, having previously served as Governor-General of Canada. He became the sixth Earl of Kerry, and following his father's death, inherited an estate in Ireland of over 48,967 ha primarily in Co. Kerry. From 1871 he began to create the now famous gardens at Derreen House in SW Kerry, and, apart from his time in Canada and India, spent the summer months every year planning and expanding the gardens at Derreen.

Derreen today is an exotic jungle of towering firs, lush tree ferns and giant rhododendrons, the latter raised from seeds brought back from India by Lord Lansdowne and from the gardens at the Royal Court at Kathmandu.

The Writers Building, headquarters of the many clerks and administrators of the British East India Company in India. Calcutta, now Kolkata, still retains much of its colonial architecture, from the days when it was the capital of British India.

Dromana, Co. Waterford. This remarkable gate lodge is the only example of Brighton architecture in Ireland and reflects Ireland's interest in India, a fellow colony, during the Victorian era. At that time thousands of Irish men and women were based there in various administrative and military roles.

Perhaps the greatest icon reflecting Ireland's interest in Indian culture is the remarkable Indian Gate Lodge that once marked the entrance to the Dromana Estate near Villierstown in Co. Waterford. Unique in Ireland, the origins of this building are uncertain, though one story has it that the structure was originally constructed of wood and papier-mâché by tenants from the estate, to greet the owner, Henry Villiers-Stuart (1803–1874) and his Austrian wife, Theresia Pauline Ott, on their return from honeymoon in Brighton in 1826. There they had admired the Royal Pavilion and were so enchanted with the new gate lodge at home at Dromana that they had it rebuilt in more durable materials.

Another theory is that the building dates from slightly later, 1849, so it may have been a project intended to provide employment to the destitute of the area in the aftermath of the Great Famine.

The builder of the gate lodge, Henry Villiers-Stuart, later ennobled as Lord Stuart de Decies, was keen to improve the conditions of the Irish, and used politics in India, another colony, as an example of how Ireland should be administered.

Elected to Parliament in 1826 as a supporter of Catholic Emancipation, Villiers-Stuart, in his maiden speech to the House of Commons, reminded his fellow MPs that Britain had come to terms with religious minorities in other parts of the empire, commenting that 'If the mild attentions of a sister for the future might effect what the harsh dominions of a mistress has hitherto failed to do,

what enables us to rule the seventy or eighty millions of our Indian territories, with comparatively more ease than the seven or eight millions [of Ireland]?'

His comparison between the conditions of religious tolerance in Ireland and India was not unusual for the time, and the last of the Penal Laws were overturned in 1829, bringing to Ireland the same religious tolerance practised in India by the East India Company, and later by the British Raj. Later generations of the family maintained an active interest in India, particularly Constance Villiers-Stuart (1877–1966), who moved to India with her soldier husband, during which time she collected material for her book *Gardens of the Great Mughals* (1913).

Though the architecture of the Dromana gate lodge has been described as Hindu-Gothic, it is in fact Indo-Saracenic Revival, combining Indo-Islamic elements with Gothic Revival features. The latter style is reflected in the elegant ogival windows of the two flanking single-celled chambers, the triumphal archway and vaulted ceilings, though the roofscape, like that of the North Gate to the Royal Pavilion in Brighton, is crowned by a quartet of minaret-like pinnacles and a bulbous onion dome. An exotic sight in this quiet, rural corner of Ireland, a piece of Islamic India had been transplanted and reinterpreted through Victorian eyes.

Dromana, one of the loveliest estates in the south-east of Ireland, celebrated eight centuries of continual occupation by the same family in 2015, and the gardens are being enthusiastically replanted with a range of rare exotics, particularly species from the Indian Himalaya, thus renewing historic links with the subcontinent.

The introduction of Indian plants through Irish sources

The 19th-century 'Plant Book', preserved in the archives of the National Botanic Gardens, Glasnevin, records the flow of plants to and from the gardens, particularly from India and the Himalaya. Commencing in 1834, some of the more notable donors to the gardens include Nathaniel Wallich, who sent two boxes of seeds from the Calcutta Botanic Garden in August 1835. From that period, dispatches of seeds from Calcutta became a regular event, and East India House in London sent a consignment of *Rhododendron* seeds collected in India in 1854.

Important packages of Indian seeds also came to Glasnevin via James Townsend Mackay (1775–1862), Curator of Trinity College Botanic Gardens in Dublin. Through him, in 1838, came a large collection of seeds received by the East India Company direct from Nepal, and four years later 36 packs of Himalayan seeds from John Forbes Royle (1798–1858), the eminent British botanist who had been based in India, particularly the Himalaya, between 1819 and 1831. Royle had formerly been Superintendent of the Saharanpur Botanic

Gardens (still known as the 'Company's Garden') in the Himalayan foothills of Uttar Pradesh, and in the 19th century Saharanpur was second only to the Calcutta Gardens for its contribution to science and the economy of India. The Saharanpur Botanic Gardens, particularly under the care of William Jameson (1815–1882), Garden Superintendent from 1844 to 1875, sent large regular seed consignments to Glasnevin during this period.

Seed also arrived frequently from Kew, thanks to Joseph Hooker, who retained a life-long interest in Himalayan plants following his Sikkim travels, and forwarded duplicate batches of seeds to Dublin that had been received directly from India.

Another source of new, exotic Indian plants were the many Irish men and women based on the subcontinent as soldiers or civil servants. As has been noted already, Ireland played an important role in the British Empire, particularly in India, and by the mid-18th century 13% of the British army in India was Irish, rising to 45% in 1780–81.[10] In the mid-19th century there were thousands of well-educated Irish citizens in the colonial service in India, many of whom took an active interest in the botany of the local area. They sent home orchids and seeds either to their families or directly to David Moore, and later, to his son Sir Frederick (who succeeded him in 1879).

Glasshouse ranges at the Royal (now National) Botanic Gardens, Glasnevin. Located in the north Dublin suburbs, Glasnevin still retains a position at the centre of Irish horticulture and botany, using Kilmacurragh in Co. Wicklow as its official country annexe where it grows plants that will not grow successfully on the Dublin site.

BOTANIC GARDENS, DUBLIN, 1782, W.L.

With such a large Irish expatriate population, it comes as no surprise that records at Glasnevin show the arrival of regular weekly consignments of seeds and living plants from India sent directly, or indirectly via family members, to the gardens. One of these enthusiasts, Captain Thompson, sent 406 packets of Indian seeds to Piperstown House in Co. Louth in 1876, and rhododendrons still line the drive to this early Victorian pile.

Seedlings from these consignments were raised in large numbers at Glasnevin throughout the colonial period and distributed throughout Britain and Ireland. Kilmacurragh proved to be the greatest beneficiary and some of these early introductions survive there today. Thus, in the latter part of the nineteenth century, India was to become a major source of plants for both British and Irish gardens.

Meanwhile, the gradual spread of the British East India Company's territories offered opportunities to explore and botanise in the remote hill regions to soldiers in its armies. Irishmen like Edward Madden and M. Pakenham Edgeworth (both friends of Joseph Hooker) became pioneers in the botanical exploration of India. Through Glasnevin, these intrepid explorers introduced exciting new hardy plants from the subcontinent.

Edward Madden

Lieutenant-Colonel Edward Madden (1805–1856), a soldier in the Royal Bengal Artillery between 1830 and 1849, was, without doubt, the most prolific Irish collector in the Himalaya. He left Ireland for India in 1830 and began to publish accounts of his travels during the 1840s in the *Journal of the Asiatic Society of Bengal* and in *Hooker's Journal of Botany and Kew Garden Miscellany* (the latter edited by Sir William Hooker). He was a highly competent botanist and plantsman, and what's more, he wrote well, so even today, accounts of his travels

and the plants he met, make for fascinating reading. Madden was also in contact with other plant explorers of the period, such as Hooker and Lieutenant-Colonel Sir Henry Strachey, all of whom were actively collecting across the Himalayan range at the time.

Madden spent four tours of duty in India, though it was during the latter two, between 1841 and 1850, that he introduced a wide range of plants, primarily through the botanic gardens at Glasnevin, Trinity College, Belfast and Kew. During this period, he sent no fewer than 35 consignments of seeds and plants to David Moore at Glasnevin. He collected mainly in the western Himalaya, in the mountains above Shimla, the deserts of Rajputana in north-west India and in the mountains of Nepal, sometimes employing local men to collect seeds for him. Initially he sent these to Glasnevin via his mother, who lived in Kilkenny, or his colleague M. P. Edgeworth, though later his collections went directly to Glasnevin.

Madden's seed collections were full of interesting new exotics, which proved hardy in the Irish climate. He visited Glasnevin while on home leave in 1843, and was also familiar with Kew, and on a visit there, in September 1853, studied the different forms of *Berberis aristata* with Joseph Hooker in the herbarium.

Dogged by ill health, it is remarkable that Madden found the time and energy to explore as widely as he did; indeed, he had many exciting journeys into the highest reaches of the western Himalaya, the results being seen in Irish gardens of the time. In 1845 he introduced, via Glasnevin, the lovely azure-blue flowered *Meconopsis aculeata* to cultivation, the first of the Himalayan blue poppies to bloom in Europe.

In 1856 Madden wrote to Joseph Hooker, thanking him for dedicating the genus *Maddenia* in his honour, though disputing Hooker's statement that *Meconopsis aculeata* had not reached

The Asian black bear, *Ursus thibetanus*, seen here near Darjeeling. Plant exploration was not without its dangers for Victorian plant hunters, and in Edward Madden's day bears were abundant in the primeval forests of the Himalaya.

European gardens, pointing out that it had in fact blossomed at Glasnevin for two years during the late 1840s.

His life as an officer in the East India Company took him to the remotest corners of the Himalaya, and in September 1846 Madden set off on an adventurous and pioneering expedition to the Pindari Glacier. As a soldier he had an interest in the strategics of the mountain passes, just as Hooker did when he later visited Sikkim.

Having passed through newly established tea gardens above Almora he found that the warmer valleys in the foothills were full of exciting plants such as *Curculigo orchioides*, *Osbeckia stellata* and abundant *Notholirion thomsonianum*. It was a wild, unspoilt, though dangerous landscape, and Madden revelled in such places:

> Khathee consists of some beautiful, open, and swelling lawns, closely hemmed in by exceedingly steep and lofty mountains, either covered with grass or enveloped in dark forest. On the n. w., about 300 feet below, the Pindur roars along its narrow gulley, up which, whenever the clouds cleared a little, several high snowy and black rocky peaks of the great range appeared close to hand … The place is a perfect bear-garden; we had not been an hour in the camp, before one appeared on the opposite bank of the river, feeding quietly on locusts. Messrs. Ellis and Corbett have seen half a dozen daily, and on the afternoon of the 16th bagged one of them about half a mile from camp. The mountaineers hold them in great dread and are unanimous in asserting that they not only devour sheep and goats, but even their own species when found dead. They are very fond of the mountain ash, or rowan fruit.[11]

It was on this expedition that Madden collected seeds of the elegant bamboo, *Himalayacalamus falconeri,* on the banks of the upper reaches of the Pindar River, sending them to Sir William Hooker at Kew, thus introducing it to cultivation.

One of Madden's most important finds was on the Gagar range, north-east of the hill station at Nainital. There he found several palms growing on Mount Takil, a 2,438 m high peak that took its name from the palm, also known locally as takil. He reported finding this species in great numbers with trees rising from 9 to 15 m tall and with superb crowns of large flabelliform leaves rattling loudly in the breeze.

These were at first identified as the widespread Indian *Trachycarpus martianus*, though it was described as a new species *T. takil* some 35 years later. Sadly, the forests of palms Madden encountered have long since been felled, apparently for their trunk fibre, and the Takil palm is now critically endangered and on the brink of extinction.

In August 1847, David Moore at Glasnevin received a package of palm seeds sent from Madden via Kew, this being 'the advance guard of a larger body which I will dispatch as soon as possible'.[12] The seeds were for Glasnevin, Trinity College and Belfast Botanic Gardens and the Co. Waterford nurserymen, Messrs Fennessey (Madden lived for a time in Waterford city).[13]

David Moore trialled a wide range of exotics for their hardiness throughout the gardens at Glasnevin, and in the autumn of 1870 had three palms planted out: the Chilean wine palm, *Jubea chilensis*, the Chusan palm, *Trachycarpus fortunei* (Robert Fortune's original introduction), and plants of Edward Madden's palm, *Trachycarpus takil*.[14] This is the first record of *T. takil* growing out of doors in Europe and though the plant no longer exists, it apparently proved hardy.[15]

The author in a drift of *Cardiocrinum giganteum* at Kilmacurragh. The giant Himalayan lily was one of Madden's finest introductions.

One of Madden's greatest gifts to gardeners was the giant Himalayan lily, *Cardiocrinum giganteum*, which he met frequently during his travels and of which he sent seeds to Glasnevin during the 1840s. The seeds germinated, for the first time in Europe, after a two-year long wait, and Moore had seedlings distributed across Britain and Ireland.

Madden's seedlings first flowered at the Lamorran Rectory near Truro in Cornwall in 1851, and in 1852 at Glasnevin and Edinburgh. Though generally stated to have been introduced via Glasnevin in 1848, this date is impossible. When one of Madden's seedlings flowered in Cunningham's Comely Bank nursery in Edinburgh in July 1852, John Hutton Balfour (1808–1884) had it featured in *Curtis's Botanical Magazine* (as *Lilium giganteum*, t.4673) stating 'the plant had been raised from seeds sent by Major Madden some five or six years ago', placing the introduction of seeds to 1846, again impossible.

According to contemporary accounts, the seeds of the giant lily germinated after two years and, as seedlings take at least a further six to seven years to flower, the actual date of introduction, when Madden sent seeds to Dublin, must be sometime around 1842–43. What a stir these great lilies must have caused at the time, their enormous white scented trumpets carried on 3.5 m (12 ft) tall stems.

David Moore was succeeded as Curator of the gardens by his son, Sir Frederick Moore, and Madden's giant lily was the first plant he was to learn of as a seven-year-old child – for all the wrong reasons:

My earliest recollections of the name of a plant goes back to the year 1864 in the Botanic Garden, Glasnevin, where I was born. In those days it was difficult to get anything but willow rods for staking pot plants. My father, Dr David Moore, then Director of the Gardens, used to buy builders' laths which were used to hold up plaster on ceilings and perhaps may still be so used. On wet days the men went into the sheds to split up these laths into stakes for pot plants; quite good stakes they made. My brother and I used to make very effective swords with them, by tying a piece of one across the bottom end of another, leaving about 6 inches for a handle. At the back of one of the greenhouses in a specially made border, a border which exists to-day, two or three plants of *Lilium giganteum* (which were much taller than we boys) were flowering for the first time, creating considerable excitement and anxiously watched. We knew nothing about their value or interest – we thought them worthy foes for our swords and slashed them down! We did not break our swords upon the lilies, but later our swords were broken on us, and my brother and I had good cause to remember the name of *L. giganteum*.[16]

Throughout the 1840s Madden continued to explore the western Himalaya, sending seeds to Glasnevin and thousands of herbarium specimens to Kew. Leaving his base at Almora, in Kumaon district in 1849, Madden explored the Himalayan forests above the rapidly expanding hill station at Nainital, before their great trees were felled to build houses in the new town below. He described seeing enormous veterans such as the Himalayan holly, *Ilex dipyrena*, with girths of up to 5 m, and that monarch of the Himalayan forests, *Rhododendron arboreum*, 12 m tall and with girths of up to an incredible 5 m.[17]

Nainital had been founded in 1841 for the same purpose as Darjeeling, and soon became popular with British and Irish soldiers, colonial officers and their families, as a place to escape the sweltering heat of the plains. The famous English botanical artist Marianne North visited the area in July 1878, and her paintings of the countryside surrounding the hill station, which capture the romantic beauty of the place, are on permanent display in her gallery at Kew.

In April 1850 Edward Madden sent roots and bulbs, including the lovely *Lilium wallichianum*, to the Botanic Gardens at Belfast and Glasnevin, thereby introducing Wallich's lily to cultivation. The Dublin and Belfast bulbs flowered in the summer of 1851, when a superb plate of the Belfast plant was prepared for Kew's *Curtis's Botanical Magazine* (t.4561).

Madden's seed collections continued to pour into Glasnevin at this time, many flowering there for the first time in Europe. One of his finest introductions was the wonderfully fragrant *Abelia triflora*, which he collected in the mountains above Shimla in September 1846, where he stated it was abundant and was known as '*spung*'.

David Moore received seeds of this lovely species in March 1847 and Madden's original plants still grow at Glasnevin near the fern house, where they form enormous fat-trunked, multi-stemmed trees. They first flowered at Glasnevin during the summer of 1852, and continue to do so to this day, a tribute to this great Irish explorer.

The mountains and glens above Shimla, fringed by great forests of *Betula utilis*, reminded him of the rugged mountain scenery of Glengarriff in West Cork, while the vacciniums seemed oddly similar to the fraochán, *Vaccinium myrtillus*, of his home country.

Madden revelled in this wild, unexplored landscape, offering first-hand accounts to European readers based in British India and at home in Britain and Ireland of a strange, exotic land. The following is an account of the view of the snowy mountains above the place where he had collected seeds of *Abelia triflora*:

> The Himalaya is seen to the best advantage, not at noon, but a little before sun-set, when, especially in the cold season, its whole extent is at once, and most gloriously lit up in a rose or copper colour, "one living sheet of burnished gold." Gradually the "sober livery of grey twilight" creeps up towards the loftiest peaks, extinguishes all their "bright lights" and replaces them with the deadly pale hue of a corpse; the soul of the mountain has departed; and if the speculator be contemplating the ranges north of Simla, he says or sings its requiem with the pun – *Sic transit gloria Mundi!*[18]

David Moore was keen to promote these exciting new introductions and had many sent to London to be featured in *Curtis's Botanical Magazine*. Among these was the lovely lilac-flowered *Buddleja crispa*, which Madden had collected in the mountains above Almora, the chocolate-coloured *Clematis barbellata*, and the alpine *Cassiope fastigiata*.

Sir Frederick Moore, pictured here in 1911, with his son, also Frederick. The only Irish person ever to have been knighted for services to horticulture, he succeeded his father Dr David Moore as Curator at Glasnevin in 1879.

Abelia triflora. Edward Madden's original seedling seen here flowering at the National Botanic Gardens, Glasnevin in July 2013. This species is wonderfully scented.

Clematis barbellata, one of Madden's Glasnevin seedlings. David Moore dispatched material to Kew where it was sketched by Walter Hood Fitch and reproduced in *Curtis's Botanical Magazine* in 1854.

Michael Pakenham Edgeworth

M. Pakenham Edgeworth (1812–1881), known to his acquaintances as Pakenham Edgeworth, also sent parcels of seeds to David Moore in the 1840s and 1850s from the Himalaya, through his aunt Henrietta Beaufort (1778–1865), daughter of Rev. Daniel Beaufort, a founder of the Royal Irish Academy.

The 22nd, and youngest child (by his fourth marriage) of Richard Lovell Edgeworth (1744–1817), Pakenham Edgeworth grew up on the family estate in Edgeworthstown, Co. Longford. Pakenham was a step-brother of the famous Irish novelist Maria Edgeworth (1768–1849), herself a keen gardener, who received plants from William Hooker for her garden at the family seat. From the 1830s her brother began sending her seeds from India, and these she shared with the Curator of the Trinity College Botanic Gardens in Dublin.

It seems all the Edgeworth children loved flowers. Pakenham's sister Lucy, writing in February 1829, mentions her teenage brother's creation; 'I have so pretty a garden to show you – great clearings in the wildernesses ... the circular bed, near the old elms, is Pakenham's Botanic Garden ...'[19]

M. Pakenham Edgeworth left Ireland two years later to join the Bengal Civil Service and travelled widely throughout the subcontinent and Sri Lanka until his retirement in 1855. He held a series of administrative and judicial posts in northern India, thus giving him easy access to botanise in the highest mountains of the western Himalaya on the Indo-Tibetan border. During this time, he explored the mountains above the Hindu holy town of Badrinath and Tungnath in the mountains of the state of Uttarakhand, and it was near Tungnath that he discovered the incomparably beautiful alpine primrose, *Primula edgeworthii*. Unlike many of his contemporaries Edgeworth did his own collecting, rather than training natives to do the work for him.[20]

From the hill station at Shimla, he also carried out a pioneering journey to several snowy mountain ridges including the Rohtang Pass in Himachal Pradesh. 'Rohtang' means 'pile of corpses', referring to the number of people that die while trying to cross in bad weather. Edgeworth was one of the earliest Europeans, certainly the first Irishman, to botanise on the alpine slopes of the 7,274 m Mana Peak in the same state. Several of his discoveries bear his name, including

Many of these must have been the first introductions of their kind into Europe. No record of how they fared in Maria Edgeworth's Co. Longford garden survives.

In 1846, while returning to India after home leave in Ireland, Edgeworth touched down in the British coaling station at Aden, where he became one of the first Europeans to botanise the region. During a two-hour stroll he discovered no fewer than 11 new species of plants,[21] including the west African *Acacia edgeworthii*. Joseph Hooker was next to botanise the region in December 1847, as he travelled to India, and Edward Madden collected at Aden in 1850, thus these early pioneers blazed a trail for later naturalists.

By 1850 Edgeworth had become Chief of Police in Punjab. He was a friend of Edward Madden and once brought a consignment of Himalayan seeds to Glasnevin on the latter's behalf. He is suitably commemorated by the Indian *Edgeworthia gardneri* and his collections are in the herbarium at Kew.

The Irish botanist M. Pakenham Edgeworth was another pioneering adventurer in the western Himalaya. Like Madden, he was a friend of the Hookers at Kew and of David Moore at Glasnevin.

The Edgeworth coat of arms, here pictured in the town of Edgeworthstown, Co. Longford. The family made a rich contribution to Irish science and literature.

Pakenham Edgeworth is commemorated by the genus *Edgeworthia*. Seen here is *Edgeworthia gardneri*, a shrub whose stems are commonly used in paper-making in the Himalaya.

the lovely yellow-flowered *Impatiens edgeworthii*, an annual balsam from north-west India and the mountains of Pakistan.

A list of seeds 'sent over by Pakenham' survives, and among these are spectacular flowering Himalayan endemics such as *Androsace sarmentosa* ssp. *primuloides*, *Cornus capitata*, *Daphne bholua*, *Elsholtzia fruticosa*, *Ilex dipyrena*, *Skimmia laureola* and *Prinsepia utilis* for example.

Harry Corbyn Levinge

Harry Corbyn Levinge (1828–1896), was another Indian civil servant turned plant hunter and naturalist. He was born into an old Co. Westmeath family, whose seat was Knockdrin Castle, a large neo-Gothic pile once surrounded by extensive gardens. His special interest was in ferns, and while collecting in the Kashmir Valley in September 1875 he discovered *Pseudophegopteris levingei*, a pretty carpeting fern, now grown in a number of British and Irish gardens.

The Irish National Herbarium at Glasnevin contains an extensive collection of Indian ferns collected during the late nineteenth century. Pictured below is the epiphytic fern, *Polypodioides amoena*. The specimen on the left was collected by Charles Baron Clarke on Tonglu while those on the right were gathered by Harry Corbyn Levinge on the slopes of Senchal near Darjeeling.

Pseudophegopteris levingei; this lovely fern commemorates Harry Corbyn Levinge who discovered it in Kashmir in 1875. Seen here at Kilmacurragh, Co. Wicklow.

As Chief Engineer and Secretary for the Bengal Public Works Department, Levinge travelled widely, and this is reflected in his fern collection gathered in places as far apart as Dehradun in the foothills of the western Himalaya, the Nilgiri mountains in the Western Ghats of southern India, and Sri Lanka. Ferns were not his only interest, though he collected many in Sikkim during the 1880s while based in Darjeeling as Inspector-General of Police, and in the Nilgiri mountains he discovered a balsam, named *Impatiens levingei* after him..

A spring scene at Kilmacurragh. *Rhododendron* 'Altaclarense' is reflected in the waters of the pond, and beside it, to the right, is an old tree of *Rhododendron arboreum* var. *roseum* of which there are several forms throughout the garden. Kilmacurragh benefited greatly from the work of early collectors in India.

Alas, just as he was preparing to retire to Ireland to work on his enormous collection of Indian ferns, Levinge placed the larger part of it in storage in 'Whiteley's fire-proof warehouse' in London. Perhaps he shouldn't have been so trusting; the building was shortly afterwards the scene of a devastating fire, one of the largest in London's history, and his entire herbarium was incinerated.[22] Fortunately duplicates survive at Kew and in Paris, and a large part of his surviving collection is now housed in the National Herbarium at Glasnevin, Dublin.

Having studied these, it is possible to map his collection areas in present-day Darjeeling District and Sikkim. From Darjeeling, he collected on the Lebong spur below the hill station, at Senchal, Tonglu and along the Singalila Range to Sandakphu. Further north,

in Sikkim proper, he travelled up the tropical Teesta valley, though to collect further afield, Levinge hired a Bhutia man to gather specimens in the Lachen and Lachung valleys of north Sikkim. It was probably the same Bhutia collector who found *Adiantum levingei* at Chungthang, at the head of the Teesta valley.

Levinge returned to Ireland following a long period in India and surrounded the castle at Knockdrin with a large collection of species rhododendrons, which had been raised from seeds collected by him or sent to him from India.[23] One hundred and thirty years later, a scion of his family, the botanist and garden designer, Dr Daphne Levinge Shackleton, travelled as part of our 2015 expedition, to retrace the routes taken by her forebear.

The Roxburgh Memorial at the Calcutta Botanic Garden. William Roxburgh was one of several prominent Garden Superintendents.

guests. From the third floor, where Roxburgh's vast herbarium had once been, we climbed onto the roof and peered across to the sadly derelict 19th-century herbarium building, where William Wright Smith (1875–1956) was Keeper. Smith, later Regius Keeper of the Royal Botanic Garden Edinburgh, moved here in 1907, alongside his orphaned nephew, Roland Edgar Cooper (1890–1962), who studied botany, initially in Calcutta, and later at the Lloyd Botanic Garden in Darjeeling.

In the following years Cooper botanised with his uncle in Nepal, Sikkim, Bhutan and Tibet, and returned to Scotland in 1910, when Wright Smith was at Edinburgh; Wright Smith described the collection of George Forrest, then arriving in the garden's herbarium in great numbers from NW Yunnan.

While in Edinburgh, Cooper studied horticulture, receiving lectures from Professor Isaac Bayley Balfour (1853–1922), though before completing his studies he was invited by Arthur Kilpin Bulley (1861–1942), the enterprising English nurseryman, to collect for him in the mountains of Bhutan.

Bulley was primarily interested in showy plants with horticultural potential, and botanical collection was not a major concern. Cooper's first expedition was to Sikkim in 1913, whence he travelled via Kurseong to Darjeeling, and then north-east to Gangtok where his collector crossed the border into the Chumbi valley into Tibet. Returning to Sikkim he re-traced Hooker's route up the Lachen and Lachung valleys, reaching alpine country in the Yumthang valley.

This trip yielded over 500 packs of seeds and Cooper used a Lepcha collector to assist him, continuing a tradition begun over six decades before by Hooker. He explored central and western Bhutan during 1914–15, becoming the second major plant collector to visit the region after William Griffith's famous journey there in 1837. There he made 3,958 collections, a staggering number, and he introduced many of Hooker's Sikkim discoveries to cultivation.

Cooper later joined the army, and many of his herbarium collections were sent for storage at the Royal Botanic Garden, Calcutta; sadly, kept in boxes in an outhouse, they were eaten by insects and

covered in mildew. Some of his collections proved difficult to grow at Edinburgh, as was the case also with many of the collections of the great plant hunters such as Forrest; this was partly due to the fact that so many young, skilled horticulturists were killed on the European battlefields. Had Cooper carried out his work before the Great War his collections might have gained wider popularity.

Meanwhile, during our visit, we saw in the garden's lakes, the giant waterlily, *Victoria amazonica*, which stretched its giant pads across the warm, murky waters. The botanical sensation of the 1830s and 1840s, the original plants of *Victoria amazonica* had been brought out to Calcutta from Kew by the new Head Gardener, Mr Scott, and Hooker was delighted by the fact that by the spring of 1853, no fewer than five plants were blooming there at once.

The giant South American waterlily *Victoria amazonica* stretches its enormous 1.5 m (5 ft) wide pads across the lakes at Calcutta Botanic Gardens. Seen here in November 2012.

Our group with Calcutta's great banyan tree, *Ficus benghalensis*, in May 2015. Back row, left to right: Bruno Nicolai, Derek Halpin, the author, Averil Milligan, Bruce Johnson. Front row: Daphne Levinge Shackleton, Kristin Jameson, Orlaith Murphy, Gráinne Larkin and Lesley Fennell.

Much of Calcutta's colonial architecture survives, with parts of the city resembling the 18th-century architecture of Dublin and London. In the city centre we visited the site of Fort William, the infamous 'black hole of Calcutta', the nearby headquarters and council chambers of the British East India Company and the elegant Victoria Memorial, constructed with white Makrana marble in the early 20th century in the Indo-Saracenic revivalist style, which uses a mix of British, Venetian, Egyptian, Islamic and Mughal elements.

The old Victorian herbarium building at Calcutta, once a thriving centre of science and a base for William Wright Smith and his nephew Roland Edgar Cooper.

Darjeeling — Queen of hill stations

From Calcutta we flew to Bagdogra, a small town in the northern reaches of West Bengal, which afforded us a good view of the Ganges and the Grand Trunk Road, the route taken by Joseph Hooker on the back of an elephant 164 years before. From Bagdogra we crossed the Terai to the foothills at Pankhabari. Trees of *Polyalthia longifolia* var. *pendula*, jacarandas, enormous banyans, coconut palms and teak lined the roads, their crowns laden with epiphytic ferns and orchids.

This was to be the first of four expeditions to India, undertaken by a group of us aiming to retrace the trail taken by Hooker through Darjeeling District and the Sikkim Himalaya in 1848–49. Travelling with me were eight Irish botanists, horticulturists, tree enthusiasts and keen gardeners.

During Hooker's extensive travels through this region, Sikkim consisted of all of present-day Sikkim proper, Darjeeling District and the Sikkim Terai, including Bagdogra. The same region, following Hooker and Campbell's incursions into Tibet in 1849, has now been divided into two separate areas, Darjeeling District, belonging to West Bengal, and Sikkim, which since 1975 has been a state of the Indian Union. Thus in the following chapters dealing with our modern-day expeditions, I refer to historic Sikkim as Sikkim and Darjeeling District.

The sun was setting as we travelled from Pankhabari on the edge of the tropical plains of West Bengal and made our initial ascent of the Himalayan foothills towards the former hill station of Darjeeling. Our route took us across the Sikkim Terai, which in Hooker's time had been a vast forest, although shortly after his visit most of the trees had been felled to make way for newly introduced tea. We stopped briefly in the tea gardens, instantly recognising the hills Hooker sketched when he passed though the region, and drove along the same winding road that he travelled on a pony.

The drive from Bagdogra to Darjeeling took six hours along narrow steep roads with hairpin bends, and it was night by the time we arrived. There we were met by Sailesh Pradhan, the Gangtok-based nurseryman, whom I had employed as our guide and coordinator. Sailesh is the third generation of a family of famous naturalists, his father, Keshab Pradhan, being India's leading plantsman, who was awarded the Royal Horticultural Society's Veitch Memorial Medal (VMM) in 2011.

Keshsab's father, Rai Saheb Bhim Bahadur Pradhan (1895–1975) had been Sikkim's Forest Manager during the earlier part of the 20th century, and through this had known the British plant hunters. Keshab followed his father's example, following a career in forestry, and as a youngster his talents were spotted by the king of Sikkim, who made him a Royal Protégé, and sent him to America to pursue his studies at the Yale School of Forestry.[1] A friend and advisor to the last king and queen of Sikkim, he became an authority on the orchids, primulas and rhododendrons of the Himalaya and the author of several good books on these topics.

Sailesh follows in this great horticultural tradition, running a large wholesale nursery specialising in orchids (over 150 species), *Citrus* fruits, *Hemerocallis*, heat-tolerant rhododendrons and azaleas (many of which he hybridises himself) as well as other exotics; he also organises botanical expeditions through the Sikkim Himalaya.

We had to wait until dawn to witness for ourselves the scenery that has made the hill town of Darjeeling world famous; no town in the Himalaya is more spectacularly located, with Mount Kangchenjunga, the world's third highest mountain, in the distance.

The famous Sikkim plantsman Keshab Pradhan VMM, seen here at home in Gangtok with his wife Shanti and son Sailesh. The latter acted as our guide through Darjeeling District and Sikkim.

Bagdogra, on the northern fringes of West Bengal. In Hooker's time this area was part of the Sikkim Terai and was extensively forested.

St Paul's and a pilgrimage to Brian Hodgson's bungalow

One of the most important surviving colonial buildings in Darjeeling is 'Brianstone', the bungalow that once belonged to Hooker's friend, the great naturalist, Brian Hodgson. His house is now the Rectory of St Paul's School, an exclusive independent boys' school, which has educated members of several Asian royal families. Just beside the bungalow is a pair of fat-trunked trees of *Cryptomeria japonica*, brought to Darjeeling by Robert Fortune when he visited the hill station.

The school was just a ten-minute walk away from our hotel, along a steep pathway through woods of *Alnus nepalensis* that led directly to the bungalow and school quadrangle. The gardens were immaculately kept and planted with a mix of native and exotic trees. Epiphytic plants of *Hedychium spicatum* scaled the trunk of a leaning *Cryptomeria*, while *Ilex sikkimensis* acted as a host for *Agapetes serpens* and a number of spectacular flowering orchids like *Epigeneium rotundatum* and *Cymbidium longifolium*, with pendulous, densely packed spikes of luminous lemon-yellow blossoms.

Agapetes serpens is a common epiphyte at this altitude throughout Darjeeling District and the lower valleys and hills of Sikkim. It was in full flower during the time of our return visit to St Paul's in April 2015, its long arching stems smothered in scarlet lantern-like blossoms.

The spectacular setting that has made Darjeeling so famous. Kangchenjunga raises her great snowy peaks over the hill station while the valleys below plunge several thousand feet deep into a tropical vegetation zone. November, when the post-monsoon rains have cleared, is the best time to see this magical scene. No other Himalayan town commands such an arresting view.

Discovered by William Griffith in Bhutan in 1838, it was described as a new species by the India-based, Scottish botanist Robert Wight in 1847, just before Hooker's departure for India. Its name is derived from the Greek *agapētos* meaning beloved, desirable or loveable, alluding to its spectacular appearance at flowering time.

Ilex kingiana also grew close by, forming a small tree with spectacularly large leaves and dense fascicles of ripe, red fruits. Joseph Hooker originally named it *Ilex insignis* in 1875, not realising that the name had already been applied to a fossil species. *Ilex kingiana* is hardy in the milder coastal parts of Britain and Ireland and is named for Sir George King, a former Superintendent of the Lloyd Botanic Garden. Labels on Hooker's specimens at Kew state that he collected it at Darjeeling, and it is highly likely he found it in the forests surrounding Brian Hodgson's house.

'Brianstone', the house in which Brian Hodgson once lived, framed by Robert Fortune's *Cryptomeria japonica*. Joseph Hooker stayed here and discovered a new species of fungus on the verandah.

Eton in India. St Paul's School, Darjeeling, with the backdrop of the Kangchenjunga massif. I gave a lecture about Joseph Hooker's exploits in India and his Sikkim rhododendrons in Irish gardens in the building to the right in November 2012.

One of Hooker's discoveries, the exotic *Brassaiopsis mitis*, pictured here near Brian Hodgson's bungalow. Its foliage is variable in the wild and the finest forms are spectacular.

Perhaps the most exciting plant, certainly in terms of foliage, growing in the grounds of St Paul's, was the showy *Brassaiopsis mitis*, a small tree bearing enormous palmate leaves similar to those of the rice paper plant, *Tetrapanax papyrifer*. Hooker discovered this exotic foliage plant in the lower Lachen valley in August 1849, and it has recently been introduced to cultivation although it is hardy only in the warmest, most sheltered parts of Britain and Ireland.

To the rear of the Anglican church belonging to the school, we matched the scene of the great snowy mass of Kangchenjunga, sketched by Hooker from the verandah of Hodgson's bungalow, and were interested to see that even some of the trees of *Rhododendron arboreum* included in the sketch by Hooker had survived the passage of time.

During our first visit to St Paul's in November 2012, we arrived to hear students singing 'Silent Night', and in November 2014, prayers were offered for our safe return from the Goecha La and the snows of Kangchenjunga; the influence of the Raj and the colonial era still certainly lingers in this remote and enchanting corner of India.

In 2012 I gave a lecture in the school, about Hooker and his Sikkim *Rhododendron* collections in Irish gardens. The event was packed and members of the Indian press attended, with the result that news of the Irish expedition made it into seven national newspapers. The Rector of the school hosted a reception for our group afterwards, in the drawing room of Hodgson's house, and I couldn't help thinking that this was the very room in which Hooker had planned his travels in 1848 and 1849, and that on the verandah outside he had discovered several new species of fungi.

Cymbidium longifolium, a spectacular epiphytic orchid seen here on *Ilex sikkimensis* close to Brian Hodgson's house.

Agapetes serpens, a common epiphyte in the Darjeeling area. Seen here beneath glass at the Royal Botanic Garden Edinburgh.

The view from Brian Hodgson's house framed by the rhododendrons depicted in Hooker's *Himalayan Journals*.

Around Darjeeling

Darjeeling retains much of its colonial charm and some of its Raj era architecture survives today. The population of the town nowadays numbers over 110,000, placing substantial pressure on fragile natural resources, though it still remains spectacular. The narrow streets are crowded with a range of Himalayan faces; Nepali, Tibetans, Bhutanese and Sikkimese predominate.

The best time to visit Darjeeling is after the monsoon (late October and November) when the rains and mists dissipate, the climate is beautifully warm and the panorama of immense snowy mountains is strikingly clear. We visited the area three times during the November period, and were richly rewarded with awe-inspiring mountain views, and latterly in late April, when we had a fleeting glimpse of Kangchenjunga through the pre-monsoon mists, but were rewarded by the spectacular colours of *Rhododendron* flowers in the valleys.

The town straddles steep slopes overlooking the Kangchenjunga massif and even the street-side embankments were awash with colour from both native and introduced plants. In November the skies become gentian-blue and one of Darjeeling's most spectacular introduced flowers opens its glorious blossoms. Anyone who has visited Darjeeling will forever associate the tree dahlia, *Dahlia imperialis*, with it. Native to the mountains of Mexico and central America, where it can reach 10 m tall on bamboo-like stems, it carries enormous panicles of nodding lavender blooms in late autumn.

Many of the older buildings are in a state of decay, and silky-fawn plumes of *Miscanthus nepalensis* sprang from the crumbling face of the clock tower of the town hall and from the gutters of surrounding houses. Several climbers sprawled their way through the street-side thickets including *Clematis buchananiana*, another autumn bloomer, which, was smothered in creamy-white, lantern-like blossoms. With

The Mall, Darjeeling. The hill station retains much of its colonial architecture and is a melting pot of ethnic groups.

Dahlia imperialis, a popular flower at Darjeeling and in the warmer parts of Sikkim. It flowers spectacularly in November once the monsoon season has passed.

Feathery plumes of *Miscanthus nepalensis* springing from the face of a clock tower in downtown Darjeeling.

as Inspector-General of Police. Hooker often visited the house when it was occupied by Campbell and his wife (for whom he named *Rhododendron campbelliae*), and was godfather to their daughter Josephine, who was born during the hostage crisis at Tumlong.

Campbell's house lay on a ridge below the town centre and it is still the residence of the Chief of Police for Darjeeling District. He very kindly allowed us to visit, and we had high tea on the lawns of the house beside what is said to be the very first tea plant brought to Darjeeling and planted by Campbell himself. One of my travelling companions, Helen Dillon, spotted a sizeable clump of the pretty Chinese *Primula malacoides* in bloom by the gates to the house, a relic of cultivation no doubt, and at the end of the lawn we were pleasantly surprised to find an old tree of *Magnolia campbellii*.

When Joseph Hooker arrived at Darjeeling, Campbell's magnolia was the predominant tree on the hills surrounding the little hill station. Within two decades they were all but gone, cleared to make way for ever-expanding tea gardens and to provide timber for tea chests. Today emerald green tea gardens surround the town, with millions of clipped knee-high bushes, on the steep slopes that fall into the deep river valleys beneath. Darjeeling produces the champagne of teas, but be wary: 40,000 tonnes of Darjeeling tea are sold annually, but only 10,000 tonnes are actually produced!

it we found the climbing herbaceous *Dicentra* relative, *Dactylicapnos roylei*, then bearing a heavy crop of fleshy violet-coloured seed capsules.

In many parts of Sikkim, *Bergenia ciliata* is commonly planted into embankments and through pockets in stone walls for soil stabilisation, and in the mild, damp Darjeeling climate it grows with amazing vigour, forming large cabbage-like clumps.

One of the highlights of our 2013 expedition was a visit to 'Campbell's Cottage', the house in which Archibald Campbell once lived, and later occupied by the Irish amateur botanist, Harry Corbyn Levinge, when he was based in Darjeeling during the 1880s

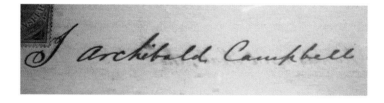

Campbell's signature. From a document we spotted in the Windamere Hotel, Darjeeling. His first name is often debated; some say he was Arthur, though this document resolves that question.

Our group at Campbell's Cottage. Formerly called 'Beechwood', the house has been modernised, though a European-style Victorian structure lurks beneath.

Primula malacoides, a Chinese native that has naturalised itself in the vicinity of Archibald Campbell's house.

Exotic spices from India and the foothills of the Himalaya, a scene from Darjeeling's colourful market.

The Lloyd Botanic Garden

The Lloyd Botanic Garden was established in 1878 by Sir Ashley Eden (1831–1887), Lieutenant-Governor of Bengal during the British Raj era. William Lloyd (then owner of Lloyd's Bank), gave 16 ha for its establishment and the grounds were laid out by the garden's first Superintendent, Sir George King (1840–1909). The gardens were established to act as an annexe to the Royal Botanic Garden, Calcutta, allowing that garden the opportunity to grow and study plants needing a temperate climate.

One of the garden's most energetic Curators was George H. Cave (1870–1965), who graduated as a Kew-trained gardener in 1895. He accepted a post at Calcutta in 1896, and in 1904 was appointed Curator at the Lloyd Botanic Garden. Over the following decades he explored and collected across much of Nepal, Tibet and Sikkim.

During his time at Darjeeling, the botanic gardens rose to prominence and Cave based his restoration of the gardens on the arrangement of the Royal Botanic Garden Edinburgh. He employed Lepcha collectors to search the mountains to the north to collect herbarium specimens that were later sent to Kew and Edinburgh, and seeds that were distributed to major botanic gardens located in temperate regions of the world.[2]

In 1909–10, George Cave and William Wright Smith from the RBG Calcutta spent a considerable time collecting in the mountains of Sikkim, where they made several discoveries, including a number of *Primula* species. Wright Smith often brought his nephew, Roland

George H. Cave, Curator of the Lloyd Botanic Garden. He employed several Lepcha men to collect for the garden and its annual seed list. He is commemorated by *Primula caveana*, which ironically is often found growing in the mouths of caves in north Sikkim.

Edgar Cooper, who was later to re-employ Cave's Lepcha collectors when he began plant hunting for A. K. Bulley in Bhutan (see p. 57).

Cave is still remembered fondly by older people at Darjeeling, and he was a colleague and friend of Sailesh Pradhan's grandfather Rai Saheb Bhim Bahadur Pradhan (see p. 59).

The Lloyd Botanic Garden supplied an exciting range of seeds on an annual basis, and in 1911 the garden's annual report stated that their staff had made an exceptional collection, particularly of hardy species of *Rhododendron* and *Primula*, along the western side of the Chola range, the mountain barrier between Sikkim and Tibet.[3] In 1926 the Curator went on a collecting expedition to Sikkim and the garden's Lepcha collectors went as far as the Nepal frontier in a quest for Himalayan seeds. In the early 20th century Glasnevin received hundreds of packs of seeds annually from the Lloyd Botanic Garden.

Darjeeling could be a difficult place to trial new plants. The rainfall varied from year to year from an annual high of 390 cm to a low of 220 cm, while torrential spring hail storms smashed young plants to pieces. Most of this precipitation fell during the monsoon season, which meant that European fruits such as apples and peaches failed to ripen.

The gardens were extremely popular in their heyday. For example in 1917, 16,205 Europeans and 35,730 Indians passed through the gates, to enjoy what was then one of the best-stocked gardens in

Pinus roxburghii, a superb specimen in the Lloyd Botanic Garden. It was probably planted during Cave's era as Curator.

The central glasshouse at the Lloyd Botanic Garden surrounded by colourful Mexican *Tagetes erecta*. The structure is based on Richard Turner's Great Palm House at Kew.

Rhododendron grande near the summit of Senchal. To the left, is the hill station of Darjeeling spilling down a northern spur. Above it rise the snowy peaks of the Kangchenjunga massif.

Tiger Hill and the Senchal Range

We climbed Senchal twice, firstly in November 2012, and again during the pre-monsoon mists of late April 2015. The November visit was certainly rewarding; the summit of Senchal (2,590 m) rises high above the town of Darjeeling and thus the view of the Himalayan range is greatly increased. Beneath a perfectly cloudless sky we came to admire a 250 km sweep of immense mountains beginning in the far west with a glimpse of the Nepalese peaks of Lhotse (8,501 m), Mount Everest (8,848 m) and Makalu (8,485 m), the fifth highest mountain on the planet.

Beyond there the scene continued with the Tibetan peak of Chomolonzo (7,804 m), before crossing back into Sikkim where the entire sweep of the Sikkim Himalaya from the snowy mass of Kabru on the Nepal–India border to the ice fields and glaciers on Pauhunri and the Dongkya La in the north-east of Sikkim.

The middle distance was dominated by the town of Darjeeling, just 11 km away. It spilled down the steep slopes of Senchal's northerly spur, and beyond there lay the immensely deep gorge of the Great Rangit River, where, beneath the snows of Kangchenjunga, figs, wild bananas and other tropical plants luxuriated. The upper forested slopes form Senchal Wildlife Sanctuary, covering an area of 39 km², with elevations ranging from 1,100 to 2,600 m and providing a habitat for barking deer, Himalayan black bear, leopards, jungle cats and the Himalayan flying squirrel. The avifauna is also quite rich, and Senchal is home to *Anthus hodgsoni*, Hodgson's pipit, a pretty little bird with long migration routes. The annual rainfall of the upper slopes of the mountain averages 270 cm, and two lakes supply drinking water to Darjeeling.

Senchal is the oldest protected area in West Bengal (a sanctuary was established there in 1915), and the mountain is important because it is the type locality for a large number of species collected over the past 160 years by various visiting botanists and naturalists.

A meadow of *Anaphalis margaritacea* near the summit of Senchal. Forest still caps the higher mountain slopes.

India. By the early 20th century, the Annual Report for the Calcutta and Lloyd Botanic Garden stated that 'for its size, the Lloyd Botanic Garden is one of the richest and best in India as the elevation of Darjeeling is not so high as to make the introduction of some of the lower elevation Himalayan species impossible, nor too low for the introduction of some of the choice species that find their best development at greater heights. Of special interest this year has been the Curator's success in growing species of *Meconopsis*.'[4] By then the gardens attracted 100,000 visitors annually.

We visited the Lloyd Botanic Gardens on three occasions, which gave us a good introduction to the flora of Darjeeling District, though it was sad to see how this once great garden has been allowed to slip into serious decline. In spite of this, fine trees planted in the late 19th century survive, including a magnificent chir pine, *Pinus roxburghii*, and several good trees of *Magnolia cathcartii*, an evergreen species collected by Hooker at Lebong, a mountain spur 300 m below Darjeeling. Hooker named it for John Ferguson Cathcart, the Calcutta judge, who retired to Darjeeling and had his Indian artist paint it.

The orchid house was one of the few areas that brimmed with colour, particularly native species such as the beautiful white and yellow flowered *Cymbidium mastersii*, which is commonly found in the warm damp *Rhododendron* forests below Darjeeling. The specific epithet commemorates John White Masters (c.1792–1873), Head Gardener of the Calcutta Botanic Garden between 1836 and 1838.[5]

However, one of the greatest challenges to the area is habitat degradation, and we witnessed large-scale damage over an extensive area during our visits.

We returned to Senchal in the spring of 2015, when the summit lay enveloped in a dense mist, but when much of the surrounding vegetation was in flower. One of my travelling companions, Dr Daphne Levinge Shackleton, was there to retrace the route of her forebear, the Irish pteridologist Harry Corbyn Levinge, who was based in Darjeeling (see p. 53) during the 1880s. During that time, he made a large collection of ferns from the Sikkim Terai, immediately around Darjeeling, on Senchal and on the slopes of Tonglu. To collect further afield, he employed a Bhutia man, who gathered specimens in the Lachen and Lachung valleys, near the Tibetan border. Levinge's herbarium of Indian ferns is now housed in the National Herbarium at Glasnevin and his Sikkim fern collections are full of interest, particularly those from the Sikkim Terai, since those forests have long since been felled.

One of Levinge's ferns, *Pseudophegopteris levingei*, discovered by him in Kashmir on 25 September 1875, was finally introduced to cultivation in 1981. It is a pretty little fern, of dense habit and given its ease of cultivation it has become popular among fern enthusiasts in Britain and Ireland. It was first grown in Ireland at Beech Park, Co. Dublin, the remarkable garden of Daphne's father-in-law, David Shackleton (1923–1988), who was regarded by many as the greatest plantsman of his time in Ireland.

Drepanostachyum intermedium, a common bamboo in the Darjeeling area.

Levinge's herbarium was catalogued, restored and remounted in 2014 by Dr Noeleen Smyth, a conservation botanist based at the National Herbarium at Glasnevin, and from his original specimens, I was able to establish where he and his Bhutia collector had explored. He had made a rich collection on Senchal, and having seen and studied his dried fern specimens in Dublin, it was thrilling to see their living counterparts dripping in the pre-monsoon mists, there in the Himalayan foothills where he had romped and explored a good 130 years before us.

Given its proximity to Darjeeling, much of the mountain's forested areas have been felled over the past 150 years, though patches of old growth forest remain near the summit, with vast areas of regenerating secondary forests on the lower slopes. On the summit itself the most conspicuous plant was the western pearly everlasting daisy, *Anaphalis margaritacea*, whose dried stems filled the meadows creating vast sheets of colour, and contrasting beautifully with the silky seedheads of *Miscanthus nepalensis*.

The suckering bamboo, *Drepanostachyum intermedium*, framed the view of distant snowy mountains. This species was described from a flowering shoot collected by Joseph Hooker, and while no locality is recorded for his specimen, it was probably found by him near Darjeeling where it is locally abundant.

Its bedfellows included *Osbeckia crinata*, sporting spectacular coppery autumn foliage, *Hypericum uralum*, bearing masses of delightful golden-yellow blossoms on arching stems, and *Rubus lineatus*, another handsome foliage plant with silky-hairy stems and striking five-foliate leaves that are silvery-downy beneath. Though widespread in this part of the Himalaya, it was not introduced to cultivation until 1905 when George Forrest sent seeds from China.

Though the views are obscured by heavy cloud and mist, spring is an altogether more exciting season for visiting plantsmen and women, though they should beware of leeches! Our April 2015 visit coincided with peak flowering season for rhododendrons and many other plants,

Miscanthus nepalensis, one of the most beautiful grasses of the Darjeeling area. It is lower-growing than its Chinese counterpart, *Miscanthus sinensis*.

Osbeckia crinata, an extremely handsome shrub with good autumnal foliage and showy fruits. The genus commemorates Pehr Osbeck (1723–1805), a Swedish clergyman and a student of Linnaeus, who travelled to India and China.

it survived for several years, but cold winters, which hit the coastal parts of Ireland once every three to four decades, decimated it and later replacements.[6]

Growing close by, and very similar to Lady Dalhousie's rhododendron, were occasional 1.5 m tall bushes of *R. lindleyi*, springing from rock faces and smothered in large, fragrant, white blooms. Hooker first found it near Darjeeling (probably Senchal) in 1848, though he confused it with *R. dalhousieae*, which meant the species wasn't described until 1864, when the English gardener and botanist Thomas Moore (1821–1887) studied a plant in Standish's nursery at Ascot, that had been raised from seeds collected in Bhutan by Thomas Jonas Booth (1829–1879).

Moore named this plant for Dr John Lindley, botanist and secretary to the Horticultural Society of London (now the RHS), who used his influence to have Moore appointed as Curator to the Apothecaries' Physic Garden at Chelsea. There he succeeded Robert Fortune, who had left London to smuggle tea plants from China for the British East India Company.

R. lindleyi is by no means common in European gardens and succeeds best in the mildest coastal gardens of south-west England, western Scotland and Ireland. Perhaps the finest plants in Britain and Ireland grow at Mount Stewart in Co. Down. These were raised by the Marchioness of Londonderry, Edith Vane-Tempest-Stewart (1878–1959), the noted society hostess and creator of this celebrated garden on the Ards Peninsula.

Rhododendron dalhousieae, one of the most regal members of this showy genus. Hooker thought it worthy enough to name for the wife of the Governor-General of India.

and we came back down the mountain slopes that evening drenched by the pre-monsoon showers, but having seen some of the most spectacular floral displays the eastern Himalaya can offer.

Rhododendron griffithianum is one of the rarer of the Sikkim species, though on the lower slopes of Senchal it was locally abundant when we were there; what's more, it was in flower, the first rhododendron species we were to see on this particular trip, causing us great excitement. It is variable in the wild, in both flower and bark colour. The form growing here bore pure white blossoms with large, decorative pink-tinged calyces.

It was *R. dalhousieae* that stole the show however. I had never seen this species in bloom either in cultivation or in its native haunts and it didn't disappoint; Lady Dalhousie's rhododendron was probably one of the most exciting plants we were to see on all four of our expeditions to Sikkim and was common growing on rocks on the edge of *Lithocarpus pachyphyllus* forest.

I could see why Hooker named it for the wife of a Governor-General of India, as this species is the aristocrat of the entire genus; we were fortunate to see it at its spectacular best. The individual flowers are enormous, particularly when compared to the plants carrying them, which were no more than 1.5 m tall. The lime-green flowers of the Senchal form were larger than the creamy-white ones typically found in cultivation.

Sir Frederick Moore stated that this species was not completely hardy at Kilmacurragh. When first introduced (from Hooker's seeds)

Much of the original forest cover has been removed from Senchal. Seen here is a reservoir that provides drinking water for Darjeeling. The pyramidal trees to the left are *Cryptomeria japonica,* which is widely planted around the hill station.

Arisaema erubescens, one of the most striking of the Himalayan aroids, seen here on the lower slopes of Senchal's Tiger Hill in late April, 2015.

She was one of the primary sponsors of the great plant hunters, particularly George Forrest and Frank Kingdon Ward, and the many clones of *R. lindleyi* growing at Mount Stewart today were raised from Kingdon Ward's Himalayan collections. The finest of these is now over 4.5 m tall, and, in May of each year, carries masses of enormous white, pink-tinged blossoms that scent the air around it over a considerable distance. It has a rather straggly habit, but when seen in full bloom few will debate its position as one of the very best of this showy genus.

As have seen (p. 26), Hooker had been most impressed by the cobra lilies (*Arisaema* sp.) on this mountain, and sent corms to Kew. I have seen these plants in other parts of the Himalaya and China, but never in greater numbers than around Darjeeling and in the Lachung valleys in north-east Sikkim. They grew in millions, with a vigour unmatched in cultivation, and were the most conspicuous plants in woodland on Senchal.

*Arisaema erubescen*s produced an enormous parasol-like leaf protecting a 30 cm tall cobra-like flowering spathe. Much more handsome and exotic was *A. speciosum,* with its rather sinister-looking white-striped purple spathes and enormous trifoliate leaves. I don't know why we don't grow this species more in the milder gardens of Britain and Ireland, and it was a thrill to see it *en masse* in a wild setting.

With the cobra lilies grew the delightful little daisy, *Ainsliaea aptera,* one of the least-known woodland perennials native across the entire Himalaya from Afghanistan to Bhutan. It was in Bhutan that I first saw it, in the autumn of 2014, in a forest of *Pinus wallichiana,* beneath the famous Tiger's Nest monastery near Paro, and it was a treat to re-encounter it near Darjeeling the following spring. On

The witch hazel relative, *Exbucklandia populnea*, one of Hooker's favourite Sikkim trees. On Senchal it was a common component of regenerating secondary forest.

Eriobotrya hookeriana, the second species found on Senchal. It forms a dense habited, pyramidal-shaped tree with striking new red growths.

Pilea anisophylla, a striking foliage plant that grew beneath the forest canopy in damp shade. The genus is distinguished from other members of the Urticaceae by its opposite leaves.

Senchal the upright spikes bore masses of pendent rose-tinted, spidery white flowers. *A. aptera* was discovered by John Forbes Royle in the mountains of NE India in the early 19th century; the genus commemorates Sir Whitelaw Ainslie (1767–1837), an assistant surgeon in the British East India Company and author on *materia medica*.

It was also exciting to find Wallich's milk parsley, *Selinum wallichianum*, a Himalayan perennial, which in recent years has become popular in British and Irish gardens for its broad corymbs of milky-white blossoms. It was too early in the season for flowers, though its finely dissected foliage, however rank-smelling, made a wonderfully striking contrast with the various cobra lilies.

These grew beneath a canopy of *Daphniphyllum himalayense*, *Exbucklandia populnea* (with new copper-flushed foliage) and the loquat-relative *Eriobotrya petiolata*, which was discovered by William Griffith in Bhutan. Hooker described it from Griffith's collections, and also from those he made on Tonglu. I have never seen it on Tonglu, though it is abundant on Senchal and large old trees are commonly seen in the older parts of Darjeeling. It is a spectacular foliage plant, with large bold elliptic-lanceolate leaves up to 22 cm

long. I don't know of this species ever having reached cultivation, but it would probably be far better adapted to the cool, moist, yet mild conditions of coastal Irish and British gardens, than the heat-loving common loquat, *E. japonica*.

Another species, *E. hookeriana*, was common higher up the mountain slopes, and it too has much garden potential if ever introduced. It forms a tight-habited, upright, pyramidal tree to about 10 m, bearing striking copper-red, newly emerged lance-shaped leaves. Though named for Hooker, this was yet again another of Griffith's 1836 Bhutan discoveries; Hooker found it by the temple at Sanga Choeling in January 1849.

In open woodland glades and by the sides of the track we spied numerous diminutive perennials, one of the prettiest being the endangered *Viola sikkimensis* (*V. hookeri* Thomson) ex Hook. f.) bearing white, violet-flushed blossoms, with a central venation of a similar hue. With it grew one of the prettiest Himalayan wild flowers, the blue pea, *Parochetus communis*, and *Pilea anisophylla*, a striking herbaceous nettle-relative with silvery-marbled leaves.

The latter made a wonderful contrast with *Hydrangea robusta*, then sporting freshly emerged soft hairy beetroot-coloured foliage on long brilliant-red petioles. The purple-blue lace-cap flowers are lovely in autumn, though even out of flower this is a handsome species with very striking foliage.

However much we admired the *Hydrangea* it was soon surpassed when we stumbled upon one of Hooker's most curious discoveries, the strangely alluring *Helwingia himalaica*, then bearing clusters of peculiar tiny purple-brown flowers on the upper surface of leaves. Due to a strange evolutionary development causing the pedicel to fuse with the petiole and leaf midrib, the flowers and

Hydrangea robusta, the newly emerged foliage is particularly striking. Despite the felling of much of its original forest cover Senchal still preserves a rich and very interesting flora.

Helwingia himalaica, one of the most curious of Himalayan shrubs. Hooker also recorded seeing it on Senchal.

Begonia cathcartii. Hooker named this charming species for his friend at Darjeeling, a retired Bengal judge with an interest in botanical art.

later, the fruits, appear in this rather unique position, and while interesting, the genus is probably of more botanical interest than of horticultural value.

Begonias are well represented in the flora of Darjeeling District, and it was a real treat to find the pretty little white-flowered *Begonia cathcartii* in fruit near the helwingias on the lower slopes of Senchal, and in flower, the following day, on the lower slopes of Tonglu. It was described from Hooker's dried Darjeeling specimens, though William Griffith had previously found it in Myanmar, where the Veitchian plant hunter Thomas Lobb (1817–1894) also found it in the late 1850s.

Hooker named it for John Ferguson Cathcart, whose Indian artists painted this species. Kew's botanical artist, Walter Hood Fitch, re-worked his Indian counterpart's depiction and a superb image appeared in *Illustrations of Himalayan Plants* (1855), where Hooker claimed it to be the hardiest of the Sikkim species and also that it was not uncommon in the woods near Darjeeling, a claim that holds true to this day.

Tonglu and an excursion into East Nepal — an autumn visit

With modern roads in place Mount Tonglu is now within easy reach of Darjeeling, and we visited it twice, first in the autumn of 2012, to see for ourselves the spectacular views of Everest and Kangchenjunga from the upper slopes, and again in late April 2015, when we returned during flowering season. The following is an account of our late autumn visit.

On the morning of 18 November we left Darjeeling for Tonglu, first explored botanically by Joseph Hooker in May 1848. Hooker always regarded it as one of the great highlights of his visit to India, as did we.[7] Hooker's visit put Tonglu on the map, so to speak, and it was to become a place of pilgrimage for later explorers. The English botanical artist, Marianne North, painted several plants on the mountain's summit in 1878, including the stunning blue poppy, *Meconopsis wallichii*, and this painting is still displayed in her wonderful gallery at the Royal Botanic Gardens, Kew.

Tonglu is a good two-hour drive from Darjeeling, along narrow mountain roads and into deep river valleys. *Quercus lamellosa* was one of the most common trees along our route, and its thick, gnarled moss-laden branches were festooned with several epiphytes such as the gorgeous autumn-flowered orchid *Pleione praecox*. Other common

Pleione praecox, the most beautiful autumn-flowered orchid from the eastern Himalaya. We first encountered it on Tonglu and saw it in bloom on several successive expeditions.

epiphytes included *Agapetes serpens* (see p. 61), an ericaceous shrub with a swollen turnip-like base to its lower stem (which stores water in the dry season) and the glorious *Rhododendron dalhousieae* (see p. 25), collected by Hooker on Tonglu in 1848 and introduced by him from Sikkim in 1850.

We drove mid-way up the mountain and soon entered the temperate zone where rhododendrons, to our great delight, soon appeared on the scene. *Rhododendron arboreum* var. *arboreum* grew scattered among other forest trees on the mid-forested slopes, and approaching the summit *R. arboreum* ssp. *cinnamomeum* Campbelliae Group clothed entire slopes; we were told that in spring the shades of white, pink and crimson-red create a dazzling effect that may be seen from miles away.

A view of Kangchenjunga and Everest from the same area of the summit where Marianne North painted her famous scene in September 1878. Tonglu provides the best vantage point from which to view this great stretch of the Himalayan range.

Rhododendron griffithianum, a form with rose-coloured calyces and rose-pink splashed flowers. This species is one of the parents of the spectacular *Rhododendron* Loderi Group.

R. griffithianum was also here, growing with *Sarcococca hookeriana,* the Himalayan yew, *Taxus wallichiana, Stachyurus himalaicus, Rubus lineatus, Hydrangea heteromalla* and *Hypericum hookerianum.* The latter was named not for Joseph Hooker, but for his father, Sir William Hooker. It is a widespread species in NE India and we saw it several times during our travels. Natural rhododendron hybrids also appeared on the slopes, including *R. arboreum* × *R. griffithianum,* forming plants with foliage intermediate between the parents.

Soon the aptly named *Rhododendron grande* made an appearance. I was particularly happy to see this because it is one of my favourite species at Kilmacurragh and the earliest to flower there. Our old tree, collected in this same area 163 years ago by Hooker, has now formed a glorious 12 m mound, and as I write it is smothered in great trusses of creamy-white, pink-tinged blossoms. In his magnificent work

The bark colour of *Rhododendron griffithianum* is highly variable. That of the Hooker plant at Kilmacurragh is a rich cherry-red hue, while that of trees on Tonglu is grey-green.

Rhododendrons of the Sikkim Himalaya (1849), Hooker described the form found on this mountain range as a new species, *Rhododendron argenteum.* On Tonglu and the adjoining Singalila Range plants have a plastered silvery-white indumentum on the underside of the leaves; elsewhere in Nepal, Bhutan and Assam the indumentum is fawn. The Kilmacurragh tree has brilliant silver leaf undersides.[8]

We delighted in the richness of the flora of Tonglu. More and more rhododendrons appeared on the scene, among them the delightful *R. triflorum,* by then of course long out of flower, but its peeling cherry-like bark more than compensating for the lack of blossom. Even better were the thickets of another great favourite, *R. barbatum,* with stunning mahogany bark that peeled in thin strips. Caught in the low November sunlight it made a lovely sight, and on Tonglu it grew with that other garden great, *R. falconeri,* whose leaves in places were up to 45 cm long.

Rhododendron triflorum, one of the best of the many rhododendrons that have beautiful bark patterns on Mount Tonglu. *Rhododendron barbatum* is another.

Rhododendron triflorum, first found by William Griffith in Bhutan in 1838; Hooker introduced it to cultivation twelve years later. Illustration from *Rhododendrons of the Sikkim Himalaya.*

Nearing the summit at 3,072 m, the panorama of the Himalayan range opened before us as we marched through *Rhododendron* and *Magnolia* forest. To the north-west, on the Tibet–Nepal border, rose the snow-clad peaks of Mount Everest, and the enormous snowy mountains pierced the sky as far as the Sikkim–Bhutan border, a sweep of several hundred kilometres across the greatest mountain range in the world.

On a broad, boggy, flat ridge, local farmers were busy harvesting the dried-up remains of a wild meadow that contained thousands of plants of the glorious *Iris clarkei,* another of Hooker's Tonglu discoveries, named for the English botanist Charles Baron Clarke,

Iris clarkei, discovered on Tonglu by Hooker. We spotted it on the same boggy flat where he found it in the autumn of 1848. It is still abundant there.

An hour or so later we had our passports stamped in a tiny border outpost and returned to India.

Good garden plants again appeared on the scene. *Primula capitata* carried a few late flowers and sheltered in thickets of the very charming *Gaultheria hookeri*, a low-growing, evergreen dome-shaped shrub, by then absolutely laden with masses of tiny cobalt-blue fruits. Hooker originally found this in the Lachen valley in 1849, and it was introduced to Western gardens by E. H. Wilson from western Sichuan province, China in 1907 and again in 1926 from Myanmar by Frank Kingdon Ward.

Primula capitata was one of Hooker's finest introductions. He collected it on gravelly banks at Lachen village, and at Yumthang, though William Griffith had previously discovered it in nearby Bhutan. He did however send seeds to Kew in the autumn of 1849, and young plants raised there flowered in the rock border in October of the following year. William Hooker, ever keen to promote his son's endeavours, had the Kew plants painted, and these were featured in *Curtis's Botanical Magazine* (t.4550) for December of that year, speedy progress indeed, for a newly introduced species. The general public loved the little primula, and it has remained a firm favourite ever since.

The best part of the day was yet to come however, when, descending a ridge, we entered a forest almost entirely composed of *Magnolia campbellii* with an undercanopy of thousands of *Daphne bholua*. Across this mountain range, just a stone's throw from Nepal, trees are mainly the very large white-flowered form (*M. campbellii* Alba Group) and not the pink form which is the common form in cultivation.

Magnolia campbellii is, without doubt, the finest of this noble genus of flowering trees and shrubs, and I can only imagine the glorious sight this forest must make when covered in enormous white blossoms in early spring, or the glorious scent that must pervade the woods during the same season when the many thousands of *Daphne* are in bloom.

The russet foliage of *Sorbus hedlundii* with one of Tonglu's heavily forested spurs in the background.

who also gathered it on the same mountainside in 1875.[9] It must have made a stunning show earlier that summer; the flower colour is highly variable and ranges from blue to red-purple and the markings on the falls also show considerable differences.

One of the most abundant, beautiful and conspicuous trees on Tonglu was *Sorbus hedlundii,* discovered there by Hooker in 1848. This fine whitebeam was first cultivated outside the Himalaya at Kilmacurragh in the late 19th century, having been sent there as a seedling from Glasnevin. By the time of our visit its leaves had turned russet brown and the gentlest breeze revealed the brilliant silver of their undersides. Hedlund's whitebeam has always remained rare in cultivation, though there are several trees at Mount Usher, which were raised in the 1920s from the original Kilmacurragh tree.[10]

We reached the summit of the mountain by early afternoon, and had a celebratory picnic; it was my birthday and I can think of no better place to throw a party. One of our travelling companions, Kristin Jameson, head of the famous Irish distilling family, had brought a bottle of Jameson whiskey to the summit for the purpose, and after this brief, but welcome, surprise we made our descent along a west-facing spur that took us across the border into Nepal.

We were surprised to find a shop on the summit of Tonglu. It lies just across the frontier in Nepal.

As I write, several plants of *Daphne bholua*, in the guise of that superb cultivar 'Jacqueline Postill', are in bloom here at Kilmacurragh, near our old tree of *Magnolia campbellii,* reminding me of the wild woods on Tonglu. The *Magnolia* is swelling its buds in anticipation of another marvellous show. It bore its first blossoms here in 1907 ,after a 31-year wait. The Kilmacurragh tree was planted in 1876, dating it to the third introduction of live plants by Sir George King, of the Calcutta Botanic Gardens. It is likely that it first started life in the mountains surrounding Darjeeling, where pink-flowered forms also occur, and travelled to Europe as a seedling in a tightly sealed Wardian case.[11]

The daphne belongs to a group of closely related species occurring in the Himalaya, whose bark has long been used in the production of rope and in paper making. The rather strange-sounding specific epithet, *bholua*, is derived from the colloquial name *bholu swa*, as recorded by Francis Buchanan-Hamilton, when he made the original collections in Nepal, and it appears that the name was most commonly used by the Newar tribe who live in the Kathmandu area.

Brian Hodgson described the Nepalese and Tibetan method of paper making in the *Journal of the Asiatic Society of Bengal* in 1832; the bark was boiled with wood ash for half an hour till the stems became soft. They were then pulped in a mortar with a wooden pestle, before being finally floated out in water, where the mucilage was collected in sieves and dried as sheets of paper. A final refinement was to lay out the paper and smooth it with a conch shell. English engravers in the early eighteenth century believed this *Daphne* paper afforded a finer impression than British paper of the time. It is thought that paper making reached Nepal from China during the 14 century.

We visited a factory in Kalimpong, West Bengal where paper is manufactured from the stem fibres of *Daphne bholua* and its close relative, *Edgeworthia gardneri.*

Back on Tonglu we dallied a little too long in the magnolia woods and the sun was setting by the time we began our descent; it was time to press on with our travels to Sikkim.

Tonglu in the spring season

While a November visit to Tonglu is certainly rewarding, and worthwhile, it is not the best time to see the mountain's wild flowers, so we returned in April 2015. Tonglu lay enveloped in a deep mist, the deciduous forests had just sprung back to life, and flowers carpeted the forest floor beneath.

Rhododendrons were the most conspicuous feature on the lower, densely forested slopes, particularly the gorgeous *Rhododendron dalhousieae* and occasional plants of *R. lindleyi*, but the most striking species at this altitude was *R. griffithianum*. By no means common in Sikkim, it was relatively frequent on the lower part of the mountain, where its blossoms varied from pure white, to clones (found at higher altitude) with wonderfully pink-tinted blossoms.

The colour of the bark too, is highly variable, the Hooker veteran at Kilmacurragh having a striking birch-red trunk, whereas the bark of trees on Tonglu was of a ghostly grey-green hue. Soon after its introduction to cultivation this species was used by hybridisers, crossing it with the Chinese *R. fortunei*, to form the superb *Rhododendron* Loderi Group. Sir Frederick Moore, the Irish *Rhododendron* expert, persuaded authorities at Kew to name it for Sir Edmund Loder (1849–1920), who raised it at his garden at Leonardslee, West Sussex, in 1901.

As before, cobra lilies abounded, and this time the most dramatic of these was *Arisaema griffithii*. Its broad, curled, rather sinister-looking dark-purple hooded spathe is actually very like an Indian cobra. It is probably more curious than beautiful, though it is a first-class garden plant, and we saw many forms during our travels with leaves wonderfully marbled purple-black along their mid ribs.

Arisaema nepenthoides, another dramatic woodlander, grew in great drifts among the rocks, rearing its dramatic snake-like flowers on brown and purple mottled pseudostems. First found in Nepal by

Viburnum erubescens, a form with flowers beautifully coloured rose at their extremities.

Holboellia latifolia, the most striking climber we encountered during our spring visit to Tonglu. Hooker recorded finding it there also.

Nathaniel Wallich in the early 19th century, Hooker re-discovered it on the upper slopes of Tonglu, and the plants we saw were obviously descendants of those admired by Hooker over 160 years before.

Viburnum erubescens is another shrub abundant throughout Darjeeling District and Sikkim, its flower colour varying considerably from snowy white to warm pink. It was also exciting to find *R. arboreum* ssp. *cinnamomeum* Campbelliae Group in full bloom, with lovely rounded, tightly packed trusses of reddish-purple blossoms. It formed large trees approaching the summit, above the boggy flat known to early explorers as 'Little Tonglu', and in *Rhododendrons of the Sikkim Himalaya* Hooker singled these trees out: 'On the summit of Tonglo it is the prevailing plant, and there, when in full flower, it exhibits a truly magnificent spectacle, gorgeous with scarlet heads of blossom.'[12]

Its bedfellows included *Berberis insignis*, then bearing large, dense clusters of golden-yellow blossoms, *Betula utilis* (introduced by Hooker from Sikkim in 1849), *Lithocarpus fenestratus* and masses of *Pieris formosa*, sporting copper-tinged, fiery new growths. *Rosa sericea* lit the woods with its lovely creamy-white pendent blossoms, the latter dripping from monsoon mist, and on the rocks overhead *Vaccinium retusum* carried masses of pretty urn-shaped red flowers.

Numerous vines ran through the trees, the most spectacular being the striking *Holboellia latifolia*, draped in pendulous racemes of lavender-purple fragrant blossoms. Other good woodland plants

Lost in the mist. Visiting Tonglu in late April meant trekking in the post-monsoon mists. It made for a magical atmosphere.

Daphne Levinge Shackleton pictured here on Tonglu where she re-traced the route taken by an earlier member of her family, Harry Corbyn Levinge.

included *Paris polyphylla*, and, nearby we found, much to the delight of my travelling companion Daphne Levinge Shackleton, *Hymenophyllum levingei*, a delicate filmy fern discovered by her forebear Harry Corbyn Levinge during the 1880s.

Nearing the summit, we encountered entire groves of *Enkianthus deflexus*, a curious form with purple-brown flowers, growing alongside flowering trees of *Rhododendron falconeri*. Hooker based his original description of the species on material from the summit of Tonglu, stating the flowers were white. I always found this strange since all plants in cultivation are a wonderful warm creamy-yellow, but the trees we saw on Tonglu (descendants of the Hooker trees) were in fact an off-white shade. Hooker collected seeds of *R. falconeri* from the

Lagyap valley in east Sikkim in November 1849, though the species (which he named for Hugh Falconer of the Royal Botanic Garden, Calcutta) had previously been introduced to cultivation by Colonel William Henry Sykes (1790–1872) in 1830.

The forests on the summit of Tonglu were felled by farmers in the late 19th century and we traversed open moorland before crossing to the Nepalese side of the mountain. Though this is the drier side of Tonglu, shadowed as it is from the Indian monsoon, we soon entered a dense mist, where only the silhouettes of moss-covered trees were visible. On their trunks were masses of one of Hooker's most beautiful Darjeeling discoveries, the gorgeous white-flowered *Coelogyne corymbosa*, an epiphytic orchid that proved abundant throughout our travels.

The author and Thupden Tsering with an almost white-flowered clone of *Rhododendron falconeri*.

Gathering fodder on Tonglu. Grazing material is always in short supply in this heavily forested area.

Coelogyne corymbosa, we found it growing commonly as an epiphyte or on rocks in the mist-shrouded forests on Tonglu. It is an easy species to cultivate in a cool glasshouse.

Kalimpong — Land of Kings

In November 2014, having spent the previous ten days trekking to the eastern flank of Kangchenjunga, we returned to warmer climes and travelled to Kalimpong, a large bustling bazaar town, 18 km as the crow flies north-east of Darjeeling. Kalimpong sits on a steep ridge at 1,247 m, overlooking the Teesta River, with views towards Kangchenjunga and the Himalaya. The etymology of the town's name means 'Land of Kings'.

In the early 18th century, Bhutan claimed this area from the Sikkim Rajah. The region was then sparsely populated by native Lepchas and by migrant Limbu and Bhutia tribes. Following the Anglo-Bhutan war, the area was ceded to the British East India Company, who promptly began to develop what was then a hamlet into a hill station along the lines of Darjeeling. Kalimpong was of enormous strategic and economic value, lying close to the Nathu la and Jelep La, both major mountain passes leading towards Tibet, offshoots of the ancient silk route.

Kalimpong developed into an important trading centre and in the following years, French, Irish and Scottish missionaries founded schools there; many of these still exist today, and students from all over South-East Asia come to study there, contributing enormously to the town's economy. Like Darjeeling, relics of its former colonial past persist including several European-style buildings and churches.

Kalimpong has long been known for its excellent nurseries, which specialise in orchids and palms, and produce staggering quantities of ginger and cut flowers. The climate of the area is warm-temperate, with annual temperatures ranging from a summer high of 30°C to a minimum of 9°C. Due to the area's proximity to the Bay of Bengal, the monsoon season (between June and September) is severe, causing enormous landslides that very often cut the town off from the rest of India and envelop it in a sea of fog and mist.

We stayed in an upmarket area, on the same road as the residence of the Queen Grandmother of Bhutan (the only Queen Grandmother in the world), Ashi Kesang Choden Wangchuck (*b.* 1930). Curiously, just a few weeks beforehand, we had trekked in the mountains of west Bhutan where we had seen her namesake, the magnificent *Rhododendron kesangiae,* a recently named tree-like species, endemic to Bhutan, where it grows in fir and hemlock forest.

Trachycarpus latisectus — a palm on borrowed time

Our reason for visiting Kalimpong was to study the last wild population of the Windamere palm, *Trachycarpus latisectus,* perhaps the rarest of the Indian palms and a plant with a fascinating history. Kalimpong belongs to Darjeeling District, historically one of the most heavily botanised areas of India, yet all the early explorers missed identifying this palm as being new to science.

The palm gains its common name from the famous colonial-style Windamere Hotel, on Observatory Hill, in central Darjeeling. Established as a boarding house for English, Irish and Scottish tea planters, it was converted into a hotel shortly before the Second World War.

It was the Scottish botanist Henry Noltie who first spotted a pair of trees by a side entrance to the Windamere Hotel, during the Royal Botanic Garden Edinburgh Sikkim Expedition (ESIK) in 1992. Noltie noticed that the palms here, unlike the more commonly encountered Chinese *Trachycarpus fortunei* (which is also planted around Darjeeling), had trunks quite bare of trunk fibre and they were also markedly different to the Indian *T. martianus,* which had been previously claimed to be native to this region of the Himalaya.

On reading a paper about the discovery of this rare palm, I brought a group in October 2013, to see the type specimens at the Windamere Hotel, from which the species was formally described.

Colonial comforts. The Windamere Hotel, the grand old lady of Darjeeling. We visited for High Tea though our purpose was really to study the living 'type' specimens of T*rachycarpus latisectus.*

The Windamere palm, *Trachycarpus latisectus*. The tree pictured here is one of a pair at Darjeeling's Windamere Hotel from which the species was described.

They are handsome specimens, about 8 m tall, and were planted in the early 1970s, having been supplied by a nursery in nearby Siliguri. We noticed several other specimens of a similar height and age planted by colonial buildings nearby.

Trachycarpus latisectus is quite a distinct palm. As previously mentioned, it bears no trunk fibre. The trunks are smooth and grey, and the large fan-like leaves are more shallowly divided than those of *T. fortunei*. Another notable characteristic is the wide leaf segments, some of which are fused for almost their entire length in groups of 2–4.[13]

During the time of the British Raj in India, the range of species of *Trachycarpus* was not fully understood. This species had been collected by earlier plant hunters and botanists, and as earlier stated, it was obviously misidentified as the better-known *T. martianus*, an allied species from Nepal and the Khasia Hills in Meghalaya State, India.

However, a small population of *T. martianus*, from which the English botanist C. B. Clarke gathered specimens in the late 19th century, has been recently rediscovered by an Indian botanist at the village of Runbing (Runbong or 'Rungbong' of Clarke) near the Temi tea plantation in south Sikkim. *T. martianus* is therefore an extremely rare plant in Sikkim and is only found growing wild west of the Teesta River, its most western point of distribution.

The true *T. latisectus*, on the other hand, is found to the east of the Teesta at Rissisom in Darjeeling District, West Bengal. The earliest mention of the genus in this area was made by James Sykes Gamble (1847–1925), an English botanist, who became Director of the British Imperial Forest School at Dehradun. In his magnum opus, *A Manual of Indian Timbers* (1881), Gamble wrote of *Chamaerops martiana*: 'The writer has once found small plants of what is probably this palm on Rissoom (Rissisom), near Dumsong, beyond Darjeeling, at 6,500 feet elevation.'[14] Gamble's *Chamaerops martiana* (an old name for *Trachycarpus martianus*), was actually the elusive *T. latisectus*, and unknown to him he had discovered a new species that would have to await description for over a hundred years.

Armed with this information, we departed Kalimpong on Friday 14 November and drove 25 km east of the hill station, to visit the last remaining wild population of this extremely rare West Bengal endemic. The original forest cover of this area had obviously been felled in the past and had been replaced by naturally regenerating trees such as *Alnus nepalensis* (a pioneer species that aggressively colonises areas following deforestation), *Saurauia napaulensis*, *Leucosceptrum canum* and *Exbucklandia populnea*.

Alnus nepalensis appears to be a good host for epiphytic orchids, and the trees by the roadside carried enormous numbers of *Pleione praecox*, *Pholidota imbricata* and one of my great favourites, the lovely *Coelogyne cristata*. How exciting it is, coming from the cold-temperate climate of Ireland, to see these orchids in the prime of health and *en masse* in their wild habitats, compared to the occasional cosseted glasshouse specimens seen in the botanic gardens of Glasnevin, Kew and Edinburgh.

Leaf detail of *Trachycarpus latisectus*, a young specimen at Mirik Busty, West Bengal.

On the brink. The Windamere palm faces a sad fate. Just about a dozen specimens survive in a tiny area of the Relli River valley. The species faces certain extinction unless action is immediately taken.

Another interesting tree in the Kalimpong area was the spectacular pink-flowered Himalayan cherry, *Prunus cerasoides*. I've seen this species throughout Sikkim, bearing its lovely blossoms in November, but in the Kalimpong region a high proportion of trees bore snow-white flowers. What a pity then, that it cannot cope with the rigours of the British and Irish climate. On the forest floor, the Indian tree fern, *Cyathea spinulosa,* was common and there formed fine specimens; often their upper trunks were hidden by a skirt of dead fronds.

From the roadhead we hired a number of ancient Land Rover-style vehicles. All were several decades old, though eminently suited to the rough gravel tracks that took us down the steep Dumsong Range of hills towards the small village at Mirik Busty. From this tiny village we headed further downhill on foot, much to the curiosity of the local inhabitants, who must rarely see Europeans in those parts.

The climate, despite the lateness of the season, was beautifully warm and we passed tiny houses with gardens full of colourful flowers and vigorous healthy vegetables, including peas, in full flower. Below there, we were led to an enormously deep valley, drained by the Relli River at its base.

The scene before us was both shocking and terribly sad. We stood on an extremely steep stone ridge at 1,400 m, from which virtually all the surrounding vegetation had been stripped by local farmers for feeding their livestock. A few stunted or re-sprouting bushes of *Pittosporum floribundum*, wild mangoes, *Mangifera indica* and the spurge relative, *Phyllanthus reticulatus* were all that remained of a once thriving forest.

Even more shocking was the remaining number of palms: less than a dozen, and we were told by a guide from the nearby hamlet, that a tree had been cut the previous year by a local resident for use as a house post. To study them closely required a dangerous scramble

Many species of *Trachycarpus* favour cliff-side habitats, though the Windamere palm has lost its surrounding forest cover and few of the remaining trees continue to produce fruit.

up a steep, rocky cleft, with little, and at times no, vegetation to cling to, and nothing to break a fall. My heart sank. The few remaining palms looked very unhappy, most were stunted, and none appeared to be producing fruits. Once they would have sheltered from the hot, baking sun in a damp forest, soaked by the Indian monsoon and made humid by its proximity to the Relli River in the valley deep below.

Now that this forest has been destroyed, the last remaining wild trees of the Windamere Palm are doomed, baked dry in the scorching heat. So far, no effort has been made to conserve these rare palms, and it seems as though *Trachycarpus latisectus* may go the way of the dinosaurs.

It is said that later this century the loss of our planet's biodiversity will happen with a cascading effect. This came to mind as I stood on the ridge overlooking these few sad palms. People are desperately poor in this area, and sadly, at the end of the day, people come before plants. Habitat destruction and excessive seed collection may yet bring *T. latisectus* to the end of its line, but I hope someone fights its cause. Ideally the area should be protected, become a reserve and be re-afforested with species that formerly grew on the ridge; once a canopy is established, young plants of the Windamere palm might yet again thrive. If the Lloyd Botanic Garden at Darjeeling was to be reinstated as a fully functioning botanic garden, it could play a vital role in protecting the flora of Darjeeling District, which, with a burgeoning population and other threats that come with global warming, is under ever-increasing pressure.

The good news is that there still, just, time to save the Windamere palm; *T. latisectus* is cultivated in the surrounding area, and certainly we saw robust specimens, heavy in fruit by local houses. It is also planted occasionally in Kalimpong and, as previously mentioned, there are old trees in Darjeeling. Thus there may still remain populations with a good deal of genetic diversity in cultivation. From these, young plants might be raised for repatriation into the wild. The Windamere Palm is now on the brink of extinction; let us hope that the Indian authorities realise its importance and act before it is too late.

CHAPTER 5

A PIONEERING ADVENTURE IN EAST NEPAL

In the autumn of 1848, Lord Dalhousie, as Governor-General of India, contacted the Sikkim Rajah, Thsudpud Namgyal (1785–1863), requiring him to grant Joseph Hooker honourable and safe passage through his realm. Dalhousie's request met with an uncompromising refusal, so pending further negotiations, Archibald Campbell wrote instead to the Nepal Rajah requesting permission for Hooker to explore the Tibetan passes to the west of Kangchenjunga, in east Nepal.

Colonel Charles Thoresby, the East India Company's British Resident at the Nepal Court, also used his influence with the then Prime Minister, Jung Bahadur (1816–1877), and permission was granted. In October 1848, a guard of six Nepali soldiers and two officers were dispatched from Kathmandu to Darjeeling, to escort a jubilant Hooker to any part of east Nepal he wished.

Hooker's ambitious and pioneering route, previously unexplored by Westerners, was to take him up the valley of the Tamur (Tambur) River, which originated on the glaciers of Kangchenjunga. From there he would explore the icy, easternmost of the Nepal passes into Tibet: Walungchung (Wallanchoon) and the Nango La (pass) above Kambachen (Kanglachem), which would bring him to the loftiest parts of the eastern flank of Kangchenjunga.

Sunset across the Himalaya. The summit of Tonglu falls into darkness as dusk casts its last rays across Kangchenjunga and Pandim. Hooker re-crossed the western ridge of Tonglu in October 1848 to reach the easternmost valleys of Nepal.

Most of the arrangements for the six-month expedition, including organising equipment and provisions, were carried out by Hooker's friend at Darjeeling, Archibald Campbell.

Since much of the terrain that Hooker planned to cover was unknown, taking pack animals was not an option and the entire expedition was to be carried out on foot, with a large team of porters to carry provisions. Up to 30 men were needed, but sourcing these was not without difficulty. Although the Lepchas were fine porters, they were unsuited to life in snowy mountainous areas and unenthusiastic about working outside Sikkim, particularly in such a warlike country as Nepal.

The Nepalese, residing as British subjects in Darjeeling, were mostly refugees from their own country, and were terrified to return, lest they be claimed by feudal lords. To employ other ethnic groups from the hotter, low elevations was out of the question, and therefore, the only choice remaining was to employ Bhutanese run-aways living in Darjeeling. This group, mainly comprised of slaves or criminals, was accustomed to high altitudes, their only fear being to return to Bhutan.

Girardinia diversifolia: nettles rarely come as handsome as this. We met with it several times during our travels and on one occasion had a very good soup made from it.

Hooker's expeditionary team numbered 56 men, consisting of himself, his servant, John Hoffman (who managed the team, thus allowing him to concentrate on the scientific aspects of the trip), and the porters. Two of the latter carried Hooker's tent and equipment, which had been supplied by Brian Hodgson, and four men carried his scientific equipment, his bed, a box of clothes and his books and papers. Seven porters were engaged simply to carry the huge volume of paper needed for pressing and drying plants. The Nepalese guard required two porters of their own, and his interpreter and chief Lepcha collector also had a man each.

Brian Hodgson, realising the potential of the region for zoological specimens, also sent a bird and animal shooter, collector, and stuffer and, with their ammunition and equipment, another four porters were required. Three young Lepcha boys, already in Hooker's service, were brought along to climb trees and change plant-papers. The party was completed by fourteen Bhutanese porters, laden with food, consisting mainly of rice with ghee, oil, chilli peppers, flour and salt.[1]

Hooker himself carried a small barometer, a large knife and trowel, notebook, sketchbook, telescope, compass and several other instruments. He kept two or three Lepcha lads close to him at all times, and they carried his vasculum, geological hammer, bottles and boxes for insects, and thermometers, as well as a sextant and artificial horizon, azimuth compass and stand. Hooker was trained in cartography, and planned to produce the first comprehensive map

of east Nepal and Sikkim, outlining passes into Tibet, much to the approval of the British Navy who were funding his travels in India.

Hooker and his team departed from Darjeeling on 27 October 1848, accompanied for a good distance by Archibald Campbell, who saw them off. The most direct route would have been across Tonglu, but due to the threats of the Sikkim government Hooker chose a more secretive track across the Ghoom (Goong) ridge, a western prolongation of Senchal, that eventually led to Tonglu, and ultimately into Nepal.

Six days later Hooker reached the flat flank of Tonglu, just 213 m below the western summit, which he had explored the previous May. By this time, the behaviour of the Bhutanese porters had become impossible; they continually plundered the expedition provisions, and neither their head man nor the Gorkha soldiers had any authority over them. Hooker considered dismissing them all, but was relieved of the decision when they made off of their own accord.

The views from the summit of Tonglu, in early November, were spectacular. By then the post-monsoon rains had long since cleared and the skies were cloudless and gentian-blue. He described the scene as follows:

> In the early morning the transparency of the atmosphere renders this view one of astonishing grandeur. Kinchinjunga [sic] bore nearly due north, a dazzling mass of snowy peaks, intersected by blue glaciers, which gleam in the slanting rays of sun, like aquamarines set in frosted silver. From this the sweep of snowed mountains to the eastward was almost continuous as far as Chola … following a curve of 150 miles, and enclosing the whole of the northern part of Sikkim, which appeared a billowy mass of forest-clad mountains. On the north-east horizon rose the Donkia mountain (23,176 feet), and Chumulari (23,929 feet). Though both were much more distant than the snowy ranges, being respectively eighty and ninety miles off, they raised their gigantic heads above, seemingly what they really were, by far the loftiest peaks next to Kinchinjunga [sic]; and the perspective of snow is so deceptive, that though 40 to 60 miles beyond, they appear as though almost in the same line with the ridges they overtopped.
>
> Beyond Junnoo [sic], one of the western peaks of Kinchinjunga [sic], there was no continuous snowy chain; the Himalaya seemed suddenly to decline into black and rugged peaks, till in the far north-west it rose again in a white mountain mass of stupendous elevation at 80 miles distance, called, by my Nepal people, "Tsungau".[2]

Tsungau is better known today as Mount Everest.

The western summit of Tonglu was tree-less, and the entire flank for the next 305 m down had been cleared by fire for pasture and flocks of black-faced sheep grazed it.

On the western Nepalese flank of Tonglu lay the Myong valley, a beautiful, warm, fertile region; at its lower reaches, 32 km south-west of the frontier, was the military fort of Ilam. The Myong, on the western side of the steep Singalila range, is dry, as the Indian monsoon sheds its waters on the Sikkim side, particularly on the eastern side of Tonglu. Hence the contrast between the deep, damp forests of the Sikkim side of the frontier, and the hot, dry valleys across the border in Nepal.

Characteristic trees of the sun-baked slopes were chir pine, *Pinus roxburghii,* and *Quercus semecarpifolia,* an evergreen tree, often with holly-like foliage, that can be common on the warmer, dry slopes of the Himalaya and western China.

The villages of the region were surrounded by fields of rice, buckwheat and maize, and teams of men, women and children were busy harvesting the latter. Gigantic stinging nettles, *Girardinia diversifolia,* with wonderfully architectural leaves, flanked the edges of these fields.

Towards Walungchung Gola

From the town of Ilam, Hooker and his men headed in a northerly direction along the road to Walungchung Gola (in present-day Taplejung District). The setting sun in this part of the Himalayan foothills can be spectacular, and, climbing a ridge over his campsite beyond Ilam, he was transfixed by the scene:

Alex Slazenger, Head Gardener at Powerscourt, seen here in the Deer Park at Kilmacurragh with the British and Irish girth champion of *Abies spectabilis.*

While my men encamped on a very narrow ridge, I ascended a very rocky summit, composed of great blocks of gneiss, from which I obtained a superb view to the west-ward. ....

The firmament appeared of a pale steel blue, and a broad low arch spanned the horizon, bounded by a line of little fleecy clouds ... below this the sky was a golden yellow ..... as evening advanced, a sudden chill succeeded, and mists rapidly formed immediately below me in little isolated clouds, which coalesced and spread out like a heaving and rolling sea, leaving nothing above their surface but the ridges and spurs of adjacent mountains. ...... As darkness came on, and the stars arose...... I quitted with reluctance one of the most impressive and magic scenes I ever beheld.[3]

In the evenings, Hooker labelled his plant collections, wrote up his journals, plotted maps, and took observations. As he retired to bed, one of his Nepali soldiers entered his tent, spread a blanket on the ground, and slept there as his guard. In the mornings, his collectors were busy changing the plant papers in the herbarium presses, while he explored the surrounding countryside, and the entire team were on the move again before 10 a.m.

Travelling northwards, they crossed a knoll, 2,835 m high, commanding a spectacular view to the west of the Tamur River, the plains of India, and the course of the Koshi River, while to the north, the summit of mountain ridges bristled with dark forests of the beautiful Himalayan fir, *Abies spectabilis*.

In the valley below lay villages surrounded by tidy fields, with crops of golden mustard and purple buckwheat in full flower, yellow rice and maize, green hemp, pulses, radishes, barley and brown millet, a colourful patchwork beneath an intensely blue sky. In the depths of the valley, skirting raging torrents, grew oranges, bananas and sugar cane.

Also in the valley, the primeval forests were impressive, with trees of enormous girth. Hooker measured one old veteran, *Toona ciliata*, with a girth of 9 m at 1.5 m above ground level. *Prunus cerasoides*, the spectacular autumn-flowered Himalayan cherry, lit up the woods with its spectacular pink blossoms and the skirts of the forests were fringed with dense thickets of blue-flowered *Strobilanthes*.

On 13 November, Hooker's party reached the east bank of the Tamur (Tambur) River, at its junction of the Khawa, in an immensely deep gorge. There it formed a gushing torrent, larger than that of the Teesta, with racing waters of a pale, sea-green, muddy colour. Across the river lay the main road from Ilam to Walungchung Gola, which he rejoined by crossing the Tamur in 9 m long canoes hewn from the hollow trunks of *Toona ciliata* trees.

The little village of Chintam, with its pretty wooden or wattle and mud thatched houses, lay 610 m above the level of the river, and afforded grand views of the upper reaches of the river, raging through a tremendous chasm, flanked on either side by well-cultivated hillsides.

Just beyond Chintam, at the bazaar town of Mywa Gola, Hooker was overtaken by a messenger, carrying letters from Archibald Campbell, stating that the Sikkim Rajah had denied any responsibility for the refusal of Lord Dalhousie's request, and authorised Hooker to return to Sikkim to explore regions of his choice. The messenger was a high-ranking official from the Sikkim Court, a Lepcha called

Looking north from the village of Chintam through the valley of the Tamur River. Trees of *Abies spectabilis* clothe the higher slopes.

Meepo, who was to join Hooker many months later in Sikkim and proved a useful companion, guide and occasional collector.

The vegetation of this part of the Tamur valley was tropical and banyans thrived in the baking heat of day. Malaria was rife, and Hooker (who had received a medical training) treated several inhabitants suffering from attacks of fever. Orchids abounded, and it was in the hot, lower valleys of east Nepal that Hooker collected *Esmeralda cathcartii*, a spectacular epiphytic orchid with chocolate-brown flowers. It also grew around Darjeeling, where William Griffith's collectors had initially found it during the late 1830s. Joseph Hooker successfully introduced living plants to the Royal Botanic Garden, Calcutta, though those he sent in Wardian cases to Kew perished on the long ocean routes back to England.

The Walungchung road continued along the west bank of the Tamur, and above Mywa Gola the valley contracted into a gorge and the river became a turbulent rapid. It was on the banks of the Tamur that Hooker discovered the pretty little blue-flowered *Strobilanthes tamburensis*, a species that later collectors would also find in the damp forests of north-east India, Bhutan and neighbouring Tibet.

A change in ethnic groups also occurred at this point, the Limbu and Khas tribes being replaced by Tibetans, and Hooker passed small parties of the latter on their way to the markets at Mywa Gola. The Tibetans made picturesque groups, herding silky black yaks, weighed down with up to 118 kilos of salt, besides a rattling miscellany of pots, pans, kettles, stools, churns, bamboo vessels, and often, buried beneath the lot, a rosy-cheeked baby, sucking on a lump of cheese-curd. Following shortly behind were flocks of sheep and goats, each carrying two small bags of salt, an important trade item in these isolated regions.

Edgeworthia gardneri was common in this warm valley, forming large, leggy evergreen shrubs with waxy, cowslip-coloured, beautifully scented blossoms. Further down the valley, its stems were used in the manufacture of paper. By the river Hooker saw, for the first time, large prayer wheels enclosed in wooden houses and turned by the force of the water. On the cylinder was inscribed the Sanskrit mantra *Om mani padme hum* or 'Praise to the jewel in the lotus'. He had entered Buddhist country.

The vegetation of the Tamur River valley assumed a more temperate aspect as Hooker gained altitude. His sketch here includes trees of the Himalayan hemlock, *Tsuga dumosa*.

Esmerelda cathcartii. Hooker sent enormous quantities of live orchids to Kew, though many perished on the long journeys to Europe. His dispatches to Calcutta fared better.

In places dense forest skirted the banks of the Tamur, and on the forest floor, on damp, mossy rocks, he found the lovely violet-purple flowered annual gesneriad *Didymocarpus albicaulis*, a species that he would later describe from his Nepal and Sikkim collections and from those made by William Griffith in Bhutan. Around Darjeeling (where it is nowadays rare) the dried leaves of this species are often burned to produce fragrant smoke.

At 2,438 m, he saw the first of the temperate conifers, *Tsuga dumosa*, the Himalayan hemlock, which occupied a belt about 305 m beneath forests of silver fir, *Abies spectabilis*. The forests in these deep gorges were obviously ancient; the largest trees of the *Tsuga* towered

Berberis insignis. Hooker collected it in east Nepal and met with it again in Sikkim. Pictured here on Tonglu in April 2015.

particularly the low-growing, aromatic *Rhododendron anthopogon* and *R. setosum*. The community was the largest village of the Walungs, who spoke a language similar to Tibetan, and shared a cultural affinity with the Sherpa people.

The slopes above the village had long since been cleared of forest, for house building and fuel, and timber was exported into nearby Tibet. The few trees that remained were draped with the long threads of the curious lichen, *Dolichousnea longissima*, whose long strands (which can measure well over a metre), when dyed yellow with the leaves of various *Symplocos* species (including one of his discoveries, *S. dryophila*), were used by the Tibetans of this region for decorating their hair.

The village was on a flat area on the east side of the river and consisted of about a hundred well-built wooden houses, each accommodating several families, and ornamented with enormous vertical prayer flags held on wooden poles. Just above the village was a large, long, two-storied convent, painted scarlet and sheltered by a grove of *Juniperus recurva*. Stupas, prayer wheels and mani walls lay scattered through the village. The hillsides were studded with bushes of deep green rhododendrons, and scarlet barberries and withered yellow roses, in autumn garb.

In the valleys surrounding the village he was to discover the lovely primrose-yellow flowered *Rhododendron campylocarpum*, a common shrub that grew just above the tree line in Nepal and in neighbouring Sikkim, from where he would later introduce it to cultivation.

Herds of yaks grazed around the village. An invaluable creature in such a seemingly hostile landscape, yaks were grazed in summer pastures as high as 5,182 m, while in winter local farmers kept them

to 46 m in places, with girths of up to 6 m. In the same area, close to the banks of the river he found *Berberis insignis*, one of the finest of the Himalayan barberries. He would see it again during his Nepal travels in the Yalung valley, and in Sikkim at Chungthang. Hooker did not discover the species – that honour must again go to William Griffith who found it in Bhutan during the 1830s – he did, however, introduce it to cultivation in 1850 from seeds collected in Sikkim.

Hooker's party reached the old Sherpa village of Walungchung (also known as Olanchung) Gola on the evening of 23 November 1848. Perched at 3,220 m, it occupied an open, steep-sided part of the valley, and the rocky sides of the surrounding mountains were covered in a luxuriant layer of *Berberis*, juniper and rhododendron,

Rhododendron campylocarpum, one of Hooker's Nepal discoveries. It is also native to Sikkim, where in places it covers entire mountain slopes.

Walungchung village, Hooker's sepia sketch. The scene depicts Hooker seated on a boulder, sketching as his Lepcha porters look on. His tents are pitched by the village where prayer flags flutter above wooden houses. Trees of *Juniperus recurva* flank the large convent, while dark fir forests scale the steep mountainside.

below 2,438 m, to avoid the snows. It was the most useful beast of burden in the Himalaya, and the wealth of these Tibetan farmers was determined by the number of yaks in their herds. The yak supplied milk and cheese, their hair was spun into ropes or woven into tent coverings, while the flesh of calves, according to Hooker, was far superior to veal. Though impervious to cold, the yak cannot live beneath elevations of 2,134 m, where it soon succumbs to liver disease and other ailments.

It was in the yak meadows above the village that Hooker discovered the gorgeous *Pedicularis trichoglossa*, a perennial lousewort up to 60 cm tall, which in summer bears densely packed racemes of blackish-purple to rose-purple blossoms. The louseworts are a characteristic part of the alpine flora of the Himalaya, where they grow as hemiparasites on grass and are pollinated by bumblebees. In the same meadows he collected the tufted grass, *Poa tibetica*, a species first found by the French botanist Victor Jacquemont (1801–1832) in western Tibet two decades previously.

Not long after his arrival, Hooker was met by the village headman who was keen to know the explorer's plans. The headman, with good reason, disputed the Nepal Rajah's authority to allow Hooker permission to visit the Walungchung Pass. The inhabitants made a good living from the salt trade over the route and were keen to keep details of the revenue raised a secret from their Rajah.

Hooker's passport said nothing about the passes,[4] his men were without suitable clothing, and his provisions were running short. The headman swore the passes had been deluged with snowfall, and had been impassable since October. With that, Hooker took the headman to one side, and pointed out the herds of yaks, which earlier that day had arrived laden with salt from Tibet, and therefore insisted on a guide, provisions and snow-boots for the next day, letting it be known that any impediment would be reported to the Rajah.

To the Tibetan passes

On 25 November, the headman grudgingly relented, and Hooker and his men headed north-west up the valley, soon crossing the tree line above the silver firs and into thickets of evergreen rhododendrons, *Rosa sericea, Juniperus indica*, spiraeas, stunted Himalayan birch, willow, mountain ash and dwarf *Lonicera*.

What surprised Hooker more than the sheer area occupied by the rhododendron thickets, was the number of species, which he quickly differentiated by the shape of their seed capsules, the shape of leaves and the indumentum on leaf undersides. All had long since blossomed, though he gathered a rich harvest of seeds, later sent to his father, Sir William, at Kew.

Above 3,658 m the valley became a wild, open landscape with steep mountain slopes broken by narrow gullies leading to blue patches of glacial ice. At this point the vegetation was stunted and chiefly composed of enormous sheets of dwarf, highly aromatic rhododendrons, *R. anthopogon* and *R. setosum* and the ling-like *Cassiope fastigiata*, which gave a heather-like appearance to the slopes.

Alpines abounded, including the diminutive *Cyananthus hookeri*, carrying tiny purple-blue, bell-shaped blossoms over mounds of densely hirsute grey foliage. Above there he found another plant destined to carry his name, the lovely yellow-flowered *Corydalis hookeri*, a high-altitude species restricted to parts of the Tibetan Plateau and the eastern Himalaya.

Hooker's sketch of a large yak in its typically high-altitude Himalayan setting. This useful beast of burden is still employed to ferry goods across snow-laden passes.

Rhododendron setosum, a high-altitude Himalayan species. It generally co-habits with *Rhododendron anthopogon*.

By the time they reached 3,962 m, the ground was frozen solid and Hooker began to feel the first symptoms of altitude sickness. He and his men camped there, and after such a tough climb, he had hoped to get a good night's sleep, though he learned before retiring that the Walungchung headman was determined that he would receive no more provisions on his return. Perhaps Hooker had gone too far in his bullying and cajoling. To remain at these altitudes without food was impossible and with this worry, he couldn't sleep.

Corydalis hookeri, photographed in north-central Nepal between Yak Kharka and Leder, at around 4,200 m.

The approach to Yangma Gola. Hooker's sketch is one of many preserved in the remarkable archives at the Royal Botanic Gardens, Kew.

The following morning dawned fine and clear; the route up the steep valley passed the vast mass of a blue glacier and fields of snow choked the gullies. Beyond there the snows lay deep, it *had* snowed heavily in October, and Hooker and his party battled on for a further 6 km, suffering all the way from the effects of altitude. The pass itself, buried in deep snow, was a low saddle between two rugged ridges, with a single cairn draped in prayer flags. There, at 3.30 p.m. on 26 November 1848, Hooker glimpsed his first view of Tibet, a land absolutely forbidden to Western travellers. From the snowy pass, wracked with headache and breathlessness, he observed range upon range of snowy mountains, being among the very first Europeans to do so.

The Yangma Valley

Hooker returned to Walungchung through thickets of dwarf juniper, rhododendron and *Cassiope*. Exhausted, he refused to meet the headman until the latter agreed to open a bazaar the next morning for his men to buy food. If he refused, Hooker told him, he would be reported to the Nepal Rajah.

At this point Hooker also arranged for the main part of his group to return to Sikkim, so that, with a smaller remaining team, he could visit the Kanglachen Pass (the Kang La or Khangla Deoral), a steep mountain gap to the north-west, perched at 5,883 m. From there he would cross two other major mountain passes; the Nango La (4,776 m) and the Khang La (Kanglanamo pass) to reach Dzongri (Jongri) in west Sikkim.

The following day, having retained 19 men and purchased a week's supply of food at the bazaar, Hooker set off again. At the junction of the Tamur and Yangma Rivers, he bade farewell to the men returning to Darjeeling with his collections. Continuing on, he followed the banks of the Yangma, along a bad road, where in places he was forced to climb ladders and ascend planks lashed to the faces of precipices, and ford torrents by means of plank bridges.

The little town of Yangma Gola consisted of about 300 stone huts, nestling under the steep south-facing flank of a flat-topped terrace. In tiny fields the inhabitants grew crops of barley, wheat, potatoes, radishes and turnips, at an altitude of 4,237 m. Several species of grasses abounded, including the feathery *Anthoxanthum hookeri*, which the Irish plant collector Augustine Henry was later to rediscover in the mountains above Mengzi in southern Yunnan, China.

Above the village Hooker entered a glaciated landscape, which brought back memories of his Antarctic Expedition with Captain James Clark Ross. He speculated as to how such enormous lateral moraines existed in ancient times, so far down the mountain slopes beneath those existing higher up at the time of his visit in late November 1848. Climbing a gigantic wall of rocks, he observed the view ahead as the sun began to set:

A superb view opened from the top, revealing its nature to be a vast moraine, far below the influence of any existing glaciers, but which at some antecedent period had been thrown across by a glacier descending to 10,000 feet, from a lateral valley on the east flank. Standing on the top, and looking south, was the Yangma valley up which I had come, gradually contracting to a defile, girdled by snow-tipped mountains, whose rocky flanks mingled with the black pine forests below. Eastward the moraine stretched south of the lateral valley, above which towered the snowy peak of Nango, tinged rosy-red, and sparkling in the setting sun: blue glaciers peeped from every gulley on its side, but those were 2000 to 3000 feet above this moraine …

Hooker's sketch of the skull of a Tibetan argali, dated May 2 1848.

I had never seen a glacier or moraine on land before, but being familiar with sea ice and berg transport, from voyaging in the South Polar regions, I was strongly inclined to attribute the formation of this moraine to a period when a glacial ocean stood high on the Himalaya, made fiords of the valleys, and floated bergs laden with blocks from lateral gulleys, which the winds and currents would deposit along certain lines.[5]

In contrast to this line of thought, Hooker had more mundane concerns – his provisions were running low and there was no food to be procured from the village except for a little watery milk and some potatoes. Worse than that, there had been some confusion as to what actually remained in stock:

My private stock of provisions – consisting chiefly of preserved meats from my kind friend Mr Hodgson – had fallen very low; and I here found to my dismay that of four remaining two-pound cases, provided as meat, three contained prunes, and one "*dindon aux truffes!*" Never did luxuries come more inopportunely; however the greasy French viand served for many a future meal as a sauce to help me to bolt my rice, and according to the theory of chemists, to supply animal heat in these frigid regions. As for my people, they were not accustomed to much animal food; two pounds of rice, with ghee and chilis, forming their common diet under cold and fatigue.[6]

And so, on a luxurious diet of turkey with truffles, prunes and rice, Hooker and his team continued towards the pass. On the banks of the Yangma, they came suddenly upon a flock of the enormous Tibetan argali, *Ovis ammon* subsp. *hodgsoni*, the largest known mountain sheep. It was named for Brian Hodgson in 1841, by the zoologist Edward Blyth (1810–1873), curator of the museum of the Royal Asiatic Society of Bengal.

These were spectacular animals, quite unlike conventional sheep, being as tall as a calf, very long legged, and with immense curling horns. Desperately in need of food, particularly fresh meat, Hooker regretted sending his gun back down the mountain with Brian Hodgson's bird stuffers. The Tibetan argali is primarily an inhabitant of the Tibetan Plateau, rarely seen below 4,267 m and Hooker was to see it later again at 5,486 m near the Kanglachen Pass.

Ancient moraines in the Yangma valley. For Hooker, who had visited Antarctica a decade previously, it must have been exciting to have returned to a glaciated landscape.

Rhododendron hodgsonii. Fitch's superb plate that appeared for the publication of the species in *Rhododendrons of the Sikkim Himalaya.*

Kambachen and back to Sikkim

Descending the Yangma valley, Hooker then turned his attention towards Kambachen, a lofty peak ascending to 7,802 m, and the glaciated valleys that surrounded it. Crossing the Yangma River, he discovered a new species of rhododendron, growing in dense thickets on a rocky moraine. It was a 'big-leaved' tree-like species, with wonderful flesh-coloured bark. Hooker stated that its hard, durable wood was used to make cups, spoons, ladles and yak saddles for the Tibetan market, while its leaves were employed as platters and as plates to line baskets carrying the mashed pulp of *Arisaema* roots, while presents of home-made butter and curd were always wrapped in its large, handsome leaves.

He was obviously impressed and later named it *R. hodgsonii* for his friend Brian Hodgson. On these mountain slopes it grew with mountain ash, maples, silver fir, birch and juniper. Hooker was astonished at the beauty of its foliage, some of its leaves reaching 41 cm long, while the bark, which had fallen on the ground beneath, was as delicate as tissue-paper, and of a pale flesh colour.

Four decades later Hooker delighted in seeing many of his rhododendrons acclimatised in Cornwall, in the garden of a fellow Antarctic explorer, General Sir John Henry Lefroy (1817–1890), who had sailed on both the *Terror* and the *Erebus.* Hooker's rhododendrons

grew with great vigour in the mild, damp Cornish climate, and the sight of them enthused him to write to Brian Hodgson's wife:

> Tell Brian with my love that I saw, in Cornwall, many, many plants, of the *Rhod. hodgsonii* in the open air, 6 feet across and more, with leaves a foot long … They were planted in the woods and throve luxuriantly. There were also noble plants of *falconeri, aucklandii, argenteum, barbatum* and others – together with *hodgsonii* forming regular shrubberies, as if natives of the soil.[7]

From Kambachen, meanwhile, Hooker headed towards the Nango La (Nango Pass), which crossed the saddle of a mountain at 4,776 m. Once above the tree-line, he found himself in an alpine landscape, colonised by *Potentilla fruticosa* var. *arbuscula,* dwarf *Lonicera* and *Polygonum vacciniifolium.* In the distance, directly to the north, was the peak of Nango , while to the east, 1,524 m below, the view across the stupendous Kambachen gorge opened out, with a distant view of immense sheets of rock, snow and ice, piled in great confusion along the base of Jannu (Junnoo), a spectacular 7,710 m peak, and across the Mirgin La (Hooker's Choonjerma).

On those hostile alpine slopes plants were reduced to tiny stature, and interesting species abounded, particularly small cushion-like arenarias, woolly saussureas and thick-rooted umbellifers.

Beyond there, near the lower slopes of Nango, he first encountered a new, unknown species of larch, previously collected by Griffith in Bhutan. In east Nepal it formed a small tree 6–12 m high and Hooker later named it *Larix griffithii,* in honour of its original discoverer.

Joseph sent seeds of this tree from Nepal to his father at Kew, though it has resisted all attempts to coax it into Western gardens, soon succumbing to larch woolly aphid and late frosts, which is a shame, and a great loss. In recent times a young tree, from the Royal Botanic Garden Edinburgh, grew on the Bleach Green at Kilmacurragh, where its chief attraction was its vivid purple juvenile cones. Sadly, it soon died, and this tree can still be seen at its best only in Nepal or Sikkim.

The tiny village of Kambachen, though perched at 3,469 m, lay deep in a snowy valley bounded by soaring grey cliffs, on a flat terrace above a raging glacier-fed river. Hooker considered it the

Ancient moraines in the Kambachen valley, a wood engraving from *Himalayan Journals.*

The Mirgin La (Hooker's Choonjerma pass). His sketch of the taller peaks in the far distant right is thought to be the first Western depiction of Mount Everest. A dense cumulus of clouds creates a sea-like effect between the taller peaks and ridges.

wildest, grandest and most gloomy gorge he had ever visited, finding it extraordinary that anyone should live in such a cold, barren landscape. The villagers were extremely kind and provided a guide for the Mirgin La, which led into the Yalung (Yalloong) valley, the most easterly in Nepal.

Hooker pitched his tents between enormous ancient moraines, among dwarf rhododendrons, prostrate juniper and the withered remains of an alpine meadow full of primulas, gentians, anemones, potentillas, campanulas and pedicularis; earlier that summer it must have presented a kaleidoscopic scene. Some of his alpines from the mountains above here proved to be new, including the tiny yellow-flowered *Saxifraga cordigera*.

Rhododendrons continued to appear in great numbers, changing in species as the party gradually gained altitude. Perhaps the most exciting species he found in the higher valleys was the blood-red flowered *Rhododendron thomsonii*. Above the tree line it formed great thickets with its polished red-brown bark. It is one of the showiest of all the Himalayan *Rhododendron* species, and Hooker named it for his life-long friend and travelling companion in the Khasi Hills of Assam in north-east India, Thomas Thomson.

Hooker reaped rich rewards in terms of discovering new species. In the higher valleys, on the edge of dark fir forests he found the beautiful yellow-flowered *Rhododendron wightii*, a small shrubby tree that he later collected in Sikkim, from where he introduced it to cultivation in November 1849. Sheep and goats grazed the surrounding valleys and Hooker observed that they were able to feed on the foliage of *R. thomsonii* and *R. campylocarpum*. On the roots of the latter species he found a previously unknown root holoparasite, *Boschniakia himalaica*.

The views from the Mirgin La were spectacular, and to the north the steep and majestic peak of Jannu rose in dismal grandeur. Hooker described the scene:

Rhododendron thomsonii, Walter Hood Fitch's re-working of Joseph Hooker's field sketch. One of the finest plates in *Rhododendrons of the Sikkim Himalaya*.

The setting sun can be a spectacular feature in the Himalaya when conditions are right. Here, on the Nepalese side of Tonglu, the sun sets across east Nepal. A Buddhist stupa dominates the foreground.

Evening overtook us while still on the snow near the last ascent. As the sun declined, the snow at our feet reflected the most exquisitely delicate peach-blossom hue; and looking west from the top of the pass, the scenery was gorgeous beyond description, for the sun was just plunging into a sea of mist, amongst some cirrhi and stratus, all in a blaze of the ruddiest coppery hue. ....
I have never before or since seen anything, which for sublimity, beauty, and marvellous effect, could compare with that I gazed on that evening from [the] Choonjerma pass.[8]

Hooker made his descent from the Mirgin La by moonlight, reaching the Yalung valley ahead of his porters, many of whom were suffering badly from frostbite. The valley ran between the peaks of Jannu and Kabru (7,339 m), a notable peak on the Sikkim border. A sharp descent brought him into a dense forest of *Abies spectabilis* and *Tsuga dumosa*, with fine trees of *Rhododendron falconeri* up to 12 m tall and with leaves 48 cm long. Below there he re-encountered the Himalayan yew, *Taxus wallichiana*, covered in red berries and a tree of *Rhododendron barbatum* in full flower in the depths of December!

It was here that Hooker discovered the epiphytic *Rhododendron camelliiflorum*, clinging high up in the topmost branches of giant forest trees, forming part of a secret, aerial flora, mostly hidden from the prying eyes of this determined Victorian plant hunter. On the forest floor beneath, he found the dwarf *Primula gracilipes*, a lovely species bearing crowded masses of pink-purple blossoms with a yellow eye. It was described from collections later made in Sikkim by George Cave, the Curator of the Lloyd Botanic Garden in Darjeeling, though Hooker's Nepal collections were the first made by any Western plant collector.

Travelling south, he and his team followed a parallel course to the Singalila Range, that great dividing ridge between Nepal and Sikkim, and they gradually left the mountains and descended into a warm-temperate forest dominated by *Quercus lamellosa*. Soon Mount Singalila (3,686 m) came into view, offering hope of finding a pass into Sikkim, and, as if to welcome him back, Kangchenjunga raised her great peaks.

Beyond there his route followed the Kabeli (Khabili) River valley, and down the course of its tributary, the Iwa River, bounded on either side by forests of *Pinus roxburghii*, dwarf date-palms and trees of the Indian gooseberry, *Phyllanthus emblica*. Ticks swarmed in the surrounding jungle, getting into his bed and hair, and even attaching themselves to his eyelids at night. On 14 December 1848, Hooker and his men camped on a narrow spur, in deep, dense forest at 2,835 m and ate the very last of their provisions. The following morning, he crossed the 3,100 m Chiwa Bhanjyang Pass (Islumbo Pass) on the Singalila range and returned to Sikkim, having blazed a pioneering trail through the mountains of east Nepal.

Rhododendron thomsonii, a superb clone with red calyces, seen here at the Royal Botanic Garden Edinburgh. One of the most beautiful of all the Himalayan *Rhododendron* species.

CHAPTER 6

MEETING THE RAJAH

The Sikkim side of the Chiwa Bhanjyang Pass was broad and grassy, with a scattering of dwarf bamboo, stunted bushes of *Rosa sericea* and several *Berberis* species, all covered in mosses and lichen. Hooker and his men crossed the pass early on the morning of 15 December 1848. The surrounding vegetation was coated in a film of ice, a harsh wind blew from the south-east and heavy fog lingered on the highest slopes.

Their descent back into Sikkim was down a steep slope into the valley of the Kalek (Kulhait) River (a tributary of the Great Rangit), and the following day they dropped into the lower, warmer reaches of the valley, and crossing the river, reached the village of Lingdam

(Lingcham) below the Buddhist monastery of Sanga Choeling (Changachelling). It was pitch dark by the time of their arrival and the village headman or kaji (kajee of Hooker) sent out a party with torches to bring them in, honouring them with a musket salute and bringing gifts of milk, eggs, fowl, bananas and local murwa beer; fresh rations that were no doubt gratefully received.

Hooker was told that a messenger with letters was waiting for him at Yuksam (Yoksun), a village three days' march further to the north, since he was expected to have crossed over the Khang La (Kanglanamo Pass) to the north-west of that region.

One of the most famous places on Sikkim's *via sacra* is the monastery at Tashiding with its many white-washed stupas. Hooker visited here following his return to Sikkim from east Nepal.

The Kalek River valley as seen from the lower slopes of Mount Maenam. Crowning the helmet-shaped hill in the centre of this scene is the Buddhist monastery at Tashiding.

He also heard that Archibald Campbell was *en route* to meet the Rajah at Samdong (Bhomsong Samdong) on the Teesta River, a place where no European had previously set foot. This was surprising, as the Rajah had previously eschewed any communication with representatives of the British East India Company, and had absolutely refused to allow any agent of the Governor-General to cross into his territory.

Campbell, being that agent, had long tried to meet with the Rajah, though with no success. The British India authorities felt hard done by, having reconquered the kingdom from the Nepalese, returned it to the Rajah, and, by Treaty, bound themselves to support and protect his territory from the Nepalese to the west, Tibetans to the north and the Bhutanese to the east.

Following the Anglo-Nepalese War, the terms of the Treaty of Titalia (1817) guaranteed to the monarch his absolute right as ruler of Sikkim, though in return the British were granted trading rights and permission to travel through his dominion as far as the Tibetan frontier. Added to that, British India assumed the title of Lords Paramount of Sikkim, a title that in theory allowed the East India Company a predominant influence in the State. Hooker

found the Rajah's lack of communication arrogant, though given the Company's colonial ambitions and annexation of independent Indian kingdoms during the 18th and 19th centuries, his cautious silence was perhaps understandable.

Hooker considered the valley of the Kalek River one of the finest in India. It ran for 28 km from the Chiwa Bhanjyang Pass to the Great Rangit River, and along its length were some of the oldest monastic establishments in Sikkim. The valley was bounded on either side by steep-sided spurs, with mountains rising from 2,134 to 2,438 m. Sixteen kilometres of the lower part of the valley was inhabited, with several villages occurring on spurs 610 m above the river.

Three days after his arrival at Linchyum, Hooker received a letter from Campbell, asking him to travel towards the Teesta River and meet him at Samdong, where he was waiting on the west bank. He set off on 20 December, accompanied by the village headman, a Lepcha, who was travelling to pay his respects to the Rajah. Along the way Hooker was treated with great kindness by village people who showered him with gifts and hospitality, since the Durbar had ordered that he be treated with respect.

Having crossed the Great Rangit River by means of a simple bamboo bridge, Hooker set up camp on its east bank in a thick tropical jungle. He and his companion then headed east, making a steep ascent of the mountain range dividing the Great Rangit from the Teesta. From there he obtained a fine view of the conical wooded hill of Tashiding (Tassiding), rising abruptly from a deep river gorge, and crowned by temples and dozens of white-washed stupas. Beyond, on successive ridges, were the ancient monasteries of Pemayangtse (Pemiongchi) and Sanga Choeling, while in the distance, to the north, lay Dubdi (Doobdi), its thatched temple rising from a wooded flank of Kangchenjunga.

It was on this ridge that he bought a little black puppy, a cross between a Tibetan Mastiff and the Sikkim hunting dog. He was just a few weeks old, a bundle of fur, and became a faithful companion during the coming months. From there Hooker's route was along the

Hooker named his dog 'Kinchin' after Kangchenjunga, the guardian deity of Sikkim. In a letter to his sister Elizabeth Hooker, he sketched his companion, outlining how clever he was, and how he pinched food from the campside kitchen.

Begonia xanthina, Booth's original introduction that appeared in *Curtis's Botanical Magazine*. Its stems are edible.

from Linchyum still travelled with him, and gave him much local information about these plants. Several species of nettles grew on the roadside, and from their fibres the mountain people made bowstrings, or thread for sowing and weaving, while others were cooked in soups.

Begonia xanthina, a new yellow-flowered species, was abundant, and its acid-tasting, fleshy stems were used as an accompaniment to pork, just as apple-sauce is used in Europe. This *Begonia* was encountered again shortly afterwards by Thomas Jonas Booth in Bhutan, who sent tubers to his uncle, Thomas Nuttall (1786–1839), in Lincolnshire. There it flowered in July 1852 and was featured as t.4683 in *Curtis's Botanical Magazine*.

The ridge was rich with temperate plants familiar to Hooker from the mountains surrounding Darjeeling. It ran as a long lofty saddle connecting Mount Maenam (Mainom) to the north with that of Mount Tendong to the south. He stopped to sketch an enormous mani stone, 9 m long, with carved Sanskrit text. The following morning, after a heavy fall of snow, he was rewarded with a view of Maenam, rising close to his campsite, in a series of rugged peaks, crested with dark firs, all dusted in snow.

From the base of Maenam, Hooker travelled east, crossing the Rangpo River valley, and from there headed north through the little villages of Brom (Broom) and Lingmo (Lingo), where the Dewan had a pony waiting for his use. Hooker rode into Campbell's campsite on the west bank of the Teesta, in an orange grove close to where pineapples and pomegranates were also cultivated. It was three months since they had last seen each other; Campbell was delighted to see the safe return of his friend and congratulated Hooker on the success of his expedition to Nepal.

Even in December it was warm in the depths of the Teesta valley. The vegetation here was wholly tropical, with several species of palms like the Himalayan dwarf fishtail palm, *Wallichia oblongifolia*, whose leaves were (and still are) used for thatching and making brooms. It

An enormous mani stone near Mount Maenam. This sketch by Hooker's was later reproduced as a wood engraving in *Himalayan Journals*.

principal road to Tumlong (Tumloong), then the capital of Sikkim, where the Rajah's palace was situated. Along the way he met dozens of Buddhist lamas, returning to the monasteries he had just viewed, from the marriage of the Rajah's eldest son.

This was a significant occasion for the lamas, who were trying to thwart the efforts of the Dewan (Prime Minister) Duniya Namgay, whom Hooker believed was corrupt and had undue influence over the Rajah and his family. The eldest son, being a Buddhist monk, carried only spiritual authority. Temporal authority devolved on the second son, who was heir apparent of Sikkim. This arrangement had been thrown into chaos, however, by the sudden death of the heir, and an illegitimate son of the Rajah was favoured by the Dewan. The monks from the monasteries further west foiled the Dewan's plans by gaining a special dispensation from Lhasa, for the eldest son to leave his monastery and marry a bride brought from Tibet. The marriage feast had lasted for 18 days.[1]

On a pass, on a steep ridge dividing the Great Rangit from the Teesta, Hooker made a rich collection of plants. The headman

Old Kashmir cypresses among the stupas at Tashiding. Hooker's sepia sketch was later engraved for *Himalayan Journals*.

grew alongside the prickly rattan-cane, *Calamus erectus*, and screw-pines with forked trunks crowned with immense saw-edged leaves and large inedible fruits, and enormous grasses, forming a thick, dark jungle, with pink-flowered balsams below. The cane bridge crossing the gushing torrent had been deliberately cut, in distrust of Campbell and Hooker, and a raft was thus the only means of crossing the river.

Soon after his arrival at Samdong, Hooker was sent gifts from the Rajah, and, that evening, they were visited by the cause of their difficulties in Sikkim — the Dewan. The Rajah's Prime Minister, a Tibetan of Lhasan birth, was related to one of the regent's wives and Hooker described him as 'a mere plunderer of Sikkim', establishing his own family in fine estates and trading for his own benefit. The Rajah, at this stage, was 63 years old (not 70, as Hooker stated), and absorbed more in spiritual activities than the day-to-day affairs of his kingdom. The Dewan, popularly known as the *pagla diwan* or the 'mad Prime Minister', regarded Hooker and Campbell with repugnance, and the feeling was mutual.

Campbell, for his part, was determined to gain the ear of the Rajah, in the hope of establishing closer relations, though when the audience did take place, the conversation was short and constrained, and totally unproductive. Compared to the Indian Maharajahs, the Sikkim monarch lived in relative poverty. Hooker described the Durbar as follows:

> The audience chamber was a mere roofed shed of neat bamboo wattle, about twenty feet long; two Bhoteeas in scarlet jackets, with bows in their hands, stood on each side of the door, and our chairs were carried before us for our accommodation. Within was a square wicker throne, six feet high, covered with purple silk, brocaded with dragons in white and gold, and overhung by a canopy of tattered blue silk, with which material part of the walls also was covered. An oblong box (containing papers) with gilded dragons on it, was placed on the stage or throne, and

behind it was perched crossed-legged, an odd, black, insignificant looking old man, with twinkling upturned eyes; he was swathed in yellow silk, and wore on his head a pink silk hat with a flat broad crown, from all sides of which hung floss silk. This was the Rajah, a genuine Tibetan, about seventy years old. On some steps close by, and ranged down the apartment, were his relations, all in brocaded silk robes reaching from the throat to the ground … A few spectators were huddled together at the lower end of the room, and a monk waved about an incense pot burning juniper and other odoriferous plants. Altogether it was solemn and impressive; as Campbell expressed it, the genius of Lamaism reigned supreme.[2]

The Rajah did not speak Hindustani, so their interpreter was a small, rosy-cheeked Lama by the name of Chebu (Tchebu). The Lama was loyal to the Rajah and his son, though in common with the other Lamas, he utterly detested the Dewan, and wished for better relations between Sikkim and Darjeeling. He was also the only person in the Sikkim court who could speak both Tibetan and Hindustani, leaving the distrustful Dewan feeling very uncomfortable indeed. The Dewan was as anxious to hurry over the meeting as Campbell was to protract it, and Hooker realised from the outset that nothing concrete would be gained from such an uneasy gathering. As a signal for departure, white scarves were thrown over their shoulders, as is the custom of Sikkim, Tibet and Bhutan, and the meeting was brought to an abrupt end.

Though the meeting was anything but productive, the audience brought about good relations with the Lamas who were in high spirits following the marriage of the Rajah's son, who also understood the Dewan's underhanded ways.

Ascent of Maenam

With this business over, Campbell and Hooker decided to spend the following days exploring nearby Mount Maenam and to visit the *via sacra*, the valley where Hooker had just entered Sikkim from Nepal, above which were Sikkim's most important monasteries and ecclesiastical buildings. Campbell planned to return to Darjeeling after this, while Hooker would head north to explore the southern flank of Kangchenjunga.

They departed Samdong on Christmas Day, accompanied by two of the Dewan's officers. Before leaving the Teesta valley they stopped to measure some enormous trees of *Terminalia myriocarpa*, one of which had a trunk of 14 m around its buttresses.

They stayed that night at Namphak (Nampok) and began the ascent towards Maenam the following morning, passing the Yangang (Neongong) monastery, built in 1787 and now under restoration. It was at Yangang that Hooker came across one of Nathaniel Wallich's discoveries from Nepal, the very graceful bamboo, *Himalayacalamus falconeri*. For a bamboo, this species flowers on a very regular basis. It was introduced to cultivation by the Irish botanical explorer, Lieutenant-Colonel Edward Madden, who sent large quantities of seeds to Kew in 1847, which were distributed throughout Europe. Seedlings raised from this introduction flowered in 1875–76, and although all these plants subsequently died, they produced seeds and a new generation was raised, and flowered between 1903 and 1908.

On the lower slopes of Maenam many of the trees were broadleaved evergreens, typical of the subtropical and warm-temperate zone, and it was in these damp forests Hooker discovered *Actinodaphne sikkimensis*, a small tree with yellow timber that was used by carpenters at Darjeeling, where it also grew in warm valleys below the hill station.

At the level of Darjeeling (2,134 m), they encountered snow, indicating a colder climate than that of the hill station (where none had fallen), and 305 m further up, plants common on the summit of Senchal appeared, with thickets of *Rhododendron hodgsonii* and the lovely *Primula petiolaris*, then covered in masses of stemless purple blossoms.

Above there the path ran along the ridge of a south-east facing precipice, allowing views of the Rajah's palace at Tumlong and the superb snowy peak of Chola mountain in the distance. At 2,743 m shrubby rhododendrons abounded, with mountain ash, dwarf bamboo and superb trees of *Rhododendron falconeri* that towered from 12 to 15 m high. Beyond there the snow lay deep, so Hooker and Campbell laid their tents on a thick layer of rhododendron twigs, bamboo and moss and camped 244 m below the summit, in a wood of *Sorbus hedlundii*, *Magnolia campbellii*, *Rhododendron* and bamboo.

That night, the temperature fell to –5°C, but scrambling through the snow the following morning, they eventually reached the summit, making the little wooden gompa (temple) there their quarters. Although the summit was a broad plateau, the depth of snow prevented them exploring further afield, and the forests of *Abies densa* were so tall that the only views were from the temple: the panorama of snowy mountains was breathtakingly beautiful.

Still busy compiling details for his map of Sikkim, Hooker took a complete set of angles and panoramic sketches from the summit, before returning on 28 December to Yangang. The following morning at daybreak, they were roused by the solemn sounds of drums, gongs and trumpets (made from human thigh-bones) emanating from the little monastery above.

On the saddle of a mountain ridge to the west of Yangang, Hooker found a bamboo, that, having bloomed, was covered with heavy panicles of dark, long grain, similar to rice. The inhabitants of this mountainous region boiled it to makes cakes, or fermented the grain to produce a crude beer. This was the fabulous Himalayan blue bamboo, *Himalayacalamus hookerianus*, whose new culms are a luminescent shade of blue, with hints of red and purple at the stem joints.

Looking west from the summit of this ridge, he and Campbell could see no fewer than ten monastic complexes, with their temples, villages and farms, perched on spurs and peaks varying from 914 to 2,134 m and commanding splendid prospects. Beyond there the men pitched their tents near the village of Lingdam, where he had first camped on re-entering Sikkim from Nepal. The following morning, in order to reach the great monastery of Tashiding, they descended into the valley of the Great Rangit River, through a tropical forest full of palms, vines, pepper, wild bananas and screw-pines, all generously clothed with epiphytes. They had finally reached the *via sacra*.

The middle temple at Tashiding. A group of Lepchas, with yaks, appear in the foreground.

Sikkim's sacred route — the *via sacra*

The steep, conical hill of Tashiding divided the Great Rangit River from its principal tributary, the Rathong (Ratong). This Campbell and Hooker crossed by means of a bamboo cane bridge, and climbed 853 m to the top, where, among palms and wild bananas, was a small flat area, crowned with temples and stupas.

Tashiding occupied an imposing position, at the very heart of Sikkim, with immensely deep river valleys on either side. The monastery was over two centuries old at that time, and during the Nepal wars had been plundered of its great treasures. Given such turbulent neighbours, all the monasteries along the *via sacra* secured

Tashiding, the southern temple. Sikkim's temples had thatched roofs until the turn of the 20th century.

The temple interior at Pemayangtse. The walls are covered in murals and the heavy timber beams are decoratively carved.

views of one another; thus from Tashiding the famous temple at Pemayangtse could be seen towering overhead to the west, while to the north-west Dubdi rose on a high, forested spur.

Entering the complex, Hooker and Campbell first visited the Lama's houses, surrounded by small, well-kept gardens, leading to three large, elaborate temples raised on stone platforms, and finally a square walled enclosure facing south, containing 25 enormous white-washed stupas, the largest of which, standing 8 m high, was consecrated to the memory of the Rajah's eldest son.

Between the stupas had been planted what Hooker thought to be the Chinese funeral cypress, *Cupressus funebris*, but was in fact the Kashmir cypress, *C. cashmeriana*. He described them as very old trees, probably as old as the monastery itself. To the Lepchas, Bhutias and Tibetans it was the '*Tchenden*' and its aromatic timber was burned in the temples.

The monks of Tashiding belonged to the red hat sect, and the principal Lama, dressed in a yellow-flowered silk robe, gave Hooker and Campbell a history of the introduction of Buddhism into Sikkim. They spent the evening studying the interior of the temples. Hooker was impressed by the solemnity of the place and viewed the colours of the interior through an artist's eye:

> The effect on entering these cold and gloomy temples is very impressive; the Dugang in particular is exquisitely ornamented and painted, and the vista from the vestibule to the principal idol, of carved and coloured pillars and beams, is very picturesque. Within, the general arrangement of colours and gilding is felt to be harmonious and pleasing, especially from the introduction of slender white streaks between the contrasting masses of colour … It is also well worthy of remark that the brightest colours are often used in broad masses, and when so, are always arranged chromatically, in the sequence of the rainbow's hues, and are hence never displeasing to the eye. The hues, though bright, are subdued by the imperfect light; their countenances of the images

are all calm, and their expressions solemn. Whichever way you turn, the eye is met by some beautiful specimen of colouring or carving, or some object of veneration. The effect is much heightened by the incense of juniper and sweet-smelling herbs which the priests burn on entering, by their grave and decorous conduct, and by the feeling of respect that is demanded by a religion which theoretically inculcates and adores virtue in the abstract, and those only amongst men who practise virtue.[3]

The morning of 1 January 1849 was beautifully bright, and so from Tashiding, Joseph Hooker and Archibald Campbell crossed the Rathong River and headed towards Pemayangtse, which lay to the west.

The monastery sat on a ridge at 2,159 m, the same elevation as Darjeeling, and was one of the oldest and most important religious establishments in Sikkim. Just below the monastery were the ruins of Rabdentse (Phieungoong), once the capital of Sikkim, though following repeated attacks from Nepal, the Rajah and his court had moved the capital further inland to Tumlong. The views from the main temple were stunning, and it was possible to see every vegetation type in a single glance, from the tropics to the poles:

> The view of the snowy range from this temple is one of the finest in Sikkim; the eye surveying at one glance the vegetation of the tropics and the Poles. Deep in the valleys the river-beds are but 3000 feet above the sea, and are choked with fig-trees, plantains and palms; to these, succeed laurels and magnolias, and higher up still, oaks, chestnuts, birches, &c.; there is, however, no marked line between the limits between these two last forests, which form the prevailing arboreous vegetation between 4000 and 10,000 feet, and give a lurid hue to the mountains. Pine forests succeed for 2000 feet higher, when they give place to a skirting of rhododendron and barberry. Among these appear black naked rocks, rising up in cliffs, between which are gulleys, down which the snow now (on the 1ˢᵗ January) descended to 12,000 feet. The mountain flanks are much more steep and rocky than those at similar heights on the outer ranges, and cataracts are very numerous, and of considerable height …[4]

A doorway leading to the interior of one of the temples at Tashiding.

Hooker's tent pitched by a giant stupa beneath the temple at Pemayangtse, the most important Buddhist monastery in Sikkim.

By pine forests, Hooker generalised, meaning forests of conifers typical of the region, such as *Abies densa*, *Tsuga dumosa* and *Picea spinulosa*. This was exactly what he had come to the Himalaya to see; having previously explored a polar flora, Hooker now wished to explore a more tropical flora, and here, on a single mountain slope, he gazed on every extreme.

Like Tashiding, the main temple building stood on a platform, and the interior was beautifully decorated, though Hooker found the designs coarser than those of Tashiding. Pemayangtse was famed throughout Sikkim for its antiquity and for being the residence of the Head Lama, appointed or ordained in Tibet. Though bound to chastity, exemptions in favour of Lamas of rank, power or wealth, were granted by the supreme pontiffs, both in Tibet and Sikkim, and Hooker claimed that he constantly found children around the lamaseries, who were invariably called nephews and nieces.

Snow lay on the ground around the main temple building so the two explorers descended the hill and camped in a glade below, among stupas and mani stones; Hooker fell asleep that night to the warbling of nightingales. The following morning, having explored the ruins of the former royal court at Rabdentse, it was time for Campbell to leave:

The ruins of the former capital, Rabdentse, are clearly visible from Pemayangtse. Hooker explored them during his visit to the region.

Here I bade adieu to Dr. Campbell, and toiled up the hill, feeling very lonely. The zest with which he had entered into all my pursuits, and the aid he had afforded me, together with the charm that always attends his companionship with one who enjoys every incident of travel, had so attracted me to him that I found it difficult to recover my spirits. It is quite impossible for anyone who cannot from experience realise the solitary wandering life I had been leading for months, to appreciate the desolate feeling that follows the parting from one who has heightened every enjoyment, and taken far more than his share of every annoyance and discomfort; the few days we had spent together appeared then, and still, as months.[5]

In rather gloomy mood Hooker returned to the monastery at Pemayangtse, and spent the remainder of the day sketching the temple, chatting with the Lamas and drinking butter tea, for which he had finally acquired a taste.

The next stage of his journey was ambitious. He planned to travel along the southern spur of Kangchenjunga, north to Dzongri (Jongri), which lay five days away, on the road to the long deserted pass to the Khang La (5,054 m), by which he had previously intended to enter Sikkim from Nepal, before he discovered the route up the Yalung valley towards the pass was impracticable.

From the temples he descended to Chonpung (Tchonpong) to buy sacks of rice for his party, noticing on the slopes above the village enormous groves of the paper-yielding *Edgeworthia gardneri*, with rounded heads of deliciously scented yellow flowers.

Crossing the Rathong, at that point a wide, furious torrent, he and his men camped on a gravel area above the river in dense thickets of bamboo, wild bananas, enormous grasses and rank-scented *Artemisia indica*. Interestingly, a hoar-frost covered the spot that night, coating the wild bananas, and other plants normally confined to the hot-houses of Europe, without any apparent damage.

Hooker's sketch of the timber beams at Pemayangtse records just how colourful the temple was in the mid-19th century.

Yuksam — Sikkim's ancient capital

Hooker reached Yuksam (Yoksun), the very last village on his route, before delving into the wilds as he made his journey into the mountains. Yuksam lay deep in an immense valley and occupied a warm, sheltered flat area, with the valley of the Rathong River plunging 549 m on its western edge. Hooker botanised there on his arrival, and again for two days when he returned back down the mountains. The surrounding flora was subtropical, and around the immediate environs of the village he collected plants of the wild sugar cane, *Saccharum spontaneum*, the spectacular Himalayan coral tree, *Erythrina arborescens* and curiously, the Chinese *Tetrapanax papyrifer*, which had only recently been confirmed as the plant whose pith was the source of Chinese rice-paper.

Another *Aralia* relative, with handsome, bold foliage, *Brassaiopsis aculeata* (syn. *B. hookeri*), was lopped and used for fodder in this part of Sikkim. Grass for pasture was a great rarity in this region of the Himalaya and several large-leaved figs, particularly *Ficus auriculata*, and the 'gogan', *Saurauia napaulensis*, a small tree with large elliptic leaves up to 36 cm, were pollarded on a continual basis to provide fodder for livestock.

Yuksam was one of the first areas settled in Sikkim, and was the kingdom's first capital. Its name meant 'the meeting place of the three wise monks' from the three great Lamas who travelled from Tibet in 1641, and were the means of introducing the first Tibetan Rajah or Chogyal into the kingdom of Sikkim. Chogyal means 'religious king' or 'king who rules with righteousness' and the Chogyal's rule had very strong ties with the Lhasa theocracy and to successive Dalai Lamas.

It was a pretty place, with scattered hamlets and charming lanes winding across gentle hills and copses. By the ruins of temples Hooker found several enormous Kashmir cypresses, with straight, erect trunks, harbouring masses of epiphytic ferns and orchids, while pendulous lichens, like old man's beard, draped from the upper branches. Nearby was a small circular lake, buried deep in the woods and fringed by wild plants of *Camellia kissi*.

Yuksam offered up several new plants; on the forest floor grew sheets of *Pellionia heteroloba*, a dense, low-growing, perennial nettle relative that was found across much of South-East Asia by later collectors. Such is the case also with *Leycesteria glaucophylla*, another of Hooker's Yuksam finds, which is nowadays known to be native in Nepal, Tibet, Bhutan and Myanmar. Perhaps his showiest discovery from this region is the lovely *Strobilanthes oligocephala*, a decumbent subshrub bearing blue (occasionally white) flowers in autumn.

The hills and lanes immediately above the village were fringed with bamboo, scarlet flowered *Erythrina arborescens* and the pretty *Osbeckia crinata*, whose showy blossoms reminded him of the hedge-roses of home. By one of these lanes he found a large enclosure, and in it, beneath a grove of Kashmir cypresses, an enormous stupa, the Great Stupa of Norbugang (Nirbogong), at 12 m, the largest of its kind in Sikkim.

Facing it was a long stone altar 4.5 m long, the famous Coronation Stone on which the first Rajah (Chogyal) of Sikkim was consecrated by the three Tibetan Lamas with the holy waters from the Kuthock Lake, which Hooker had just explored. The enormous Kashmir cypress

The temple at Dubdi sat on a broad, paved platform and was dominated by an enormous Kashmir cypress. Its crown was heavily laden with epiphytic shrubs, ferns and orchids, an irresistible scene for a young botanist.

behind the Coronation Stone was believed to have been associated with the ceremony. Behind, lay a 'Song-boom', a small hollow, conical stone building, used for burning juniper branches as incense.

In the evening Hooker ascended to the Dubdi monastery, along a broad path that ran up the steep, forested slopes above Yuksam. There, 305 m above the village, on a broad, partially paved platform, stood two temples, surrounded by several more Kashmir cypresses. Hooker (see chapter 7, p. 108) singled one of these out for special mention, and sketched it; it was perhaps the oldest tree in Sikkim.

The temples at Dubdi were remarkable for their heavily two-storied porticos, though the interiors were in a near ruinous state. Hooker found the view of Yuksam from here fascinating; the village lay on an enormous plain, in a part of Sikkim where there was hardly an acre of level land, and he tried to understand the manner in which this landscape had been formed.

Another sacred site near Yuksam was Khechopari (Catsuperri) Lake, which was sacred to both Buddhists and Hindus. From Yuksam, this meant a journey west across the Rathong River valley and climbing again on to a steep spur on the opposite side. A Lama from the Khechopari monastery on the hills above the lake joined the group, and Hooker's men took on an air of great solemnity as they approached the water.

Khechopari, the largest lake in Sikkim, lay hidden in dense forest at 1,841 m; the lake itself was around 457 m across, bordered by a wide marsh of *Sphagnum* moss Along its banks the Lamas had planted what Hooker took to be *Rhododendron barbatum* and *Berberis insignis*, transplanted there from the mountains above. The *Rhododendron* was in fact a natural hybrid between *R. barbatum* and *R. arboreum* ssp. *cinnamomeum* Campbelliae Group, *R. × lancifolium* (see Appendix 1).

Hooker crossed the marsh to the edge of the lake across a crude path constructed from decaying logs, at times plunging up to his knees in bog. The Lama had brought with him a boat-shaped piece of bark, a little juniper incense, and a match-box, with which he

lit the juniper, put it in the boat, and launched it across the lake as an offering. Though it was dead calm, the little boat shot across the lake, whose surface was soon covered in a thick layer of white smoke. Hooker was greatly impressed by the scene, and his men even more so.

That evening he pitched his tent by the village of Thingling (Tengling), a few miles east of the lake, and the following morning crossed the Rimbi (Rungbee) stream before ascending a steep spur to the 17th-century Sanga Choeling (Changachelling) monastery.

The Coronation Stone at Yuksam. It was on this same altar-like structure that three Tibetan Lamas crowned the first Chogyal or Rajah of Sikkim in 1641. The structure to its rear was used to burn incense, though it has not been sketched in scale with the Coronation Stone.

The ridge on which the monasteries of Pemayangtse and Sanga Choeling were built was extremely narrow, and was traversed by the famous 'via sacra', connecting the two establishments. It was a particularly picturesque walk, past temples and stupas encrusted with lichens and mosses; to the north the snows of Kangchenjunga rose above the surrounding forests, while to the south Senchal and Darjeeling loomed in the distance. Hooker took numerous bearings from there, and from Pemayangtse, which later proved to be of enormous value in completing his map of Sikkim.

The temples and stupas of Sanga Choeling crowned a rocky, flat-topped eminence on a narrow ridge, their roofs and spires soaring above groves of *Rhododendron arboreum* and towering bamboo. On the summit was a fine two-storied temple, surrounded by the ruins of others.

The temple was being restored at the time of Hooker's visit, and monks were busy re-painting murals on walls of the principal chamber. The monks had been trained as artists in Lhasa, and the paints and pigments had also come from there. Admiring the finer details of the freshly painted murals, Hooker was amazed by one in particular:

> Amongst other figures was one playing a guitar, a very common symbol in the vestibules of Sikkim temples: I also saw an angel playing on the flute, and a snake-king offering fruit to a figure in the water, who was grasping a serpent. Amongst the figures I was struck by that of an Englishman, whom, to my amusement, and the limner's great delight, I recognised as myself. I was depicted in a flowered silk coat instead of a tartan shooting jacket, my shoes were turned up at the toes, and I had on spectacles and a tartar cap, and was writing notes in a book. On one side a snake-king was politely handing me fruit, and on the other a horrible demon was wreathing.
>
> A crowd had collected to see whether I should recognise myself, and when I did so, the merriment was extreme. They begged me to send them a supply of vermilion, gold-leaf, and brushes; our so called camel's-hair pencils being much more superior to theirs, which are made of marmot's hair.[6]

The interior of the monastery at Pemayangtse with butter lamps and silk scarves. Hooker found the temple interiors curiously similar to the Roman Catholic church.

Sikkim lamas, one holding a prayer wheel. Hooker's sketch was later engraved for *Himalayan Journals*.

Hooker was then entertained to butter tea, and the Head Lama, a jolly, fat man, came to share breakfast, followed by numerous children; 'nieces and nephews' he told Hooker, who found them suspiciously like him for such a distant relationship. From the temples at Sanga Choeling he and his men descended into the valley of the Kalek River, from there making at brief stop to visit the village headman at Lingdam, who pleaded with Hooker to get him a pair of spectacles, so he too could look wise. He told Hooker that his had attracted respect throughout Sikkim, and after all, he had been wearing them, in the temple murals at Sanga Choeling.

On 19 January, having travelled south for several days, Hooker and his men ascended the Tukvar (Tukvor) spur to the hill station at Darjeeling, which they found practically empty of Europeans. Brian Hodgson had departed for a shooting expedition in the Terai, while Archibald Campbell was on official business on the Bhutan frontier. Darjeeling lay silent, and the surrounding forests had that dismal look of winter, though Hooker's Lepcha porters rejoiced at being home again.

CHAPTER 7

OUR BUDDHIST PILGRIMAGE – THE 'VIA SACRA'

November is undoubtedly one of the best times to visit Sikkim. By then the post-monsoon rains have cleared, the skies above are gentian-blue and there is still a staggering variety of exciting garden-worthy plants in bloom.

Modern-day Sikkim is one of India's smallest states, covering an area of 113 km long by 64 km wide, with undulating topography and huge contrasts in altitude. Over 82% of this area is classified as forested and almost half the flora and fauna is protected in National Parks and Sanctuaries. Modern-day Sikkim is one of Asia's greatest biodiversity hotspots, with 4,458 species and intraspecific taxa of flowering plants. Of these there are 515 species of orchids, 30 primula species, 38 rhododendrons, and over 500 ferns and fern allies.

As we have seen in the previous chapter, Sikkim's great floral wealth was first made known to the Western world by Joseph Hooker, when he visited the region in 1848–49. In November 2013, I led a second successive group of Irish botanists, horticulturists and tree enthusiasts to retrace Hooker's historic route. During this expedition we concentrated on the regions unvisited during our previous autumn trip in 2012, in particular the mountains and valleys of west and central Sikkim, the areas of Lagyap and the Nathu La in the extreme east, and Darjeeling and nearby Senchal and Tiger Hill in West Bengal.

Prayer flags and peaks, the scenery around the small town of Ravangala. This part of Sikkim is in the heart of Buddhist country.

West Sikkim — big tree country

On 15 December 1848, following his tour of east Nepal, Joseph Hooker crossed a mountain pass on the Singalila Range into west Sikkim.[1] A few days later he joined Archibald Campbell at the small town of Samdong on the Teesta River, and the two men explored the nearby Buddhist monasteries before trekking to the summit of nearby Mount Maenam, a densely forested peak rising to 3,263 m.

Our 2013 expedition was primarily based on the routes taken by Hooker and Campbell, and we started from the small town of Pelling, 2,085 m, famed for its expansive views north towards Kangchenjunga. On the morning of 10 November we set off along west Sikkim's famous *via sacra*, on which this Himalayan State's oldest and most famous golden-roofed Buddhist monasteries are perched.

Built in 1697, Sanga Choeling (Changachelling), the oldest of Sikkim's gompas, was the first monastery on our itinerary and was within walking distance of our base. It appeared on the horizon just as Hooker described it:

> Changachelling temples and chaits crown a beautiful rocky eminence on the ridge, their roofs and spires peering through groves of bamboo, rhododendron … the ascent is by broad flights of steps cut in the mica-slate rocks, up which shaven and girdled monks with rosaries and long red gowns, were dragging bamboo stems, that produced a curious rattling noise.[2]

The bamboos referred to were probably the culms of *Dendrocalamus sikkimensis*, used for scaffolding while they restored the interior of the main monastery building. Among the many freshly painted murals in the interior, Hooker had been highly amused to see a depiction of himself, though sadly this mural had been painted over by the turn of the twentieth century.

The little monastery at Sanga Choeling. Hooker found his image painted in a mural on one of its interior walls.

The monastery at Sanga Choeling sits on a narrow steep ridge. Its gold roof was visible for miles around.

Our visit to Sanga Choeling, like Hooker's, was on foot, along a dirt track and up the same flight of steps. In the far distance we could see the summit of the hill: an enormous mass of chlorite slate, flat-topped and crowned by the ancient gompa, whose roof of glittering gold reflected brilliantly against an intensely deep-blue cloudless sky.

The flora of these lower hills and valleys is warm-temperate, and one of the most dominant trees in the area was *Alnus nepalensis*, an autumn-flowered species, carrying masses of pendent catkins. Its trunk hosted several superb flowering plants of the epiphytic *Dendrobium longicornu*, a glorious orchid with beautiful sprays of white blossoms with a golden-brown labellum. Our interest however,

Silk prayer flags caught in a breeze beneath an intensely blue sky at Sanga Choeling.

Dendrobium longicornu seen here growing epiphytically on *Alnus nepalensis* below the Sanga Choeling monastery.

paniculata, a wide-spreading species with large, lax, terminal panicles of rose-purple blossoms. In the surrounding forests large shrubs of *Mahonia acanthifolia* bore stout, densely packed racemes of rich yellow flowers above handsome pinnate leaves. The young foliage of this species is of a beautiful salmon-red hue and it is equally common around Darjeeling; alas it is suited only to the mildest areas of Britain and Ireland.

One of the most exciting plants we were to find beneath the monastery was *Astragalus stipulatus*, a woody perennial up to a metre tall, bearing large, handsome pinnate leaves, reminiscent of those of the South African *Melianthus major*. It is a stunning plant, especially when in flower, bearing densely packed racemes of yellow pea-like blossoms, from the axils of leaves, whose petioles are slightly hidden by large highly decorative green stipules. Discovered by Francis Buchanan-Hamilton in Nepal in 1802, it is reported to have been cultivated in England as early as 1822, though I know of no one growing it today. The astragali are generally regarded as being of botanical interest only, though *A. stipulatus* is a show-stopper and certainly impressed our group.

was diverted to a grove of young trees of *Exbucklandia populnea*, with a heavy canopy of bright, shining poplar-like leaves and distinctive fleshy stipules. The last time I had seen this tree in the wild was on the Daweishan Range in SE Yunnan, China, in 2005 and I was struck by the similarity of the flora of the two regions. On the edge of a forest dominated by *Quercus thomsoniana* and fine fruiting trees of *Magnolia doltsopa*, *Rhododendron arboreum* formed superb tall trees with broad spreading crowns laden with epiphytic orchids, ferns, *Agapetes serpens* and various *Vaccinium* species.

By the track I was pleased to find one of my favourite shrubs, *Luculia gratissima*, carrying an impressive display of intensely scented, almond-pink flowers in rounded corymbs. One of the most abundant shrubs in the warmer interior valleys of Sikkim is *Oxyspora*

Luculia gratissima, one of the showiest Himalayan shrubs. In Sikkim it is generally confined to the warmer, low-lying valleys. It is powerfully fragrant.

Astragalus stipulatus, an exciting foliage plant, though if ever introduced to cultivation it most likely will be suited only to the mildest coastal districts of Britain and Ireland.

Impatiens stenantha. The balsams are richly represented in the flora of the Sikkim Himalaya and this is a group that fascinated Hooker, even into old age.

The most important of all Sikkim's monasteries is the gompa at Pemayangtse, established in 1705 and visited by Hooker and Campbell on New Year's Day 1849. Set on a hilltop at 2,060 m, the views of immensely deep subtropical river valleys and soaring snow-capped distant peaks are magnificent. As Hooker pointed out, it is possible to see every vegetation type from this famous temple, from fully tropical to alpines beneath the ice fields. In all, the altitudinal distance of the views from Pemayangtse, between the low-lying valleys and the highest peaks is a staggering 7,620 m. Before leaving the monastery area we visited a small flat piece of ground beneath the main temple, where, 164 years previously, among the giant stupas, Hooker had pitched his tent and fallen asleep to the song of nightingales.

With it grew *Gleichenia longissima*, an apt epithet for a fern whose fronds were over 4 m long. In damp shade *Impatiens stenantha* colonised shattered slate at the base of a sheer cliff. Hooker was an expert on this taxonomically difficult group, and he described this pretty yellow-flowered balsam from material he collected in the Lachen valley in May 1849 and from an earlier collection made by one of William Griffith's collectors in the Khasi Hills of Meghalaya (Assam) in north-east India. In his correspondence with Charles Darwin, Hooker estimated that there were probably about a hundred species in India. Nowadays we know the figure to be twice that. In India the balsams have a high degree of endemism, with a very localised pattern of species distribution, thus exhibiting an extraordinary level of diversity within the genus.

The approach to the gompa at Pemayangtse. The stupa to the left is the second, upper stupa that appears in Hooker's sketch. The first, lower stupa, still exists.

Pemayangtse, the most important of all Sikkim's Buddhist monasteries. The monks in the foreground are seen raising a flagstaff.

Khechopari Lake, another place visited by Hooker during his travels in Sikkim. The place possesses a magical atmosphere.

The little town of Yuksam sits on one of the few areas of flat ground in west Sikkim. To the right is the valley of the Rathang River.

Yuksam and Dubdi — Giant Kashmir cypresses

From Pemayangtse our travels took us north-west to Khechopari Lake, a tranquil stretch of water sacred to Buddhists and to the Lepchas. We visited at dusk, with enough light to see the densely wooded hills reflect in the lake's tranquil, mist-shrouded waters. The atmosphere was magical and the place certainly has an aura about it. Prayer flags lined the shores of a lake full of rainbow trout (fishing is strictly forbidden) and we left in the dark beneath enormous trees of *Castanopsis tribuloides*.

From Khechopari Lake our route followed the course of the Rathong (Ratong) River to the tiny village of Yuksam, which is situated on a flat valley floor and sheltered by an amphitheatre of pyramidal, heavily forested hills and snow-capped mountains. The flora there is warm-temperate and on the edge of the village was a copse containing Roxburgh's fig, *Ficus auriculata*, a small tree with large, bold, broadly ovate leaves and small cauliflorous fruits springing directly from the stem. Its bedfellows included *Rhododendron arboreum* and *Engelhardtia spicata*, the latter draped with handsome pendulous spikes of clustered three-winged nuts.

Yuksam, the first capital of Sikkim, was where the trio of Tibetan holy men crowned the first Chogyal (divine ruler) of Sikkim in 1641. The coronation stone, sketched by Hooker in January 1849,

The Coronation Stone at Yuksam. Sketched by Joseph Hooker, the first Chogyal of Sikkim was crowned on it in 1641. Behind it is one of several Kashmir cypresses that grow on this historic site.

Our group at the Great Stupa at Yuksam. Front row, left to right: Neil Porteous, Kristin Jameson, Tracy Hamilton, Margie Phillips, Helen Dillon, Robert Wilson-Wright, Thomas Pakenham. Centre row: Sailesh Pradhan, Adam Whitbourn, Paul Stewart, Philip Quested, Billy Alexander. Back row: Bruce Johnson and the author.

High above Yuksam, on a forested ridge, is Dubdi (the hermit's cell); established in 1701, it is one of the oldest monasteries in Sikkim and is surrounded by a pretty garden. A 40-minute walk along a steep, stone-paved pathway through old-growth forest brought us to the summit on which the gompa is located. The forests there were dominated by *Docynia indica*, *Magnolia velutina*, *Alnus nepalensis*, *Toona ciliata*, *Spondias axillaris* and *Castanopsis indica*, the latter forming enormous, fat-trunked 30 m tall trees.

Ilex excelsa is the common holly of this area and was smothered with crowded clusters of white blossoms. The kiwi-fruit relative, *Saurauia napaulensis* is commonly found across much of the Himalaya and western China, and around Dubdi and Yuksam it formed small trees with large elliptic leaves. Beneath the forest canopy was a medley of superb garden plants of which *Edgeworthia gardneri* was abundant, and we were delighted to find several plants in bloom. With it grew fruiting shrubs of *Dichroa febrifuga*, *Clerodendrum glandulosum* (with clusters of showy white blossoms) and the splendid fern, *Angiopteris evecta*.

Tracy Hamilton, one of my travelling companions, seen here with one of the giant Kashmir cypresses at Yuksam. Behind it is the Coronation Stone.

The Kashmir cypress at Dubdi. This is the same tree Hooker sketched when he visited the monastery. Compare to Hooker's image in chapter 6.

still exists, as does the enormous 12 m high stupa, both of which he mentions in *Himalayan Journals*. The sight that left our group in awe however, was an enormous 29 m tall tree of *Cupressus cashmeriana*, believed to be associated with the original coronation ceremony. If that's true, the Yuksam tree is over three and a half centuries old, which is likely, considering the age and size of the trees in Hooker's sketch of the pair of trees among the stupas at nearby Tashiding.

It is debated as to whether or not these weeping cypress, known in Sikkim and Bhutan as *tsenden*, are in fact *C. cashmeriana*. The tree known in European gardens as the Kashmir cypress (the tree is not native to Kashmir) was first cultivated in France and Italy; the largest tree in Europe grows on Isola Madre in northern Italy. This particular clone, which has been widely propagated vegetatively, differs from Himalayan trees in its wonderful grey-blue heavily pendulous sprays of foliage. Trees in Sikkim and Bhutan, while still very beautiful, are generally of a dull green shade, and young trees are much more cone-shaped in their habit.

This has led to the belief by some authorities that the *tsenden* of Sikkim and Bhutan is quite a different species, though this has not been accepted by conifer experts at Kew. My guess is that the lovely grey-blue European Kashmir cypress is a selection, an outstanding horticultural form.

The main monastery building at Tashiding. All the monastic complexes visited by Hooker have been modernised in recent times, though Tashiding still remains the most fascinating place on the *via sacra*.

On the edge of forest between Yuksam and Dubdi, one of my companions, Billy Alexander of Kells Bay House and Gardens in Co. Kerry, stumbled across what we would later vote 'plant of the day'. A huge, 4 m tall bush bearing enormous panicles of azure-blue flowers, it was later identified as *Spermadictyon suaveolens*, a giant subshrub widely distributed in the hot, dry lower valleys of the Himalaya; it made a great impression.

Spermadictyon suaveolens, one of the most spectacular flowering shrubs of the warmer valleys of Sikkim. It won our vote for 'plant of the day'.

Bauhinia vahlii, the camel's foot climber, formed large evergreen vines bearing distinctive, rusty-haired two-lobed leaves. On the forest edge, *Parochetus communis* sported its pretty blue pea-like flowers close to a fruiting spike of *Arisaema utile* and a showy clump of *Onychium japonicum*, a fern I have previously encountered in central China.

The monastery at Dubdi, visited by Hooker, has been modernised in recent times. We were pleased to encounter another gargantuan Kashmir cypress, the same tree evidently that Hooker sketched (see chapter 6, p. 101) in 1849 and appears in *Himalayan Journals*. He described the Dubdi trees as follows:

> In the evening I ascended to Doobdi … nearly 1000 feet above Yokson, is a broad partially paved platform, on which stand two temples, surrounded by two cypresses: one of these trees (perhaps the oldest in Sikkim) measured sixteen and a half feet in girth, at five feet from the ground, and was apparently ninety feet high: it was not pyramidal, the top branches being dead and broken, and the lower limbs spreading; they were loaded with masses of white-flowered Coelogynes and Vacciniums. The younger trees were pyramidal.[3]

Tashiding is one the most remarkable monastic complexes on the *via sacra*. Located on the summit of a helmet-shaped hill at 1,465 m, it occupies a spectacular site above the confluence of the Rathong Chu and the Rangit River, with Kangchenjunga providing a spectacular backdrop. Founded in 1641 by one of the three lamas who held the coronation ceremony at Yuksam, the present buildings date back to 1717, though their exteriors have been modernised in recent times.

Five colourful religious buildings are strung out between the monks' quarters; the four-storey prayer hall is the most elaborate of these and was chosen by the Dalai Lama in 2010 for a two-day meditation retreat. Just beyond the monastery complex is a striking compound of white-washed stupas, overshadowed by two mature

Among the stupas at Tashiding. It's said that by staring at one of these stupas, that one's sins are forgiven for a lifetime.

Among the stupas at Tashiding. The pair of Kashmir cypresses sketched by Hooker have grown fatter over the past 166 years.

Kashmir cypresses, a scene sketched by Hooker in December 1848. These cypresses, like those at Dubdi, have also survived and it felt rather remarkable to gaze on the same trees that Hooker had admired and sketched 165 years earlier. Nowadays they are enormous, sheltered by equally large trees of *Toona ciliata* and *Spondias axillaris*.

Surrounding the main monasteries were dozens of younger Kashmir cypresses, all of a typically, tight pyramidal habit. Before leaving the monastery we gave Hooker's sketch of the stupas and cypresses to a delighted crimson-robed monk, who immediately recognised the scene. Dusk was falling as we left Tashiding, to the sound of a large bell echoing through the ancient stupas, and the breathtaking mountain views gradually turned fiery amber in the setting sun.

Our group on the mid-forested slopes of Maenam. Most of the trees overhead are *Lithocarpus pachyphyllus*.

The Big Buddha at Ravangala stares on serenely. Behind are the lower slopes of Mount Maenam and the jagged peaks of the Chola range of mountains.

The Big Buddha at Ravangala and Mount Maenam

Ravangala is a small town beautifully perched on a ridge at 2,010 m, overlooking an expansive sweep of the western mountainous regions of Sikkim. Just outside the town is the stunning and beautifully serene 41 m tall golden Buddha, framed against a horizon broken by jagged snow-capped peaks that pierce a cloudless ink-blue sky.

Ravangala was our base while we climbed and botanised on nearby Mount Maenam (Mainom), at 3,263 m the highest peak within the Maenam Wildlife Sanctuary. This mountain was first explored botanically by Campbell and Hooker during Christmas 1848, and has been little visited by Western botanists since, so there was an element of anticipation to our visit.

Maenam means 'treasure-house of medicines', a reference to the mountain's rich array of medicinal plants long utilised by the Lepcha people. It has an exceptionally rich flora, with forests of giant magnolias and rhododendron, and also provides refuge for red pandas, leopards, Asiatic black bears, goral, monal pheasants and black eagles.

Mt Maenam's flora consists of several climatic types, and it was interesting to observe the altitudinal sequence of the vegetation as we climbed its slopes. From 2,100 to 2,400 m, the slopes are enveloped by broadleaved evergreen subtropical forest dominated by trees such as *Castanopsis indica*, *Magnolia doltsopa*, *Engelhardtia spicata*, *Juglans regia* and *Schima wallichii*.

Between 2,400 and 2,700 m, wet-temperate forest hosts species such as *Quercus lamellosa*, *Q. thomsoniana*, and the related *Lithocarpus pachyphyllus*, as well as *Acer campbellii* and *Magnolia campbellii* (both named for Hooker's travelling companion); many of these trees are hosts to epiphytic ferns, orchids, algae and fungi.

Moist, cool-temperate forest is finally encountered near the summit between 2,700 and 3,000 m, where the forests are composed mainly of *Abies densa*, *Sorbus hedlundii* and *Rhododendron falconeri*, while the summit area, at 3,000–3,263 m, is primarily scrubby sub-alpine forest with dwarf *Rhododendron* shrubbery, *Buddleja colvilei*, and sub-alpine herbaceous plants.

There is also an extremely diverse avifauna, with several restricted-range birds, a number of which commemorate Hooker's friend at Darjeeling, Brian Houghton Hodgson. While best remembered in Western gardens by the superb tree-like *Rhododendron hodgsonii*, in the skies over Maenam his namesakes include *Tickellia hodgsoni* (broad-billed flycatcher-warbler), *Muscicapella hodgsoni* (slaty-backed flycatcher) and *Columba hodgsoni* (speckled wood-pigeon).

Daphyniphyllum chartaceum, a handsome evergreen. It formed large trees on the lower slopes of the mountain.

first found by Francis Buchanan-Hamilton in Nepal. Native to the eastern Himalaya, Myanmar, Thailand and W China, this stunning little orchid created a spectacular show on Maenam's lower slopes, and indeed, in many of the other warm-temperate areas we visited. Literally thousands of plants clung to the trunks of surrounding trees and lit up the ancient forests with their large, showy pinkish-purple blossoms. Hooker collected *P. praecox* on the hills near Darjeeling, where it is still abundant. *Pholidota imbricata*, another autumn-flowered epiphytic orchid, was also quite abundant and carried dense, many-flowered peduncles of creamy-white blossoms.

Daphniphyllum chartaceum was another astounding large tree on these lower slopes; here it grew up to 21 m tall, with slender absolutely straight trunks clear of branches for the first 12 m. Strangely, authorities have sunk this superb species into synonymy with the more widespread *Daphniphyllum himalayense*. Anyone who has studied both species in the field will be left with no doubt that they are quite distinct.

Pholidota imbricata, a common autumn-flowered orchid in Sikkim. It was common in the trees on Mount Maenam, though this plant is photographed at Khechopari Lake.

Botanising Mount Maenam

A local forest guide led our way along Maenam's slopes, along a narrow mud track overlooking Ravangala's golden Buddha. Along the lower slopes we walked alongside a network of gushing water pipes; the mountain is the only reliable perennial source of water both in this area and for virtually all of southern Sikkim.

The predominant tree here was one of Hooker's Darjeeling discoveries, *Lithocarpus pachyphyllus*, a large evergreen oak-relative that was common in the 19th century on the mountains surrounding Darjeeling and near the summit of Tonglu. These formed enormous, straight-trunked trees and we arrived as they were shedding their enormous fruits.

These ripe fruiting spikes posed two dangers during our visit to Maenam; firstly, their sheer weight, falling from 30 m overhead (several of our party had near misses) and secondly, the fact that they are a favourite food of the local Asiatic black bears. Descending the mountain that evening, our guide sang and chanted, keeping our group close together, in a bid to prevent a bear attack. Such are the dangers of modern-day plant exploration!

With the *Lithocarpus*, in places, grew *Quercus lamellosa*, which, as previously mentioned, is a superb host to epiphytes, and we spotted numerous orchids flowering in these high-rise aerial gardens. The showiest of these was the superb autumn-flowered *Pleione praecox*,

Lindera neesiana, one of the prettiest autumn-flowered trees in Sikkim, is relatively common at these altitudes in Sikkim, and formed small trees, covered with dense umbels of small yellow flowers. *Maesa chisia,* another small understorey tree or bush, also appeared on the scene. It is common in the eastern Himalaya, and on Maenam bore short, stout racemes of small green-white blossoms during our visit.

We continued our journey uphill, determined to reach the temperate zone as soon as possible. This area, at 2,335 m, showed signs of past fire damage in several places. Here *Acer campbellii* formed superb column-shaped specimens to 27 m tall, and by then its foliage had assumed butter-yellow autumnal hues, which contrasted brilliantly with the cloudless blue sky. It was pleasing to see such venerable specimens of Campbell's maple on the very slopes he had climbed so many years before. Originally discovered by William Griffith's collector in Darjeeling, Hooker later recollected it, and named it in 1875 for his friend, the Superintendent of the hill station at Darjeeling.

Schefflera rhododendrifolia, one of the most exotic foliage plants we found on Mount Maenam where it formed multi-stemmed trees.

Acer campbellii, Campbell's maple, formed enormous trees on Mount Maenam. It was wonderful to see it on the very mountain that Archibald Campbell climbed with Joseph Hooker during Christmas 1848.

Hooker described it as a tree to 15 m tall, though it reaches far more in the more remote valleys of Sikkim. He introduced it to cultivation, but may have collected seeds at low elevations, as the early introductions proved tender, succeeding only in coastal gardens in Britain and Ireland.

At 2,475 m the forests were almost entirely composed of *Acer campbellii, Magnolia campbellii* and *Rhododendron grande*, the latter soaring to an astounding 21 m tall. I was particularly pleased to see such giants, as it is one of my favourite rhododendron species and at Kilmacurragh we grow one of Hooker's original seedlings. This giant cultivated specimen, generally considered the finest in Europe, by no means matches the scale of the colossal trees of Mount Maenam.

Other trees intermixed in the woodland included *Acer sikkimensis* (up to 15 m tall), *Schefflera rhododendrifolia* (similarly tall, though multi-stemmed), *Alangium platanifolium* (a common understorey tree) and *Gamblea ciliata* (a small tree in the Araliaceae with striking five-foliate leaves). First found by Joseph Hooker on the higher slopes of Tonglu, it was described as a new genus as a compliment to James Sykes Gamble, a forester based in India. Though rare in cultivation it will appeal to avid collectors of fine foliage plants. *Polygonatum cathcartii*, one of Hooker's Lachen discoveries, grew epiphytically on the trunks of many of these trees.

Sorbus hedlundii, a mammoth tree on the upper slopes of Mount Maenam. This striking Himalayan whitebeam was first cultivated at the National Botanic Gardens, Kilmacurragh. Sadly, it never reaches similar dimensions in European gardens.

We had a late morning picnic on the summit of Mount Maenam and re-traced our route back down the mountain reaching the base early that evening with plenty of daylight to spare.

Common understorey shrubs here included *Senecio scandens, Viburnum erubescens, Berberis insignis* (sometimes epiphytic) and *B. hookeri* (discovered by Hooker on the upper slopes of Tonglu) and *Hypericum hookerianum. Daphne bholua* was abundant, forming 4 m tall shrubs, some already in bloom and varying in shades from pure white to pinkish-mauve.

The Lardizabalaceae is an interesting relict family of woody climbers and shrubs distributed in Asia and South America. On this mountainside two genera belonging to this group grew side by side: *Holboellia latifolia* formed enormous vines that raced through surrounding trees, while a thicket of *Decaisnea insignis* filled an entire glade, its long pinnate leaves by then ochre brown. Hooker named this striking genus for the distinguished French botanist Joseph Decaisne (1807–1882), Director of the Jardin des Plantes in Paris, who, in his day, was an expert on the Lardizabalaceae. William Griffith, who discovered this handsome species in Bhutan and originally placed it in the genus *Slackea,* was obviously equally impressed, giving it the specific epithet *insignis* (distinguished or remarkable). It is not quite as strong growing as the Chinese *D. fargesii,* and bears golden-yellow fruits.

Above 2,800 m *Rhododendron falconeri* appeared on the scene, some specimens up to 18 m tall and with enormous girths. The dominant tree at this level was *Sorbus hedlundii,* and trees here were up to 21 m tall and obviously very old. Hedlund's whitebeam was introduced to cultivation by Thomas Thomson, when two consignments of his Sikkim seeds reached the Royal (now National) Botanic Gardens, Glasnevin in Dublin in 1858 and 1859.

The gompa on the summit, built at the request of the last Queen of Sikkim. The moon has just emerged in the skies above it.

Mount Maenam doesn't reveal its greatest secret – its spectacular views – until the absolute summit is reached. It certainly makes the climb very worthwhile, with views extending as far as the Chola Range on the Sikkim–Tibet border.

Our travels in west Sikkim began at the small town of Pelling. Seen here is the view from our base there, with Mexican *Cosmos bipinnatus* and *Tagates erecta* beneath the peaks of Kabru and Kangchenjunga.

Sir Frederick Moore (who somehow erroneously referred Thomson's seedlings to *Sorbus thomsonii*) sent seedlings to Thomas Acton at Kilmacurragh, where they fruited for the first time in cultivation.[4] This is probably the finest of the Himalayan whitebeams; in autumn its leaves turn russet and are retained on the trees for a long time. Alas, it doesn't seem to reach the same enormous dimensions in cultivation as it does in its Himalayan homeland.

As the forest thinned out, the altitude began to affect our breathing, and we climbed slowly through thickets of dead bamboo (*Drepanostachyum intermedium*). At 3,050 m higher-altitude rhododendrons appeared, for example *R. arboreum* ssp. *cinnamomeum* Campbelliae Group, *R. triflorum*, *R. barbatum* and the superb *R. thomsonii*. Approaching the summit *Abies densa* dominated the scene, and in the subalpine meadows the silvery plumes of *Calamagrostis emodensis* blew on the wind among scattered clumps of *Cotoneaster glacialis* and two of Hooker's discoveries, *Vaccinium nummularia* and *Gaultheria hookeri*; the latter was carrying pea-sized cobalt-blue fruits. Reaching the top, we were rewarded by one of the most stunning scenes in Sikkim: a panoramic view of the snow-clad Himalaya from Kangchenjunga to the great Chola Range in the east. There we lunched by a tiny Buddhist monastery (built at the request of the last Queen of Sikkim), before retracing our route down the mountainside, with vultures circling overhead.

CHAPTER 8

TO THE SNOWS OF KANGCHENJUNGA

On 7 January 1849 Hooker and his men departed from Yuksam on an expedition to Kangchenjunga. He knew that deep snows would prevent him from reaching the higher slopes, so he decided to try and reach the lower limit of perpetual snow that descended in one enormous sweep from 8,534 m to 4,572 m.

His route took him across the flat lands around Yuksam, and then due north along the precipitous east flank of the Rathong River, following the rough track through dense subtropical forest. This proved to be tough going, with other rivers cascading down from the mountain slopes above, and Hooker and his men risked their lives to cross these torrents using bamboo stems to steady their way.

The trail he was following was an old salt smuggling route, which crossed the Khang La into the Yalung valley in east Nepal, where salt fetched enormous sums; in theory, this pass had been closed since the Nepal wars. The closure of the pass actually had more to do with the common practice of human trafficking, rather than the sale of salt, the authorities at Kathmandu being understandably determined to prevent Sikkim people kidnapping children and slaves from their territories.

Hooker's route continued along the east bank of the steep-sided Rathong River, where immense landslips, having swept the forests into the raging torrents below, exposed the white bank of gneiss and

The turquoise waters of the Rathong River in a gorge above Yuksam. Hooker followed a track to the right of the river to reach Dzongri. We followed the same route in 2014.

Hooker's winter box, *Sarcococca hookeriana*, seen here at the Royal Botanic Garden Edinburgh. One of the finest Himalayan shrubs for the winter garden on account of its handsome evergreen foliage and powerfully fragrant blossoms.

Mans Lepche

From the upper part of the valley, Hooker and his team headed north-west, crossing the raging waters of the river, and began the ascent of Mans Lepche (Mons Lepcha). He camped again at 2,636 m on the steep slopes at Bakhim (Buckeem), in a great forest of rhododendrons, *Lithocarpus pachyphyllus*, *Tsuga dumosa*, *Abies densa* and *Taxus wallichiana*.

Snow lay on the ground above 2,438 m and the night was cold and clear. Following yak tracks in the snow, he continued his ascent the following day, surrounded by dwarf juniper and rhododendrons. At 3,353 m, snow lay deep in the silver fir woods and his porters waded through it with great difficulty.

The summit of Mans Lepche, at 3,987 m, was a broad plateau, fringed by thickets of *Rosa sericea*, *Berberis*, *Rhododendron anthopogon*, *R. setosum* and dense colonies of the Himalayan heather, *Cassiope fastigiata*. At this altitude, the sun was so powerful that when Hooker reached the summit, he was so warm that he walked barefoot across the frozen snow (which he found preferable to wet socks). The temperature at the time was –1.4°C, later falling to –17°C and the snow sparkled with broad flakes of hoar-frost under a full moon, which was so bright that Hooker could continue taking observations into the night.

granite. Their campsite that night lay at 2,033 m, among plants he had hardly expected to see so close to the snows of Kangchenjunga: a rich mix of oak, maple, birch, rhododendron, daphne, *Jasminum, Arisaema, Begonia,* gesneriads, pepper, figs, *Menispermum,* wild cinnamon, bananas, ginger lilies, vines, ferns and epiphytic orchids.

Higher up the valley the surrounding flora took on a more temperate aspect. It was while travelling up the valley on 7 January 1849 that he discovered *Sarcococca hookeriana*, a winter box that would later become popular in gardens on account of its exquisitely scented, small white blossoms.

The oak relative *Lithocarpus pachyphyllus*. It is still common in the valleys below Mans Lepche and Dzongri.

The view from Mans Lepche. In the centre foreground is the cone-like granite mass of Kabur, dominated to the rear by the pyramidal-shaped Kangchenjunga, while to the right is Pandim with its great black cliff face.

The following morning the sun rose high in the sky, allowing views of the plains of India, 225 km to the south-east. The mountain was spectacularly located and the setting must have reminded Hooker of his tour in east Nepal.

From his vantage point on Mans Lepche, Hooker once again marvelled at being able to discern extreme vegetation types in a single glance:

> The view to the southward from Mons Lepcha, including the country between the sea-like plains of India and the loftiest mountain on the globe, is very grand, and neither wanting in variety nor in beauty. From the deep valleys choked with tropical luxuriance to the scanty yak pasturage on the heights above, seems but a step at the first *coup-d'oeil*, but resolves itself on closer inspection into five belts: 1, palm and plantain; 2, oak and laurel; 3, pine; 4, rhododendron and grass; and 5, rock and snow. From the bed of the Ratong, in which grow palms with screw-pine and plantain, it is only seven miles in a direct line to the perpetual ice. From the plains of India, or outer Himalaya, one may behold snowy peaks rise in the distance behind a foreground of tropical forests; here, on the contrary, all the intermediate phases of vegetation are seen at a glance. Except in the Himalaya this is no common phenomenon, and is owing to the very remarkable depth of the river-beds.[1]

The plantain Hooker refers to is the wild Indian banana, *Musa balbisiana*, one of the ancestors of the common cultivated bananas, alongside *Musa acuminata*. Of oaks, Hooker was referring to the evergreen oak-relatives *Lithocarpus elegans* and *L. fenestratus*, and the true oaks, *Quercus glauca* and *Q. lamellosa*. No pines occur in this region. In this case Hooker was referring to *Abies densa*, *Juniperus recurva*, *Larix griffithii* and *Tsuga dumosa*.

From Mans Lepche Hooker and his team of collectors and porters headed in a north-westerly direction to Dzongri (Jongri), along an open, bare mountain spur, past the dry beds of pools and the odd thicket of dwarf junipers and alpine rhododendrons, which broke the monotony of this harsh moorland landscape.

After a two-hour walk they came within sight of Dzongri, a wide, open alpine plain, backed by the cone-shaped peak of Kabur, lying to the north and east in an amphitheatre of soaring snow-clad mountain peaks. Two stone huts were the only signs of human presence on this bleak spur; Hooker gave up one of these to his men, while he made a make-shift tent outside. Dzongri was a summer grazing ground for the inhabitants of Yuksam, but was wholly deserted during Hooker's visit in January 1849. The ground was so frozen around the campsite that it took four of Hooker's porters, working with hammers and chisels, several hours to dig a hole just 41 cm deep to sink a ground thermometer.

Hooker spent the afternoon botanising, though when the sun rose, the resinous scent of the foliage of *Rhododendron anthopogon* and *R. setosum* was oppressive, particularly since any form of exertion at this altitude brought on throbbing headaches. Mosses, one of his specialist interests, were few, though crustaceous lichens abounded, and their specific epithets like *tristis* (dull), *gelida*, *glacialis*, *arctica* and *frigidus* gave an indication of their origin, their vivid colours or their weather-beaten aspects. He was particularly pleased to stumble across *Rhizocarpon geographicum*, the map lichen, named not to indicate its

Pandim, seen here from the valley of the Prek Chu. Its black cliff face makes it one of the most distinctive mountains in north-west Sikkim.

Rhododendron lanatum, one of Hooker's Dzongri discoveries. It has particularly beautiful foliage, with the undersides of leaves lined with a woolly brown indumentum. Hooker also introduced it to cultivation.

global distribution across the highest mountains of the world on both sides of the equator, but for its curious map-like patterns, which its yellow crusts form on rocks.

Of the many plants Hooker discovered on the steep mountain slopes above Dzongri, perhaps the most beautiful was the stunning blue-flowered *Corydalis ecristata*, a species he initially mistook for the better-known *C. cashmeriana*, though his plant was far smaller, not reaching much more than 3–5 cm high. It was collected, with permits, at Dzongri, by the 1983 Alpine Garden Society Expedition to Sikkim, who described the plant they saw 'as a treasure, appearing above the tuft as clusters of sapphires set against the grey boulders. Stunning and startling – and tantalising, for as we reached out later to pick the still green seed pods, they burst at the first touch.'

On the spur above Dzongri Hooker found enormous blocks of gneiss, which he correctly guessed were transported by pre-existing glaciers, during a period when the planet had a cooler climate. There the boulder-strewn moorland was cut with masses of little pools, then dry, though some of large circumference.

It had been freezing hard all day, even with a powerful sun shining, and by evening a north wind set in. At sunset, the moon rose, illuminating the snow-capped peaks while Hooker sat watching the weather anxiously. Dzongri was particularly exposed to heavy snow drifts and the track leading there was difficult to find. During the night the temperature fell to –11.5°C .

The following morning, he set off to explore a rocky valley on the western flank of Kabur, where he managed to make a good collection

The alpine plateau at Dzongri. The vegetation at this altitude is mainly dwarf thickets of *Juniperus indica* and two dwarf *Rhododendron* species, *R. anthopogon* and *R. setosum*, both of which are highly aromatic.

of plants, though he failed to ascend the mountain. Temperatures remained beneath freezing all day, with a piercing, damp south-west wind that numbed two of his porters so badly that they needed assistance to return to the campsite.

By 11 a.m. a thick fog forced them to retreat, followed all day by heavy snowfall, and that night an Arctic-like cold set in. Hooker was greatly worried about exposing a party of porters to so great a danger at such a cold season. The snow continued to sweep down all night and his heart sank, his fire declined and his little puppy Kinchin, who had spent the early part of the day running about playfully in drifts of snow, began to whine and crouched beneath his master's woollen cloak.

During the night the roof of Hooker's make-shift tent began to bow beneath the weight of snow. Dreading the thought of suffocation, he propped it with sticks, placed a tripod over his head, and slept soundly with his dog at his feet. By the next morning 60 cm of snow had fallen, and immense drifts of snow obscured his route back down the mountainside. His minimum thermometer had fallen to –15.7°C during the night, and he was glad that no harm had come to his men. The next morning his porters loaded their provisions without a single complaint, seeking protection for their eyes from the freshly fallen snow. Hooker provided some with pieces of a crape veil, others made shades from brown paper, or from the hairs of yaks' tails, while the Lepchas simply loosened their pigtails and combed it over their eyes. It was a difficult descent, and it was dark by the time they reached their base at Bakhim.

The black mass of Kabur, with the snow-capped Kabru to its rear. The mountain to its right is Kangchenjunga.

The following day the party halted by the banks of the torrential Rathong River, frozen at its edges, at the foot of Mans Lepche. *Rhododendron grande* grew on its banks and Hooker was surprised to see a beautiful filmy fern, which he compared to the Irish Killarney fern, *Trichomanes speciosum*, abundant in the region. The same day, at the same altitude (2,179 m), he gathered no fewer than 60 species of ferns; this lush vegetation lay just a day's march below the frozen landscape at Dzongri. Returning down the river, he and his men reached Yuksam that evening. No Western explorer had ever ventured so deep into the wilds of Sikkim, or had come so close to the snows of Kangchenjunga.

The mountains and deep valleys below Dzongri receive enormous annual rainfall, creating cloud forest-like conditions. Filmy ferns are still abundant in the region. Pictured here is a species of *Trichomanes*, which smothered the stems of *Rhododendron lanatum* on the approach to Dzongri from Tshoka.

The map lichen, *Rhizocarpon geographicum*, seen here on a rock near the Goecha La, a mountain pass above Dzongri leading to Kangchenjunga.

The Dak bungalow at Pankhabari. This image appeared as a wood engraving in *Himalayan Journals* and was based on Hooker's original sketch (see image in chapter 2, p. 13).

A trip to the Terai

Hooker's collections for 1848 amounted to a staggering 80 loads, carried by as many porters to the foothills, where carts were waiting to carry them for five days to the banks of the Mahananda (Mahanuddy) River, which flows into the Ganges, from where they would be finally shipped to Calcutta.

At the end of February 1849, he left Darjeeling to join Brian Hodgson at Tentulia (Titalya) on the plains. Snow lay around Darjeeling, so a trip to warmer climes was a welcome distraction. Trees of the oak-relative, *Lithocarpus pachyphyllus*, surrounding the hill station were enjoying a mast year, and its enormous, densely clustered acorns were by then falling in such numbers that it was hardly safe to ride downhill.

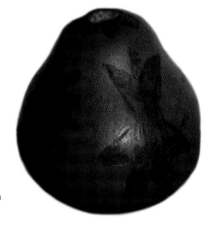

A Lepcha snuff box made from the fruit of a bottle gourd, *Lagenaria siceraria*. Hooker sent several items from Sikkim for his father's collection of economic plants at Kew.

The plains of Bengal were obscured by a dense haze, partly due to fires in the Terai forests, which filled the air as far as the foothills with charred particles of grass stems. *Vitex quinata* was common on the roadsides in the lower tropical valleys and the perfume from its cymes of white blossoms lingered in the warm, still air. Hooker stayed that night in the Dak bungalow at Pankhabari, and woke the next morning to the call of innumerable birds and the humming of enormous bees; these sounds were strangely absent from the Himalayan region.

It was in the forests of the Sikkim Terai that he found Hooker's morning glory, *Argyreia hookeri*, an herbaceous climber bearing large pink-purple flowers. The pileas, nettle relatives, are common understorey herbaceous plants in the subtropical and warm-temperate forests of Asia, and *Pilea hookeriana,* found near streams and damp areas in the Terai forests, was named in his honour in 1856.

The tropical forests of India have two main periods of flowering: summer for the great majority of species, and the second period in winter. The forests below Pankhabari were full of colour as Hooker trekked down the slopes, particularly from members of the acanthus family. Colour was also provided by the spectacular orchid trees *Bauhinia purpurea* and *B. variegata*, and *Dillenia indica*. The latter is commonly called the elephant apple, since its fruits are much favoured by the megaherbivores of the Terai. The most spectacular tree by far was the red silk cotton tree, *Bombax ceiba*. Old specimens were particularly handsome, with fat, smooth-barked, cylindrical stems with enormous buttresses at their base. It was one of the commonest trees there and the forest canopy was transformed into a sea of red by its showy blossoms.

Hooker made good progress across the wooded plains and gravelly slopes of the Terai, which extended from the base of the foothills at Pankhabari to the Dak bungalow at Siliguri, where he stopped briefly. From there he followed the course of the Mahananda River for a further 28 km to Tentulia. Kinchin, Hooker's dog, soon became exhausted by the stifling heat of the plains, and travelled on the saddle of his master's pony, licking his swollen paws, and ultimately falling fast asleep, in spite of the motion.

Joseph Dalton Hooker, a portrait of the explorer when he was 32 years old in 1849, giving a good indication of what he looked like during his Himalayan travels.

The Terai offered up new plant species, including *Lindernia hookeri* and the shrubby spurge relative *Glochidion daltonii*, which were later to bear his name. The latter is a useful plant since its bark, leaves and fruits contain copious tannins. Some of Hooker's other Terai discoveries commemorate notable contemporaries, for example, *Litsea kingii*, named for George King of the Royal Botanic Garden, Calcutta (who also later collected it in Sikkim). King became a regular correspondent of Hooker, sending him regular consignments of seeds and living plants of Indian endemic species.

After leaving the Terai at Siliguri, the woods disappeared and were replaced by towering tropical bamboo and occasional trees of banyans such as *Ficus benghalensis* and the sacred peepul, *F. religiosa*, while betel-nut (*Areca catechu*) was commonly planted around villages.

At the Dak bungalow at Tentulia, overlooking the Mahananda River and the plains beyond, Hooker received a great welcome from Brian Hodgson, who listened avidly to the stories from Hooker's Nepal and Sikkim expeditions. That evening, 27 February 1849, an earthquake struck, with a hollow rumbling sounding, as Hooker put it, like a wagon passing over a wooden bridge.

On the morning of 1 March, Archibald Campbell arrived at the bungalow, and Hooker and Hodgson travelled with him, on the backs of elephants, along the boundary of Sikkim and British India, as far as the Michi River, which marked the frontier with Nepal to the east. The Nepal side of the Terai was noticeably deforested of its large trees, with a scattering of various species of *Diospyros*, *Terminalia* and *Symplocos,* and the spectacular flame of the forest, *Butea monosperma*, which was in full bloom. Associated with the arrival of spring in West Bengal, its showy flowers are used to treat leprosy and skin diseases and are also said to be efficacious in birth control.[2] Tentulia, a town under Campbell's supervision, held occasional bazaars and fairs, and Hooker had arrived in time to see one of these take place:

> Shortly after Dr. Campbell's arrival, the meadows about the bungalow presented a singular appearance, being dotted over with elephants, brought for purchase by the government. It was curious to watch the arrival of these enormous animals, which were visible nearly two miles across the flat plains; nor less interesting was it to observe the wonderful docility of these giants of the animal kingdom, often only guided by naked boys, perched on their necks, scolding, swearing, and enforcing their orders with the iron goad. There appeared as many tricks in elephant-dealers as in horse-jockeys, and of many animals brought, few were purchased.[3]

From Tentulia, Hooker and Hodgson set off across the savannah, in an easterly direction towards the Teesta, to see where it exited from the mountains onto the plains of India. On the second day they arrived at Jalpaiguri (Jeelpigoree), then a large straggling village by the Teesta, where they waited for the arrival of their elephants.

The people of the area were Koch (Cooches), of Mongoloid origin, who replaced the Mech tribes of the Terai. Hooker and Hodgson were greeted by the Dewan (minister), who ruled the district for the Rajah, a young boy of ten, whose estates were tied up in a difficult trial of succession, being pursued by another son of the late Rajah. Hooker got on well with the Dewan, who promised to arrange elephants for his trip to the Teesta.

The Terai, a narrow band of land along the outer foothills of the Himalaya, had been formed from alternating beds of sand, gravel and boulders brought down from the mountains by the many rivers and streams that formed wide alluvial fans when they reached the plains. Consisting of marshy grasslands, savannah and forest, the Terai soils were light and dry, and much favoured by the ubiquitous sal trees (*Shorea robusta*). The area was notoriously malarial and feared by European and Indian settlers alike, although a few ethnic groups like the Mech tribe seemed impervious and eked out a living there.

In addition to the rivers, the area was well watered by the Indian monsoon, and the rich savannah grasslands made good grazing for buffalo and humped Indian cattle, driven from the banks of the Ganges 483 km away. Unfortunately, once there, many cattle were carried off by tigers.

This brief visit to the princely state presented Hooker with the opportunity to attend a Durbar (court) of a royal family. Eastern courts had long been marvelled at for colour, pomp and ceremony, though by the mid-19th century Indian courts were subject to the British East India Company, and as Hooker lamented, were shorn of much of their former glory. The palace was entered through a guard house, where the Rajah's Sepoys, in European uniform, stood guard alongside troops from the East India Company, lent to the young boy for his protection against the several tempestuous pretenders to his title.

Hooker arrived at the palace during the spring Holi festival, the Hindu festival of colour, where those attending the court chased and pelted each other with balls of coloured powder. He was not at all amused, though it is obvious in his description of the event that he was not aware of its significance. The Holi celebrations continued and his second meeting with the young regent went better:

> On the following evening the elephants were again in waiting to conduct us to the Rajah. He and his relations were assembled outside the gates, mounted upon elephants, amid a vast concourse of people. The children and Dewan were seated in a sort of cradle; the rest were some in howdahs, and some astride on elephants' backs, six or eight together. All the idols were paraded before them, and powdered with red dust; the people howling, shouting and sometimes quarrelling. Our elephants took their places amongst those of the Rajah; and when the mob had sufficiently pelted one another with balls and dirty red powder, a torchlight procession was formed, the idols leading the way, to a very large tank, bounded by a high rampart, within which was a broad esplanade round the water.
>
> The effect of the whole was very striking, the glittering cars and barbaric gaud of the idols showing best by torch-light; while the white robes and turbans of the undulating sea of people, and the great black elephants picking their way with matchless care and consideration, contrasted strongly with the quiet moonbeams sleeping on the still broad waters of the tank.[4]

The next morning Hooker and Hodgson set off on their elephants, travelling along the west bank of the Teesta. The river was navigated by canoes up to 12 m long, cut from solid logs of sal trees, while others were built of planks, the seams caulked with the fibres of the root of *Butea monosperma*, and later again smeared with gluten of *Diospyros malabarica*. The bed of the Teesta was here over 0.8 km (half a mile) across, and as they approached the mountains, the landscape became greener, in the distance dark sal forest loomed on the horizon, and to the east rose the outer hills of Bhutan, backed by the eastern ranges of the Himalaya.

In the evening they explored the skirts of the sal forest. The fat trunks of these were often scored by tigers' claws, though these great forests had been much felled within living memory. Beyond there, they explored the Terai forests at Baikunthapur (Bai-kant-pore), once part of the estates of the Rajah of Jalpaiguri. The botanising was good in those tropical forests; *Dillenia indica* was smothered in enormous blossoms, the Indian horse-chestnut (*Aesculus assamica*) was laden with long racemes of white blossoms, and both terrestrial and epiphytic orchids provided a spectacular show of colour.

The following day they travelled north, sometimes through dense sal forest, or at other times through breast-high savannah. Hooker delighted in the coolness of the atmosphere and the beauty of the great primeval forests. Wild bananas and orchid trees, the latter fully in flower, led the way to the outlet of the great Teesta River, where, confined by two low forested spurs, it spilled its way from the foothills of the Himalaya. In April 1870, the English naturalist and big game hunter, Sir Henry Elwes, visited the same region since in those days rhinoceros, elephants and tigers were abundant in the surrounding jungles; unfortunately many of his companions contracted malaria.

On 14 March Hooker and Hodgson had reached Siliguri, where the most conspicuous object of natural history was the many species of birds, Hodgson's specialist area, and wild hogs occupied the tropical forests. *Salix tetrasperma* grew there, an odd fact, considering that willows are characteristic of colder, even arctic latitudes. At this point Hooker was tiring of using elephants as a mode of transport. Travelling on them over a long distance is extremely uncomfortable, and as Hooker was to point out in *Himalayan Journals*, drivers never reached great ages, often, in their youth, succumbing to spinal diseases, brought on by constant motion to the vertebral column.

From the forests of the Terai, Hooker and his travelling companions headed back into the foothills, following the old military road towards Darjeeling. The warmth of the tropics lay behind them, and on 20 March, enormous hailstones battered the forests, and the mountains above. Sheets of hail lay on the ground at Darjeeling for over a week, congealing into ice 30 cm thick.

Rising higher into the foothills, they soon arrived at Kurseong, where by night, a great storm with forked and sheet lightning lit up the plains far below them. Hooker reached Darjeeling on 24 March, finding that the hail had done great damage to surrounding gardens and also to Archibald Campbell's experimental tea plants.

TO

CHARLES DARWIN, F.R.S., &c.

These Volumes are Dedicated,

BY HIS AFFECTIONATE FRIEND,

J. D. HOOKER.

KEW, *Jan. 12th*, 1854.

We used the first edition (1854) of *Himalayan Journals* as our guide while re-tracing Hooker's footsteps in Sikkim. It is still the best guide available to visiting naturalists. He dedicated this superb work to his friend and mentor, Charles Darwin.

CHAPTER 9

INTO THE SIKKIM HIMALAYA

Yuksam has grown little since Hooker's visit 165 years prior to our arrival. It remains entirely unspoilt, a cluster of traditional houses nestled on a broad plain at 1,784 m, sheltered to the rear by an amphitheatre of heavily forested mountains. Hooker wrote of beautiful lanes winding through gentle slopes and copsewood, with a scattering of stupas, ruined temples and great Kashmir cypresses; little has changed. We explored and botanised along many of these lanes above the little town, particularly near the Coronation Stone, giving us a good idea of the flora of the surrounding district.

Nowadays Yuksam is the staging-post for Kangchenjunga National Park and the last outpost of civilisation before entering that wild, rugged landscape. The town was our initial base as we carried out a nine-day circular trekking trail, which would enable us to follow Hooker's route to the alpine meadows at Dzongri and further afield. Hooker visited the region in January 1849 when the mountains surrounding Dzongri had experienced heavy snow, limiting the scope of his travels. Our visit, in November 2014, coincided with the post-monsoon season, when the skies are clear and heavy snowfall is still several weeks away.

Above Yuksam roads cease to exist. To reach Dzongri and Kangchenjunga we hired a large team of porters and pack animals to carry our goods and equipment. They are seen here on a bridge between Yuksam and the Tibetan refugee village at Tshoka.

Mountain peaks in NW Sikkim. The two taller peaks are Kabru, seen to the left, and Kangchenjunga to the far right. Immediately beneath the highest peak of Kabru can be seen the cone-like mass of Black Kabur. It sits on the alpine plateau at Dzongri. We trekked from there through a gap between Kangchenjunga and Pandim to reach the Goecha La, just 5 km or 3 miles from the south-eastern flank of Kangchenjunga. To the bottom right is the immensely steep Rathong River valley where tropical plants luxuriate.

We were therefore able to explore a far larger area beyond Dzongri than was possible for Hooker. We had allowed additional time to visit some of the surrounding glacial lakes, the Dzongri La, a mountain pass overlooking the peak of Mount Kabru, and, ultimately, the Goecha La, another high pass that would take us within 5 km of the eastern flank of Kangchenjunga.

We were not the first visitors to try and fill in the blank left on Hooker's map of Sikkim. In the autumn of 1899, the British lawyer and mountaineer Douglas William Freshfield (1845–1934) explored the Kangchenjunga region. He later (1903) published a book, *Round Kangchenjunga*, on his experience, dedicating it to 'Sir Joseph Dalton Hooker – the pioneer of mountain travel in the eastern Himalaya'. Freshfield was thoroughly excited by the opportunity of exploring virgin territory:

> Maps, if caviar to the general, are, as Louis Stevenson has insisted, very suggestive to persons with proper imagination. The little map of Hooker's long retained its hold on my memory, and from time to time I felt stirred by a vague ambition to supply the missing links in the tour of Kangchenjunga. Apparently no one else, outside India, shared this ambition. The years and decades slipped by, half a century had been completed from the date of Sir Joseph's journey, and still no European attempted to penetrate the inhospitable wilderness at the back of the snowy range that faces Darjiling, [sic] still no even approximately correct representation of its glacial features was obtainable.[1]

I found it an exciting prospect to take a group of plant enthusiasts through so many climatic zones: from the benign warm-temperate climate at Yuksam to the highest alpine plants in the shadow of Kangchenjunga. We engaged a large expeditionary staff of cooks, porters, guides and ponymen to carry our tents, luggage and equipment across the various mountain passes. These men belonged to five ethnic groups: Sherpa, Nepali, Tibetan, Bhutia and Lepcha, the latter being the indigenous Sikkimese peoples from whom Joseph Hooker preferred to recruit his porters.

Kangchenjunga National Park — an introduction

Kangchenjunga National Park was established under the Indian Wildlife Protection Act in 1977, conserving an area of some 1,784 km², an area which has subsequently increased to almost double that. Falling within the Indo-Burma global biodiversity hotspot, it has the widest altitudinal range of any of the world's National Parks, with a vertical sweep of 7,336 m (24,068 ft). Thus, it contains a mosaic of several vegetation types with complicated ecosystems ranging from subtropical to alpine, creating habitats that harbour a wide range of rare and endangered species.

Mount Kangchenjunga, the highest point within the park, has long been revered as the guardian deity of Sikkim since the land was first settled by the Lepcha peoples, a belief further promoted when Buddhism was introduced to Sikkim in the 17th century. Most of the landscape of the north-west region of the park, and the entire Rathong River valley, is considered sacred by local Buddhists, and this has had a positive influence on conservation.

At 8,586 m Kangchenjunga is the third tallest peak on the planet and is considered the world's finest example of an independent mountain, having its own glacial system radiating from several summits; it is one of 20 mountains within the national park that soar over 6,000 m. The north-west is an icy landscape, with 73 glacial lakes and 18 glaciers, including the famous Zemu Glacier, one of the largest in Asia. Ninety percent of the park lies above 3,000 m while 34% is under glaciers, ice sheets and perpetual snow.

The National Park faces growing pressure from tourism however, in the form of trekking and hiking, and particularly from the large-scale removal of rhododendron and juniper by porters for fuel; lighting campfires within the park is now forbidden.

Just over 70% of the landscape lies above 4,000 m, with major river systems running in a north–south fashion. Watersheds in the southern part of the National Park facing the monsoon-laden south-westerly winds have an oceanic climate. Beyond there successive west–

east running ridges obstruct moisture-laden winds, thus creating a continental climate in the central section. The northernmost fringes of the park, particularly the upper Lhonak valley, lie in a major rain shadow and as a result have a distinctly Trans-Himalayan (cold, arid desert) climate, with floral elements found in adjacent Tibet. The Lhonak valley is physically a direct continuation of the Tibetan Plateau, with an average altitude of 5,250 m and a low annual rainfall (by Himalayan standards) of 1,334 mm.

Geologically and structurally this mountainous area falls within the region of Darjeeling gneiss. Extensive granites constitute the formation of the main body of mountains to the north while the pH varies from 5.0 to less than 6.0. The soils are quite rich and are derived from conglomerates, shales and schists, with an organic matter content of up to 10%.

With such a complicated topography, the climate of this area varies from region to region, but is dominated by the summer monsoon between May and September (July is the wettest month) and a drier winter season from November to March, with a brief spring in April and autumn in October.

Relative humidity is high throughout the year (70%) and in the cloud forest, or subalpine zone, heavy mists encourage an abundance of filmy ferns such as *Hymenophyllum* and *Trichomanes* species. During the monsoon, relative humidity remains at 100%, even in the upper valleys, so the climate of the region is best described as moist and cool during the summer months and cold and dry in winter. August is the warmest month, when temperatures in temperate areas reach 22°C. The July average for alpine areas is 12°C, while the same areas can plummet to –17°C in January.

Four distinct forest types are found within the National Park, the first being the subtropical belt, which occurs between 1,220 and 2,100 m. Average annual rainfall here varies from 1,000 to 2,500 mm, with maximum summer temperatures of 24°C. Typical trees in this region include *Aesculus assamica*, *Alnus nepalensis*, *Castanopsis tribuloides*, *Ficus semicordata*, *Cinnamomum tamala*, *Engelhardtia spicata*, *Lithocarpus fenestratus*, *Quercus lamellosa* and *Schima wallichii*.

The temperate zone lies between 2,100 and 3,100 m, with rainfall varying between 1,500 and 2,700 mm. Winters here are cold and summers damp and cool. Characteristic trees include *Magnolia campbellii*, *Taxus wallichiana*, *Tsuga dumosa* and several *Rhododendron* species, including *R. falconeri*, *R. grande* and *R. hodgsonii*.

Moving higher up, the subalpine zone is encountered between 3,100 and 4,000 m. Here the predominant tree is *Abies densa*, and other forest components include *Abies spectabilis*, *Betula utilis*, *Juniperus recurva*, *Salix sikkimensis* and *Sorbus microphylla*.

The alpine zone extends beyond the tree-line to 4,500 m, and the growing season here is short (May to September). Summers are short and cold, rainfall is low and most precipitation arrives in the form of snow. Cushion plants are a conspicuous feature, particularly *Androsace* and *Arenaria* species, while genera such as *Corydalis*, *Primula*, *Saussurea*, *Pedicularis*, *Anemone* and *Potentilla* abound.

In all, 1,580 species of vascular plants have been recorded from Kangchenjunga National Park, of which 1,463 are angiosperms (flowering plants, of which there are 4,458 species in Sikkim), 11 are gymnosperms and 106 are pteridophytes (ferns & fern allies).[2]

Many of the plants found in the park are very rare, for example *Thelypteris elwesii*, a Sikkim endemic that grows along the fringes of

Relative humidity is high throughout most of Kangchenjunga National Park, particularly where valleys run in a north–south fashion, allowing the monsoon-laden winds to permeate deep into the mountains. Many of the Sikkim *Rhododendron* species relish this moist climate, so it is no wonder they thrive best in places like Cornwall, the west coast of Scotland and coastal regions of Ireland. One of my travelling companions, Gail Gilliland, is seen here at Kabur Lam Lake as mist swells from the valley below. A minute later we were engulfed in fog.

the National Park, while *Oreopteris elwesii*, originally found by Sir Henry Elwes at Lachen (on the edge of the park), is presumed extinct in Sikkim, though it also grows in Sichuan, China.

Extremely rare endemics include *Blumea sikkimensis* (found in warm-temperate areas), *Inula macrosperma* (a species confined to alpine regions) and *Astragalus zemuensis* (named for the glaciated valley in which it was found). *Arenaria thangoensis,* meanwhile, named for the village of Thangu on the NE fringe of the park, has not been seen since its collection in 1911 and may now be extinct.

New species continue to be discovered and named from the region, and this includes the endemic orchid *Myrmechis bakhimensis*, found by Indian botanists in July 1999 on a trek between Bakhim and Dzongri; there are probably many more plants waiting to be found in the remote and little-explored higher valleys to the north-west.

The fauna of the area is equally rich, with large mammals better known from the Tibetan Plateau found here including the Tibetan wild ass, *Equus hemionus* and takin, *Budorcas taxicolor*. Visitors may catch rare glimpses of snow leopards, clouded leopards, black panthers, red pandas (the State animal of Sikkim), musk deer, blue sheep, leopards, goral and Himalayan black bears.

Thanks to the wide variety of habitats, 127 species of birds have been recorded, including seven globally threatened and restricted range species. The many high-altitude lakes are an important stop-over for migratory fowl and rarities such as black-breasted parrotbills, *Paradoxornis flavirostris*, Pallas's fish-eagle, *Haliaeetus leucoryphus* and the lesser kestrel, *Falco naumanii*, all find a home there.

Hydrology of the National Park area

Seven major river systems, namely the Churang, Lachen, Lhonak, Prek, Rangit, Rangyong and Zemu, form watershed areas within the reserve, Six different habitats are found within the watershed of the Prek River alone: subtropical, temperate, subalpine, alpine pasture, rock and snow cover. The average annual rainfall of the Prek River watershed is 2,230 mm, though due to the alignment of the Kangchenjunga massif, this varies throughout the national park.

As already explained, the south-westerly monsoon winds, laden with precipitation, are obstructed by successive west–east ridge formations which dramatically reduce precipitation, towards the north (the Tibetan border area,) so the annual rainfall decreases from 2.75 m in the south-eastern part of the park to a mere 0.75 m in the north.

The sources of many of these rivers are glaciers radiating from the summit of Kangchenjunga, particularly the massive Zemu Glacier. Many of these have receded dramatically in recent decades, and satellite imagery has shown that the Zemu Glacier has shrunk 27 m per year between 1967 and 1984.

The Himalayan fir, *Abies densa*. It is seen here approaching the tree-line, above which is a dense scrub of *Rhododendron* species and dwarf *Juniperus indica*. Above that last line of woody plants is the alpine zone that extends to snow, ice sheets and glaciers.

A bridge across the Prek Chu (Chu is Tibetan for river or stream), draped with prayer flags. To Buddhists, and to many Hindus, this is a sacred landscape. The Prek River originates on the great glaciers of Kangchenjunga.

During our visit in November 2014, our group's approach to the Goecha La involved an energetic trek along a lateral moraine belonging to the huge Onglathang Glacier above Samiti (Sungmoteng) Lake. Originating on the icefields of the nearby Kabru Dome 6,600 m, the snout of this glacier lies at 4,500 m from where the headwaters of the Prek River are found. The glacier extends to the south-east through a valley for 5 km, with an average depth of 80 m, and dominates the landscape on the approach to the Goecha La. Perched at 4,940 m, this major pass allows panoramic views onto the south-eastern flank of Kangchenjunga and onto the winding snout of the Talung Glacier.

One of the perils of global warming in this alpine landscape is the build-up of menacing bodies of water from retreating ice behind fragile walls of ice and rock. These enormous unstable dams balance on the edge of steep-sided valleys, threatening towns and villages below. The melting and subsequent retreat of the Onglathang Glacier in recent times has caused the formation of several lakes in the foreground of the glacier, which pose an enormous flood threat to the inhabitants of the lower Rathong valley; in 1994 a catastrophic glacial lake outburst occurred, causing devastating flooding as far as the Tibetan refugee village of Tshoka.

The great stupa at Yuksam. Its all-seeing eyes gazed on as we departed the village for higher elevations. The dark background is provided by a young Kashmir cypress.

Departure from Yuksam

We departed Yuksam early on the morning of 4 November, as the ponymen were busily loading equipment onto waiting pack animals. The most versatile beast of burden in the Himalaya is the dzo, a yak–cow cross, widely used, for example, on the Goecha La trail approaching Kangchenjunga and in the region of Everest basecamp. Dzo (the male hybrid) are sterile, while the females (dzomo) are fertile. As a result of hybrid vigour, they are stronger than cattle or yaks from the Himalayan region, and well adapted to carrying heavy loads. Several of our porters also carried baskets on their backs, woven from local bamboo species, while the straps were manufactured from the fibre of *Edgeworthia gardneri*. Even at altitudes of over 4,000 m, these men climbed steep, slippery slopes with weights of over 40 kg strapped to their foreheads. Our support staff freed the way for us to botanise at leisure and in comfort.

Yuksam is one of the prettiest villages in west Sikkim and despite the lateness of the season, the streets were a riot of colour. The inhabitants of this area love flowers, and every house, no matter what its size, was surrounded by favourite garden exotics. The showiest of these included *Cestrum fasciculatum*, chrysanthemums, roses, *Cyphomandra betacea* (the South American tree tomato), 3 m tall bushes of *Euphorbia pulcherrima*, the Mexican poinsettia (reminding us that Christmas was fast approaching), and the spectacular *Dahlia imperialis*, then carrying masses of large single, lavender blossoms on 4 m tall stems. Native to Mexico, Central America and Columbia (where it can grow up to 10 m tall), this striking perennial refuses to blossom in Britain and Ireland, due no doubt, to our dull summers, though it finds conditions more to its liking in the valleys of Sikkim and Darjeeling District.

A porter packs our luggage and belongings onto the backs of pack animals at Yuksam. Beyond there we were on steep forest tracks. Dzo are regularly used to carry goods through Himalayan regions, but they are ill-tempered creatures and those sharply pointed horns can do a lot of damage!

We followed along the route of the steep-sided Rathong River valley to reach our destination. Above Yuksam the surrounding vegetation is subtropical, and the large-leaved tree seen to the left is *Macaranga denticulata*.

Our primary interest lay with the surrounding local flora however. Even in Hooker's time the original forests surrounding Yuksam had long since been felled and replaced by regenerating copsewood. *Alnus nepalensis* formed pure forests on the perimeter of the village, tall slim trees supporting vigorous vines of *Trichosanthes tricuspidata*, a cucurbit bearing masses of fist-sized, fleshy red fruits.

Alnus nepalensis is planted on a large scale in many of the warmer valleys in Sikkim, including Yuksam, to act as a shade tree to cultivated plantations of cardamom, *Elettaria cardamomum*. Sikkim is the major source of cardamom in India and contributes over 90% of the country's total production. The root nodules of the Nepal alder fix atmospheric nitrogen, owing to a symbiotic association with the nitrogen-fixing bacteria, *Frankia alni*, which ultimately contributes to the robustness of the cardamom plants.

In the shade of the alders we spotted several 1 m tall plants of *Tupistra nutans*, a rare perennial member of the lily-of-the-valley family (Convallariaceae), that looks like a giant aspidistra. In late autumn plants produce 10 cm long mustard-coloured flower spikes that are a great delicacy in Indian cuisine. This handsome perennial is widely grown by villagers in the hill regions of Sikkim, and in flowering season, the inflorescences are sold in local markets alongside other vegetables. We later had the opportunity to sample some of these slightly bitter blossoms in a curry on our return to Yuksam.

Yuksam to Sachen

Our route brought us parallel to the course of the Rathong River valley, along a rough track no more than 2 m wide, on the right bank of the river, whose furious turquoise-blue waters raced and raged beneath us. It was exactly the same path Hooker had taken in January 1849 and we were eager to see what, if anything, remained from the time of his visit.

We soon entered a gorge-like landscape. It was the height of the trekking season and we met a constant stream of heavily laden pack animals, including the rather ill-tempered dzos whose long sharp horns were much feared by our group. The steep-sided mountains were clothed with dense subtropical forest, and it was wonderful to travel through such a perfectly preserved landscape.

One of the most impressive foliage plants in this region was *Macaranga denticulata*, a medium-sized tree belonging to the spurge family (Euphorbiaceae), bearing strikingly handsome large peltate leaves that contrasted beautifully with *Tetradium fraxinifolium*, a small tree carrying enormous corymbs of small red fruits set among its large, glossy pinnate leaves. I was also particularly pleased to find *Myrsine semiserrata* common in these parts. The last time I had seen it was in the mountains of central China where it formed small evergreen trees and bore large axillary clusters of red-purple berries. It's a shame this species is so rarely found in British and Irish gardens, as when happy, its autumn fruits provide a spectacular show; it is tender and best suited to coastal sites. First found in Nepal by Nathaniel Wallich in 1821, it is a widespread tree in the Himalaya, with a range from Pakistan in the west to Assam and Myanmar in the east, continuing south into China and Thailand. The tree is of great medicinal value in Sikkim and Darjeeling, where the fruits are used to treat tapeworm, colic and to act as a blood purifier.[3]

Macaranga denticulata, a spurge relative, has strikingly handsome leaves. Below it is the Indian mugwort, *Artemisia indica*, while behind is the Sikkim maple, *Acer sikkimense*.

We found *Edgeworthia gardneri* in bloom in the forests above Yuksam. Its stems are still employed in paper-making in many parts of Sikkim and Nepal.

The *Hydrangea* relative *Dichroa febrifuga* was abundant on the mountainside where landslides had cleared the forest cover. There it formed metre-high shrubs bearing terminal clusters of metallic blue berries. The specific epithet refers to the stems and bark, which are used to treat malarial fevers in the Himalaya, China and South-East Asia.

Other showy shrubs along our route included *Leucosceptrum canum*, *Zanthoxylum armatum*, *Viburnum erubescens* and *Dobinea vulgaris*. The latter is a handsome evergreen whose season of interest is extended by the large and showy pendulous white bracts that remain persistent long after the flowering season has ended. Below the track we also encountered enormous 4 m tall bushes of the Indian paper tree, *Edgeworthia gardneri*, some of which were by then in full bloom.

Several large suspension bridges took us across substantial rivers, including the lower reaches of the Prek Chu, a furious torrent that cascaded in a series of waterfalls down the mountain face and joined the Rathong River on the valley floor far below us. In 1890, J. Claude White, the British Political Officer, plantsman and explorer, based at Gangtok between 1889 and 1908, organised an expedition to cross the high Himalayan pass of Goecha La, becoming the first European to do so. His route took him past the very point where the lower Prek River cascades down a steep precipice into the currents of the

Rathong River far below. He had departed from the monastery at Dubdi earlier that day and was making good progress until the first mishap occurred:

A torrent, swollen by the heavy rain, came rushing down a perpendicular rock with an almost deafening roar right across the path, which at that point was water-worn rock and very slippery, and then leapt into an abyss below, the bottom of which I could not see. A couple of saplings about four inches in diameter had been placed across, and I had gone over in safety and was resting on an incline on the other side, when one of my coolies came up. For some reason, as he was crossing the poles, he either slipped or lost his balance, I could not see which, but he fell on the up-stream side, was immediately carried under the bridge and swept over the precipice before my eyes. All this happened in a moment, and such was the inaccessibility of the spot and so dense the jungle, it was quite impossible to do anything for the poor fellow. Some more coolies now came up, and we tried to cut away down through the dense tangle of trees and undergrowth, but this proved quite impossible, though, after an hour's work, one man managed to get down by a circuitous route, only to find that his unfortunate companion had been swept into the main torrent, and that nothing was to be seen of either him or his load. I am thankful to say that in all my wanderings in the Himalayas I have only lost one other coolie.[4]

By the Prek Chu bridge, we found a substantial clump of wild bananas, *Musa balbisiana*, the same species Hooker mentions seeing when he travelled through the valley in the mid-19th century.

Several climbers scrambled their way through this track-side medley, including the herbaceous Indian madder, *Rubia manjith* and the rarely encountered *Dactylicapnos grandifoliata*, a *Dicentra* relative whose long slender trailing stems were laden with clusters of striking cylindrical-shaped, violet-coloured fruits. This was my first encounter with the scarlet clock vine, *Thunbergia coccinea*, whose long scandent stems streamed from tall trees overhead. The fruits of this species are particularly showy, the long extended 'beaks' looking exactly like a bird's head.

Wild bananas, *Musa balbisiana*, seen here by a bridge over the Prek Chu, in the subtropical zone. In *Himalayan Journals* Hooker spoke of seeing plantains in the Rathong River valley; this is the species he encountered.

A little further on, dense tufts of *Ophiopogon intermedius* carpeted the forest floor. The canopy overhead was mainly formed by giant 30 m tall trees of *Lithocarpus elegans* and *L. fenestratus*, with a scattering of large trees of *Toona ciliata*, *Exbucklandia populnea* and *Betula utilis*, the latter with wonderfully polished mahogany-coloured bark. Many of these giants were absolutely laden with epiphytes; orchids, filmy ferns and shrubby species like *Agapetes serpens* abounded. The crowns of the biggest trees carried several tonnes of these opportunistic exotics.

The scarlet clock vine, *Thunbergia coccinea*. Its curiously beaked fruits hung from the trees overhead. Nathaniel Wallich originally found this vigorous climber in Nepal.

Our porters herd the pack animals ahead as we botanise the route behind. The campsites were generally set up by the time we reached our base each evening.

David Koning, Head Gardener at Dargle Glen, Co. Wicklow, standing beneath a tree of *Quercus lamellosa*. On its trunk are numerous epiphytic ferns and ginger lilies.

Quercus lamellosa is one of the best host species for epiphytes in the Himalaya, and in this valley it supported an aerial population that included the woody *Sorbus microphylla*, a Himalayan rowan which we had also seen growing epiphytically in western Bhutan just a week earlier.

Q. lamellosa was the predominant tree in this part of our route, and is without doubt the most striking of the Himalayan oaks. Discovered in Nepal by Dr Francis Buchanan-Hamilton in December 1802, it was later described in 1819 by Sir James Edward Smith (1759–1828), who gave it the common English name, the 'many-cupped oak'.

It is relatively common in mixed mesophytic (i.e. demanding neither particularly wet or dry conditions) montane forests in eastern Nepal, north-east India (including Sikkim), Bhutan, Myanmar, northern Thailand and W China between 2,000 and 2,800 m.[5] In his monumental *Flora of British India*[6] Hooker, who considered this the noblest of all the oaks in terms of foliage and fruits, described this species as a lofty evergreen tree, attaining 36.5 m, with trunks making girths of up to 4.6 m. James Sykes Gamble, in his *Manual of Indian Timbers* (1922), claimed *Q. lamellosa* to be 'the finest and most important tree of the forests of Darjeeling', a statement Indian foresters from that region would agree with even today.

The trees we encountered were venerable giants and were laden down with literally tonnes of epiphytes. Our host, Sailesh

Pradhan, told us that in his nursery in Gangtok, he has found its leaf mould to be the best growing medium for orchids and vireya rhododendrons available.

Commonly called *Búk* by the Lepcha people of Sikkim,[7] the ovate-elliptic, heavily serrated leaves of this species are up to 30 cm long and sometimes as much as 25 cm broad. Rare in cultivation, it was raised at Caerhays Castle in Cornwall from seeds collected by George Forrest in 1924 on the Yunnan–Myanmar border (F. 24183).[8] At its upper limits in Sikkim it grows with *Rhododendron grande* and *Magnolia campbellii*, and so should be hardy in the milder coastal regions of Britain and Ireland and milder parts of the USA (USDA zone 9).

During our visit, trees were carrying an abundance of acorns, each encased in a woody cup up to 8 cm in diameter and held on 25 cm long spikes, forming enormous clustered fruits. The cupule encasing the flattish acorns, is composed of up to ten very thin, concentric downy rings set one over the next, hence the specific epithet, *lamellosa*.

Perhaps the most impressive tree in this area was a gargantuan 40 m tall *Acer campbellii*, displaying wonderful yellow autumn foliage; this must have been a young sapling when Hooker trod the same track so many years before us.

These unspoiled forests provide habitats for a wealth of rare mammalian species, including a number of primates. We were extremely fortunate, when, turning a bend on the track, we disturbed a large troop of northern plains grey langurs, *Semnopithecus entellus*, high in the trees, feasting on the fruits of *Lithocarpus elegans*. As soon as they spotted us, most of the monkeys made a rapid retreat, swinging through the tree canopy, though two more curious males remained a moment, gazing through the branches of *Rhododendron arboreum* and *Toona ciliata*. In this part of the Himalaya, they provide a food source for leopards and tigers; no wonder then, their suspicion of our party.

The many-cupped oak, *Quercus lamellosa*, one of the most beautiful of all Himalayan trees. Its foliage is equally interesting, the enormous ovate leaves being silver beneath.

A grey langur resting in the branches of *Rhododendron arboreum*. We disturbed a troop of these primates as we botanised among the trees below.

Giant vines streamed down from those enormous trees, one of the showiest, and most common, being *Parthenocissus semicordata*, whose foliage had by then turned claret. Other climbers included several species of *Smilax, Tetrastigma, Rhaphidophora* and the wild pepper, *Piper longum*, from whose common name, pippali, the word pepper is derived. In Sikkim a decoction of the immature fruits and roots of this species is used as an antidote to snake bite and scorpion stings.

The perennial climbing gentian, *Crawfurdia speciosa* is commonly encountered in damp forest in the eastern Himalaya, Myanmar and Tibet, where in autumn its wiry 2 m long stems carry masses of gentian-blue, trumpet-shaped blossoms. One plant we encountered trailed its way through a bamboo, its flowers fading to almost pure white with violet-rose tips to the apex of the corolla.

Beneath the upper canopy lay a scattering of *Rhododendron arboreum*, while *Cinnamomum tamala* grew close by, reminding us we had yet to reach temperate woodland. The distinctive three-veined lanceolate leaves of this 'Indian bay leaf' make it easy to identify, and it has been widely used in traditional herbal medicine since ancient times; it is also an important ingredient in Indian and Pakistani recipes.

Tree ferns are a conspicuous feature of the warmer, monsoon-drenched forests of Sikkim. One of the commonest we were to encounter was *Cyathea spinulosa*, whose slender, stipe-scarred trunks can reach 6 m. There it formed small groves by the track-side, its parasol-shaped crown formed by masses of elegant 4 m long fronds. Not far away from it grew another exotic fern, *Pteris wallichiana*. This fellow throws up tall stems almost 1.5 m high, with fronds radiating from the apex, well over a metre across. Rare in cultivation, it is hardy throughout Ireland and much of Britain and deserves to be more widely grown.

The Sikkim knotweed, *Polygonum molle* (syn. *Aconogonum molle*) was abundant at this altitude (2,000 m) and is common throughout the eastern Himalaya and parts of China. This dichotomously branched subshrub bears large thyrsoid, velvety panicles of creamy-

white blossoms in autumn and though normally a lax-habited shrub of 2–3 m, here its scandent stems climbed double that height into surrounding trees.

Our first campsite lay at a place called Sachen (2,179 m), where we pitched our tents beneath trees of *Prunus nepalensis* (then carrying small dark-purple fruits) and the camellia relative, *Eurya acuminata*, whose branches were liberally wreathed with dense fascicles of small white, bell-shaped flowers.

The plant that stole the show however, was a mahonia with a much confused taxonomic history. *Mahonia acanthifolia* was described in 1831 by the Scottish botanist and plant collector, George Don, who based the species on two dried specimens deposited in the East India Company herbarium.

It was first collected in Nepal in 1821, by Nathaniel Wallich, and again by Edward Gardner (*b*. 1784), the first East India Company British Resident at Kathmandu, sometime between 1816 and 1829. Don made use of the manuscript name (*Mahonia acanthifolia*) coined by Wallich and written on labels on the East India Company's specimens.[9]

In his original description Don made comparisons with another species, *Mahonia napaulensis*, stating that *M. acanthifolia* grew taller, with a greater number of leaf pairs, but then stated that perhaps his new species was the same as *M. napaulensis*; thus began the confusion between these two rather distinct Himalayan shrubs.

The climbing gentian, *Crawfurdia speciosa*. Nathaniel Wallich named the genus in 1826 in honour of John Crawford (1783–1868), the Scottish surgeon employed by the British East India Company as a colonial administrator.

Our porters and ponies beneath *Mahonia acanthifolia* by our campsite at Sachen.

Hooker, a famous taxonomic 'lumper', however, ruled in 1855 that both species should be united under *M. napaulensis*, probably as a result of the scarcity of specimens in European herbaria at the time. *Mahonia acanthifolia* is however a very distinct species, with longer leaves with up to 11 pairs of leaflets (3–7 pairs in *M. napaulensis*) and with longer, showier racemes of deep yellow blossoms that appear in autumn, while those of *M. napaulensis* are borne in spring.

Mahonia acanthifolia was everywhere abundant around our campsite at Sachen, forming enormous evergreen mounds over 8 m tall, all bearing masses of exquisitely fragrant blossoms carried in long, extremely dense-flowered racemes, above glossy, handsome pinnate foliage.

That night, having pitched our tents among the mahonias, we fell asleep listening to the roar of the Rathong River in a valley deep below. Overhead nocturnal particoloured flying squirrels, *Hylopetes alboniger*, filled the forests with their night-time calls, as they glided from tree to tree.

The root holoparasite *Balanophora dioica* was one of the strangest and most intriguing plants we met with during our travels in west Sikkim. A male pollen spike is seen to the rear, while the female structures are fatter and more rounded. This parasite causes a distortion to the roots of trees it grows on, and the timber as a consequence was greatly valued by Lepcha wood turners in the past.

Sachen to Tshoka (Choka)

Early the next morning we broke camp and spent the day trekking to the Tibetan refugee village of Tshoka (2,956 m), by way of the small hamlet of Bakhim. This should have taken just a few hours, but so good was the botanising *en route* that it was late evening before we reached our next campsite. The scent of the mahonias was glorious as we departed that morning, their golden flowers like beacons in the dark woods.

High up in the trees enormous clumps of the showy autumn-flowered epiphytic orchid, *Pleione praecox*, made a stunning show with its masses of violet flowers. *Luculia gratissima*, common around Yuksam, extended as far as Sachen; a native of the Himalaya, Vietnam, Myanmar, Thailand and Yunnan Province in W China, it forms a tall, spreading shrub up to 3 m, bearing corymbs of beautifully scented pink blossoms.

A little further on, on a steep bank beneath a *Lindera* species, we stumbled across one of the most exciting plants we were to encounter that day. *Balanophora dioica* is an achlorophyllous root holoparasite (the plant we found was obviously growing on the *Lindera*), native to the eastern Himalaya. Male and female flowers are carried separately from a fleshy underground rhizome. Fifteen species of these curious parasites are native to the Old World tropics and are known to be parasitic on 75 species of trees and shrubs. Of these, *B. dioica* is the rarest in the north-east of the subcontinent, and is also of great medicinal value, though is protected under Appendix 2 of CITES and the sale of plants is prohibited throughout India.

Gradually, as we climbed higher, the forest changed, taking on a more temperate aspect, though the Sikkim knotweed, *Polygonum molle*, continued to smother the forest floor in almost weed-like proportions. This species is quite rare in British and Irish gardens, though it has been known to escape into the countryside; with global climate change, it would be wise to keep an eye on this knotweed, in case it should become an invasive pest.

Within an hour of setting off, warm-temperate elements of the surrounding flora, e.g., *Ficus auriculata* and *Lithocarpus elegans*, disappeared, having reached the upper limits of their habitat. Deciduous trees became more abundant, and several maples appeared: *Acer campbellii* (carrying masses of epiphytic *Agapetes incurvata* along its trunk and in its crown), *A. sikkimensis* and *A. pectinatum*, the latter two forming small to medium-sized trees. *Prunus nepalensis*, the Nepal cherry, formed slim upright trees of about 10 m, providing a support for the climbing *Hydrangea anomala*, common in these dense, damp woods. The form growing here appeared to be particularly good, with corymbs of fertile flowers, surrounded by extra-large peripheral sterile white blossoms.

I have to admit to a fondness for the birthworts (*Aristolochia* spp.), that curious group of primitive perennials and woody climbers. Griffith's birthwort, *Aristolochia griffithii*, one of the most spectacular of the Himalayan species, was common in this region and romped its way through surrounding thickets. It is a beautiful plant, with large cordate leaves and large, rather sinister-looking purple and mustard-yellow blossoms in the classic shape of a Dutchman's pipe. These are followed by long sausage-like fruits. Hooker collected specimens in the lower Lachen valley, on which the species was partly described, though it had been found previously in Bhutan by its namesake, William Griffith. Ten days earlier we had seen it in mountain forests above Paro in western Bhutan, so it was exciting to re-encounter it here in Sikkim.

Senecio scandens completed this trio of climbers, rambling through *Zanthoxylum acanthopodium* and *Elaeagnus parvifolia*, and while it is colourful, I find this plant slightly weedy and reminiscent of the dreaded ragwort, *Senecio jacobaea*.

I was pleased to meet with *Rhododendron griffithianum*, though in the woods we passed through it grew only as a single tree. In all my travels in Sikkim I have found this to be very scarce, never forming gregarious masses as do the other tree-like species, and certainly never making enormous specimens like the 8 m Hooker veteran that grows near the pond at Kilmacurragh.

Rhododendron arboreum, on the other hand, was far more abundant, forming mini-forests up to 23 m tall; all red-flowered forms, we were told. Another favourite, *R. grande*, formed large forests of trees 15–20 m overhead. Wherever this species grows in Sikkim you are sure to find *Magnolia campbellii* nearby. It grew freely in the surrounding forests, and our guides told us that only white-flowered forms occurred in this area, while one of our porters claimed that white-flowered trees grow only at the highest level of the tree's altitudinal limit.

The summit of Pandim seen through the autumnal foliage of *Lindera neesiana*. We had finally reached the Tibetan refugee village of Tshoka, and the forests were dominated by enormous trees of *Magnolia campbellii* and giant *Rhododendron grande*.

Part of the settlement at Tshoka. The village, though obviously very poor, has a spectacular setting, with views down the Rathong River valley and an amphitheatre of snowy mountains to its rear.

Quercus lamellosa gradually became more abundant, shedding its enormous acorns in thousands, a bounty for the many mammals in these forests. With it grew the ubiquitous *Alnus nepalensis*, and together they provided a canopy and a rich leaf-litter layer for the giant Himalayan lily, *Cardiocrinum giganteum*, whose enormous flowering spikes were covered with plump seed capsules. One never tires of seeing this woodland aristocrat in its native habitat, conditions mimicked by the gardens of Cornwall, West Cork or Kerry. Also here were an amazing medley of other good garden plants, including *Mahonia acanthifolia*, *Edgeworthia gardneri*, *Sarcococca hookeriana*, the rather aptly named *Rubus splendidissimus* and various species of *Impatiens*.

We reached the Lepcha settlement of Bakhim about midday. This scattering of forest bungalows lies about 16 km from Yuksam at an altitude of 2,740 m and is surrounded by dense forest. *Lithocarpus pachyphyllus*, one of the higher-altitude evergreen oaks of Sikkim, suddenly appeared on the scene in thousands, forming fat-trunked trees well over 30 m tall. Several of these carried epiphytic plants of *Acer sikkimense* and the superb *Rhododendron dalhousieae* high up in their branches, forming an impressive sight.

Above there, we saw woods formed entirely of *Magnolia campbellii*; what a sight they must make in early spring when covered with many thousands of large white blossoms! On their trunks we found epiphytic plants of the Solomon's seal relative, *Polygonatum cathcartii*, a pretty perennial described from material collected by Joseph Hooker in the Lachen valley in June 1849, and named for his friend John Ferguson Cathcart, who had previously collected it near Darjeeling, and arranged for it to be painted.

At this altitude *Rhododendron arboreum* is replaced by the hardier *R. arboreum* ssp. *cinnamomeum*, whose leaf undersides are typically covered with a thick, rust-coloured indumentum. It also formed entire woods, beneath which we found several good plants including

Berberis insignis (perhaps the finest of Sikkim's evergreen barberries), forming handsome dome-shaped shrubs. Its bedfellows included *Arisaema griffithii*, whose large spikes of fleshy red fruits were scattered through the surrounding thickets.

At 2,900 m we entered a mixed coniferous–deciduous layer of temperate forest, dominated by tall, straight trees of *Abies densa* and enormous specimens of the Himalayan hemlock, *Tsuga dumosa*, that must have been several centuries old. *Rhododendron grande* was particularly fine there, towering to 21 m overhead through even taller trees of *Magnolia campbellii*. *Lindera neesiana* was sporting autumn foliage, its spectacular flowering period just weeks away, and it was exciting to find enormous 8 m tall, multi-stemmed trees of *Osmanthus suavis*. Generally seen at half that height in cultivation, *O. suavis* was described from specimens collected in Bhutan by William Griffith in 1838 and from material gathered by Hooker ten years later in E Nepal and on the summit of Tonglu near Darjeeling. The specific epithet *'suavis'* meaning 'sweet' alludes to the masses of white, fragrant flowers borne in spring. It is also native to Yunnan in W China, from where it was introduced by George Forrest (1873–1932) and first flowered in cultivation by one of his sponsors, the Marquess of Headfort, Geoffrey Thomas Taylor (1878–1943), in his garden near Kells, Co. Meath, Ireland.[10]

Equally impressive were the many fine columnar trees of the Himalayan holly, *Ilex dipyrena*, towering to 15 m overhead. The juvenile leaves of this species are quite different to those of mature plants, being larger and more viciously armed. Near these hollies we found a single 2 m tall bush of *Daphne bholua* in full bloom and identical to the popular cultivar 'Jacqueline Postill'.

The views south, down the Rathong River valley are spectacular. Here the valley is framed by noble trees of *Tsuga dumosa*, the Himalayan hemlock. The light in late evening in this part of the Himalaya is magical.

One of the finest shows of autumn colour was provided by *Lyonia ovalifolia*, a small deciduous tree of about 10 m tall. At the time of our visit its foliage had assumed purple-red tints, contrasting well with the golden-yellow foliage of nearby *Lindera neesiana*. Other good trees included *Alangium platanifolium*, *Tetracentron sinense* (whose foliage was then amber-gold), *Symplocos lucida* (bearing masses of deep blue fruits) and *Magnolia globosa*, a species similar to the better-known, summer-flowered *M. wilsonii*. *Magnolia globosa* is also native to China and was another of George Forrest's introductions (1919), though it is said that plants of Indian origin are hardier than those of Chinese provenance.

Towards late evening we reached Tshoka, a tiny village established in the early 1960s by the last king of Sikkim, to accommodate refugees fleeing turmoil in neighbouring Tibet. It consists of little more than

a few houses, some stupas, and a monastery perched on a ridge, and looks south through the deep Rathong River valley towards the ridge on which Darjeeling sits. To the north lies an amphitheatre of snow-clad mountain peaks and expansive forests of *Abies densa*, from which the Tibetan refugees originally built their village. Tshoka is the last permanently inhabited settlement on this route, and beyond there we really were venturing into the wilds. We camped to the east of the village, with spectacular views towards Mount Pandim and the surrounding peaks.

The following morning we woke to a clear blue sky and fine weather, a good prospect since we were now heading into the higher mountains; very often rolling clouds and mists obstruct the most spectacular views, though we were to be lucky in the following days.

The Lepchas continue to be used as porters in Sikkim. Pictured here is one of our helpers, Dawa Lepcha (the ethnicity of many Sikkim groups is signified in their surnames). His basket is manufactured from local bamboo species while the strap is made from the bark fibre of *Edgeworthia gardneri*.

Tshoka to Dzongri

As the ponymen packed the expedition equipment onto the backs of the dzo and ponies we headed north, warmed by the early morning sunlight. Our plan was to trek to the yak herders' camp at Dzongri (3,956 m), passing through Phedang (3,650 m), where the Alpine Garden Society (AGS) had collected during their visit to this area in 1983.

From this point the forests were dominated by conifers, particularly *Abies densa* and *Tsuga dumosa*, mammoth specimens that must have been several centuries old. Rhododendrons abounded, along the edge of the track, and in open glades *R. glaucophyllum* formed dome-shaped mounds to about a metre high. Even high up on the moss-laden trunks of *Abies densa*, the epiphytic *Rhododendron pendulum* found a suitable home and was extremely common; a little further on we spotted a single mature bush of *R. barbatum* also growing epiphytically, such is the humidity of this region.

Rhododendron falconeri was everywhere abundant, forming dense impenetrable thickets alongside fine trees of *Betula utilis* and *Sorbus hedlundii*, both trees boasting spectacular golden colours. Perhaps the best autumn colour seen on the upper forested slopes was provided by *Viburnum nervosum*, a common species in this part of the Himalaya. I have previously seen it in both Bhutan and Tibet, where, as in Sikkim, its foliage turns vividly deep purple. One of its woodland companions was *Gamblea ciliata*, a small tree with handsome palmate leaves turning yellow.

At 3,100 m, *Rhododendron falconeri* was replaced by another large-leaved species, *R. hodgsonii*. It too formed dense impenetrable thickets, with beautifully peeling grey-brown bark, exposing pale flesh-coloured trunks beneath. A dense moss layer had formed underfoot, and through this, dwarf carpeting shrubs of *Gaultheria pyroloides* had suckered. At this late stage of autumn its foliage had

Our campsite at Tshoka, just opposite the centre of the small village. Large trees of *Tsuga dumosa* bristle the ridge behind.

Gaultheria pyroloides was abundant on the mossy forest floor beneath *Rhododendron* thickets. Its foliage turns burgundy-red as the days shorten and temperatures plummet in late autumn. It is not terribly common in cultivation, though there are good plants at the Royal Botanic Garden Edinburgh.

assumed a wonderful burgundy hue. Hooker discovered this species at nearby Mans Lepche and Dzongri, during his exploration of the region in 1849, and he also re-collected it in the Lachen valley. It is always a diminutive, low-growing shrub of the higher altitudes. It was later collected during the famous 1922 Mount Everest expedition, from which the National Botanic Gardens, Glasnevin received a large consignment of seeds.

Our Lepcha porters soon passed us as we botanised along the route. As previously mentioned, each of these men is capable of carrying enormously heavy loads in baskets strapped to their foreheads,with straps made from the stem fibre of *Edgeworthia gardneri*, apparently the toughest and most hard-wearing material available. They raced up the steep mountain tracks, all good heartedly, and I could see why Joseph Hooker was so fond of these gentle, friendly people.

Reaching 3,500 m, we encountered, yet again, a change of *Rhododendron* species. At this point the great silver fir woods thinned out and we entered cloud forest. There, dominant species included *Rhododendron cinnabarinum*, *R. campylocarpum* and the blood-red flowered *R. thomsonii*, all growing in great thickets, though a little higher still it was *R. lanatum* that stole the show. The plants there were obviously many centuries old (they were most certainly in their prime when Hooker passed through) and their gnarled, moss-laden stems formed a tunnel over our route, creating a scene reminiscent of a Tolkien tale. Most of their trunks were densely covered in various filmy ferns, *Hymenophyllum* and *Trichomanes* species.

Hooker named this lovely sulphur-yellow flowered species *R. lanatum*, in reference to the dense, woolly, tawny-coloured indumentum on the undersurface of leaves. He discovered it at Dzongri, stating that, of the many places it grew in Sikkim, it was most abundant there. He also introduced it to cultivation when he collected seeds on the pass at Cho La in east Sikkim in November 1849, and seedlings were raised in Britain and Ireland in the spring of 1850 from this consignment.

Our Tibetan guide, Thupden Tsering and Margie Phillips from the National Botanic Gardens, Glasnevin in Dublin, turn the corner through a natural arch of gnarled *Rhododendron lanatum*, one of Joseph Hooker's Dzongri discoveries.

R. lanatum is often a difficult plant to grow, as it suffers during dry summers, and, taking note of its perpetually damp habitat, and the mass of filmy ferns that cling to its stems, it comes as no surprise that it will thrive only in the wettest parts of Britain, Ireland and the NW coast of the USA. The foliage is exceedingly handsome, though I have found in my travels in the Himalaya that the forms growing in W Bhutan are far finer than those found in Sikkim.

Phedang marks the upper limits of the tree-line along the Dzongri trail, and where the dark forests of *Abies densa* petered out, great thickets of rhododendrons dominated the landscape. One of the most abundant of these was the yellow-flowered *Rhododendron wightii* named for Robert Wight, a Scottish surgeon and botanist who became Superintendent of the Botanic Gardens at Madras (now

The undersides of leaves of *Rhododendron lanatum* are covered in a wonderful woolly cinnamon-coloured indumentum. It is a lovely species but can be tricky in cultivation.

Chennai). On the slopes above Phedang it grew in tens of thousands, forming bushes to about 3 m high.

It is amazing to think that in the early 20th century only two plants purporting to be *Rhododendron wightii* remained in Britain and Ireland: a Hooker original at Kilmacurragh, Co. Wicklow and another at Littleworth, near Farnham in Surrey.[11]

It was the English *Rhododendron* grower, Lt-Cdr John Guille Millais (1865–1931), who first noticed that the cultivated plants at Kilmacurragh and at Littleworth (where Clara Mangles was custodian of seedlings raised by her deceased brother, the rhododendron enthusiast and hybridizer James Henry Mangles) were quite different in both foliage and flower, from the true wild species.

Plants of this misnamed '*wightii*' are still in circulation, and these generally have a lax, one-sided truss of yellowish flowers – unlike the dense, upright habit of the true species, as shown in Hooker's plate (although this illustration is not entirely accurate). The plant grown and sold under the erroneous name is generally thought to have been a natural hybrid with *R. falconeri*.

A truss of a *Rhododendron* that for many years represented *Rhododendron wightii* in cultivation. Seen here near the Palm Walk at Mount Usher, Co. Wicklow.

The complicated saga of the introduction of true *R. wightii* is too lengthy to reiterate here (see McQuire & Robinson, 2009, *Pocket Guide to Rhododendron Species*, for further reading), but it is well-documented that mature plants currently in British and Irish gardens were all raised from seeds sent to Britain from Nepal in 1971 by Beer, Lancaster and Morris (BLM 92).

Alongside Wight's rhododendron, both *Rhododendron fulgens* and *R. wallichii* were equally abundant and formed shrubs of a similar size. Scrambling through these rhododendron thickets were several plants of *Clematis montana* var. *brevifoliosa*, then covered in silky seed heads.

At 3,800 m we reached the southern spur of Mans Lepche, where the landscape opened out into grassy alpine meadows bordered by great thickets of *Rhododendron lanatum*, *R. campanulatum* and *R. wightii*. We soon found and matched Hooker's sketch of Black Kabur and enjoyed a sweeping view of the Prek Chu valley and its extensive glacial moraines.

A superb rosette of *Meconopsis paniculata* photographed on a southern spur of Mans Lepche. Even out of flower this species is highly decorative.

Rhododendron wightii, Hooker's rather misleading plate of the species from *Rhododendrons of the Sikkim Himalaya*. The inflorescence of the true plant is less tightly packed near the apex.

Meconopsis paniculata occurred here and there in the shelter of surrounding thickets. Even out of flower this species makes a stunning sight with its tight evergreen rosettes of foliage densely coated with golden-brown hairs.

Soon we reached the upper scrub layer. Yet more rhododendrons appeared as we gained altitude, a trio of dwarfs, *Rhododendron anthopogon*, followed by the deciduous *R. setosum*, and ultimately, squat plants of *R. nivale*. The first two species, particularly *R. setosum*, filled the air with a strong resinous scent. I could commiserate with Hooker who said his porters disliked the scent of this species and associated it with the headaches they suffered from trekking at such high altitudes. The last species, *R. nivale*, is one of the highest-growing shrubs in the Himalaya, and its specific epithet meaning 'growing by snow' was an apt choice by Joseph Hooker.

This completed the altitudinal sequence of the *Rhododendron* flora of this region, and having trekked through the mountains of west Bhutan just two weeks previously, I came to appreciate the astounding biodiversity of the Sikkim Himalaya. There is nowhere in the Himalaya that can match it for the sheer number of different species per square kilometre, and for a group visiting from a small island in Western Europe, with such a meagre native flora, it was exhilarating to explore.

At Dzongri we entered a truly alpine landscape. The settlement consisted of little more than a shop, a few yak herders' huts and a trekkers' lodge, and we camped on a small windswept plateau about a kilometre to the north of the village. This plateau and the surrounding area are used by farmers from Yuksam to graze their yaks between March and October. This was exactly the same campsite that Joseph Hooker and his porters had used when they passed through here in January 1849:

> After a two hours' walk, keeping at 13,000 feet elevation, we sighted Jongri. There were two stone huts on the bleak face of the spur, scarcely distinguishable at the distance of half a mile from the great blocks around them. To the north Gubroo rose in dismal grandeur, backed by the dazzling snows of Kubra, which now seemed quite near, its lofty top (alt. 24,005 feet) being only eight miles distant. Much snow lay on the ground in patches,

and there were few remains of herbaceous vegetation; those I recognised were chiefly of poppy, *Potentilla*, gentian, geranium, fritillary, Umbelliferae, grass and sedges.

On our arrival at the huts the weather was still fine, with a strong north-west wind, which meeting the warm moist current from the Ratong valley, caused much precipitation of vapour. As I hoped to be able to visit the surrounding glaciers from this spot, I made arrangements for a stay of some days: giving up the only habitable hut to my people, I spread my blanket in a slope from its roof to the ground, building a little stone dyke round the skirts of my dwelling, and a fire-place in front.[12]

Nowadays there are three stone huts at Dzongri. One of these, slightly crudely built, was constructed in the 1960s for the 12th and last king or Chogyal of Sikkim, Palden Thondup Namgyal, to house his flock of rare sheep. The other two are quite clearly the huts Hooker spoke of, and looking at the site, it was obvious that it was the first, south-west facing built on three levels, that he gave up to his porters. I hadn't expected these structures to survive, but there they were, adding to the excitement of an already exhilarating trip and adding to the sense that we were indeed in the footsteps of Joseph Hooker.

In 1977, following the creation of the National Park, these huts were burned by the authorities to reduce domestic grazing and human presence in the area, and on hearing this, the Buddhist monks at Pemayangtse petitioned the government to have them restored. To Buddhists, this is a sacred landscape and for centuries monks from Pemayangtse had made an annual pilgrimage in September to pray to the deities of the mountains. Luckily, all the huts were restored and remain a feature of the area.

Approaching the alpine plateau at Dzongri, we were thrilled to find that the stone huts Hooker mentioned in *Himalayan Journals* still existed. His men slept in this building in January 1849, while he and his dog Kinchin slept in a makeshift tent in front of it. The peaks of Tingchen Khang and Jopuno pierce the sky behind.

Our campsite at Dzongri in November 2014. Behind our mess tent is 'Hooker's hut', as we would later dub the structure. Our porters are busily preparing supper while our group retired for pre-supper drinks and the daily vote for 'plant of the day'.

It was a spectacular spot. To the north-east , the landscape was dominated by the towering snow-clad Mount Pandim (6,691 m), the jagged peaks of Tingchen Khang (6,010 m) and Mount Jopuno (5,936 m), presenting a formidable wall of icy peaks. Directly behind these lay the great granitic mass of Kabur, Hooker's Gubroo (4,810 m), and still further to the north lay the dazzling icy summit of Kabru, Hooker's Mount Kubra (7,412 m).

The view to the south was also magnificent, with successive mountain ridges as far as Darjeeling, while to the west the Rathong River valley plunged to enormous depths. Above it, on its west bank, lay the northernmost part of the Singalila Range, at this point a narrow ridge, and beyond there lay Nepal.

The area surrounding the campsite was a broad, flat expanse of moorland, rising steeply to the north, where the slopes were a patchwork of the remains of flower-filled meadows broken by dark thickets of dwarf *Juniperus indica* and great masses of the deciduous *Rhododendron setosum*, with tan-brown autumn colour.

Douglas William Freshfield used the same campsite in the autumn of 1899, and from his tent, witnessed a magical scene:

I have read somewhere that there are no sunsets or afterglows on the Himalayan heights. As far as Sikhin [sic] is concerned this assertion is the reverse of the truth. To watch a sunset from Jongri is a thing worth living for. More beautiful colours in earth and sky I have never seen. Beyond the broad shadows of the foreground the more distant foothills turned into solid waves of sapphires, the snows blushed rose-red, until the flush, slowly dying out on the lower heights, lingered last on the crests of Pandim and Kabru. Slowly the dusk deepened in the luminous sky until the moon rose, and, kissing the icy foreheads of the loftiest peaks, threw its mild radiance over the vast spaces spread out beneath us.[13]

The Dzongri La and the Rathong valley

The following morning, at 4.30 a.m., we began our trek towards the Dzongri Peak, a large cone-shaped mass about 2 km south-east of the campsite. It rose over 300 m above our base, providing panoramic views of the Kangchenjunga massif. It was a chilling experience at such an early hour, dark and silent apart from the occasional ringing of bells on our pack animals. A deep hoar frost carpeted the ground, and streams and rivulets lay frozen. We kept close together; snow leopards had killed baby yaks near Dzongri earlier that season and with our head torches we were conspicuous in the landscape. The ascent was steep, and in places slippy, and the altitude made some of us quite breathless. We went slowly, and reached the summit before the sun rose.

When it did, it presented an amazing spectacle as the soft light fell over the mountains, painting the peaks, glaciers and ice fields a shade of warm amber. We were ice-cold, though the view had definitely been worth it, and returned to the camp for a welcome breakfast, botanising along the way.

Following this, we set off again, this time in a north-westerly direction, to explore the alpine vegetation surrounding the Dzongri La (pass) and the adjacent upper Rathong valley. The Dzongri La (4,550 m), to the west of Kabur, was a two-hour trek from our base.

A gathering of snow pigeons, *Columba leuconota*, foraged for food in the surrounding thickets as we left the campsite. Though it was late in the season alpines abounded, one of the most spectacular being *Androsace lehmanii*, a carpeting species forming great cushions 90 cm across at the base of Kabur. It was ubiquitous, with, at this late season, rusty-brown foliage. With it grew one of Hooker's Lachen valley discoveries, *Diapensia himalaica*, a tiny prostrate shrub with glorious deep purplish-red foliage. Completing this medley was the Himalayan heather, *Cassiope fastigiata*, an aptly named species, with upright stems rising to no more than 15 cm; in summer it bears masses of delicate white inverted bell-shaped blossoms.

Though the flowering season was long over, we could identify the remains of many of the meadow inhabitants. The yellow-flowered *Potentilla coriandrifolia* was perhaps the most abundant and its foliage remained ungrazed by visiting yaks. Also here were the seed heads of the lovely *Pterocephalus hookeri*, a pretty little alpine I had seen previously, flowering in north-west Yunnan. These meadows must have presented a kaleidoscopic scene that summer, filled with several species of *Primula, Silene, Pedicularis, Gentiana* and terrestrial orchid species.

Trekking towards the Dzongri La, one of the most pristine, though isolated, landscapes in the Himalaya. Above Dzongri we entered a truly alpine landscape.

Androsace lehmanii, a primrose relative, seen here in autumn foliage. It formed dome-shaped hummocks beneath Kabur.

Snow pigeons foraging on the slopes above our campsite.

Higher still, close to the pass and well above the scrub layer, another curious alpine, *Arenaria polytrichoides*, formed dense cushions of tiny moss-like foliage. It was described by the Irish botanist, M. Pakenham Edgeworth, from specimens collected by Hooker on the Kongra La, a mountain pass in north-east Sikkim that leads into Tibet at 5,133 m; it is one of the highest-growing of Sikkim's flowering plants. With it grew enormous colonies of *Primula glomerata*, a great sheet of silver farinaceous foliage. This species is similar to *P. capitata* and bears masses of blue-purple flowers in a tight globose inflorescence.

At midday, following a final steep ascent, we reached the Dzongri La. It was worth every effort, and suddenly an expansive view of the upper Rathong valley opened before us, with the Rathong Glacier and the snowy peaks of Kabru piercing an intensely blue, cloud-free sky.

The Rathong Glacier descends from the west-facing slope of Kabru Dome, with its source at 6,200 m. Extending in a south-easterly fashion for over 4 km, it is 50 m deep and forms the headwaters of the Churang River, which later joins the Rathong River.

The upper Rathong valley, looking towards the glaciers of Kabru. Ponies belonging to Indian mountaineers approach.

A caravan of ponies approached the pass from the valley below. Kabru is used by the Himalayan Mountaineering Institute at Darjeeling for training, and the team had been practising for their intended ascent of Mount Everest the following year.

Hooker sadly never reached this pass, as the snow was too heavy to risk the lives of his porters; the campsite at the foot of Kabur was as far as he managed to venture.

Dzongri to Thangsing and Lamuney

It was at Dzongri, in 2014, following three expeditions to West Bengal and Sikkim, that we finally completed our task of retracing Hooker's footsteps. It was an appropriate place to fulfil this ambition, though we decided to go further still – right up to the eastern flank of Kangchenjunga and the famous Goecha La pass.

Taking us to 4,940 m, this route is considered to be one of the most challenging hikes in the Indian Himalaya, and we were more than a little apprehensive about taking on such an arduous task. Meeting some very ill German trekkers returning from the pass didn't instil confidence. Our plan was to make a slow three-day ascent, hopefully allowing for further acclimatisation, though we knew not everyone in the group would make it.

From Dzongri we continued north through the valley of the Prek River, with the great peak of Pandim dominating the scene. Our next base was Thangsing, an alpine meadow at 3,930 m, entailing a descent into fir forest again.

Our route took us past Kabur Lam, one of many sacred glacial lakes in this area. It was particularly beautiful on the day we visited, its clear waters reflecting mirror images of the snowy mountains to the north-east while myriad prayer flags strung along its western edge fluttered in the breeze. Dense thickets of rhododendrons surrounded the lake in places, particularly at the base of cliffs on its western side. These included *R. campanulatum*, *R. wightii* and low-growing hummocks of *R. nivale*, while the warm morning sun filled the surrounding area with the heavy resinous scent of *R. anthopogon* and the ubiquitous *R. setosum*. *Juniperus indica* was particularly abundant in this region, forming handsome low-growing dark mounds, with feathery foliage covered in a heavy crop of blue-black, gin-scented fruits.

On the cliffs above the lake grew several good garden plants, including one of Hooker's best-known Sikkim discoveries, *Bergenia purpurascens*, which he found in the upper Lachen valley on 7 June 1849. By the lake it grew in great clumps, its foliage at that season beetroot red. With it grew the foxglove relative *Lagotis kunawurensis*, a common alpine in the Himalaya, where its range extends from Pakistan to the mountains of south-east Tibet. In summer it produces densely flowered 30 cm tall spikes of flowers that vary in shades from white, to mauve, to intensely deep blue. With it we spied the handsome seed capsules of *Meconopsis simplicifolia*, *Silene nigrescens*, *Potentilla atrosanguinea* and several species of *Saxifraga* and *Erigeron*.

Moving on to the north-east of the lake we found two Hooker discoveries growing side by side. *Gentiana stylophora* is a giant of a plant, reaching 1.5 m tall, its stems (in summer) sporting large, handsome *Veratrum*-like pleated foliage. The yellow, green-chequered flowers are the largest in the genus and though long out of flower at the time of our arrival, I could still appreciate its stately beauty, as I

Kabur Lam Lake, one of several high-altitude sheets of water in this magical corner of the eastern Himalaya.

had been fortunate enough to see it in bloom in SE Tibet in July 2001. Hooker first found it near Dzongri in 1849. With it grew an altogether more curious plant, *Boschniakia himalaica*, a parasite similar to ivy broomrape (*Orobanche hederae*). It is native of the Himalaya and much of China where it grows on the roots of *Rhododendron* species; here it sprang from the roots of *R. wightii*.

From the lake we headed north-east towards Thangsing, bringing us to the summit of a steep boulder–covered ridge. On one of these boulders we found *Pleurospermum govanianum*, a spectacular carrot-relative, bearing (in summer) wide spreading umbels of pinkish-white flowers. In Sikkim both its foliage and seeds are used as a spice to flavour food, while in China its deep tap root is important in traditional medicine.

From the same ridge we had a sweeping view of the Prek Chu valley, dominated by Mount Pandim, with enormous tongue-like glaciers falling from the south-western flank of the mountain. As we sat on the ridge absorbing this dramatic view, a lone golden eagle glided across the valley scanning for marmots and hares beneath.

We descended into the valley below through a thicket of *Rhododendron lanatum, R. cinnabarinum* and *R. campanulatum*. Soon we were back in a dark forest of *Abies densa*, with a scattering of smaller trees such as the pink-fruited *Sorbus microphylla*, a pretty little rowan that is common in Sikkim at these altitudes. By the sides of streams *Salix sikkimensis* formed small trees, with golden-yellow autumn foliage. Many of these trees had their lower stems densely draped with the delicate fronds of several species of *Trichomanes*, indicating how constantly humid the climate of this valley is.

We crossed the turbulent waters of the Prek River by means of a simple, rough-hewn wooden bridge. This river finds its headwaters in the nearby Onglathang Glacier and ultimately joins the Rathong River. On its banks grew the most spectacular plant we were to see that day, *Myricaria rosea*, a low-growing, lax-habited, prostrate *Tamarix*-like shrub, described from material gathered in NW Yunnan (where it is also a characteristic riverine species) by the plant

Juniperus indica seen here on moraines below Pandim. At lower altitudes it can form a small tree.

Myricaria rosea was especially abundant in the upper reaches of the Prek Chu valley, so much so that its autumnal hues were recognisable from several miles away.

collectors George Forrest and Frank Kingdon Ward in the early 20th century. It had however, been previously discovered in the upper Lachen valley by Hooker in July 1849. On the banks of the Prek Chu it was abundant, scrambling in gravel through boulders, and at the time of our visit it had assumed vivid pink-red autumn colour visible from several miles away.

The following stage of our journey involved a steep hike from Thangsing to our next base at Lamuney, a campsite in a wide mountain valley at 4,145 m. Lamuney is named from the many mani or prayer stones, written on in Sanskrit, found in this spectacular valley. We reached our campsite after a long trek along the banks of the Prek River, and it was there that we began to encounter the first elements of a morainic flora.

Morainic scrub is found mainly in the glaciated valleys of north-west Sikkim, along lateral and terminal moraines between 3,900 and 4,500 m. In its lower reaches this shrub-dominated flora is very diverse, becoming sparse and stunted in the upper levels. *Potentilla fruticosa* var. *arbuscula* and *Spiraea arcuata* are the diagnostic species of this morainic scrub flora in the mid elevations. The *Potentilla* was particularly abundant above our campsite where it covered vast areas, its grey stems contrasting brilliantly with the glowing red of the *Myricaria*.

Lamuney was the most spectacularly located campsite we used, in an enormous valley, well above the tree-line, through which the Prek River cut a deep track. Our tents were dwarfed by the massive walls of the valley, and to the north-east lay the huge cliff-like face of Mount Pandim while Kangchenjunga loomed up directly to the north.

Ascending to the Goecha La via Samiti Lake

On 10 November 2014, at 3.30 a.m., we departed from our campsite heading due north towards the Goecha La and Kangchenjunga. It was cold and pitch dark, and for several days I had difficulty breathing (due to altitude); I questioned if any of us would reach the pass.

Our Tibetan guide, Thupden Tsering (who had travelled on all our previous expeditions to Sikkim), led the way slowly across the alpine moorland; we must have made a curious sight in the still darkness, slowly ascending into the glaciated Onglathang valley above us.

Perched high up in the Onglathang valley at 4,800 m, Samiti Lake is the most famous of the many glacial lakes found in this region of the National Park. Its Sikkimese name is Sungmoteng, and it is considered sacred by Buddhist pilgrims who travel from afar to worship there; prayer flags line the shore.

The lake is filled from a stream descending from Mt Pandim, while its outlet later joins the Prek River to the south, and we skirted its cold, deep waters in the darkness. Above there we reached the first of several mountain passes we were to cross that day. Famous for its early morning views of Kangchenjunga, we stopped and rested on the pass as a bracing, icy breeze, whipped through the surrounding prayer flags. We were glad of the large stone cairns that provided a small element of shelter, though it was bitterly cold, and despite being well equipped with thermal clothing, we were all thoroughly frozen.

Pandim loomed ominously above us. A furious icy gale roared off its summit, sounding like a jet engine overhead. Finally, by 5.40 a.m., the sun raised its head above Kangchenjunga, drowning it in a cold amber light.

Pandim with the boulder-filled bed of the Prek Chu in the foreground. Dark trees of *Abies densa* scale the slopes of the valley on either side.

The final push; at 3.30 a.m. we departed from base camp and in the pitch darkness headed north towards the Goecha La. Some people like sun holidays. Ours was nothing of the kind!

took us along substantial, steep-sided lateral morainic ridges of pre-existing glaciers, and finally, by those of the massive Onglathang Glacier, its lower tongue dusted with gravel and boulders.

We kept well to the track. Below us we spotted a huge sinkhole in the ice, below which we could hear the roar of a subterranean river. The soaring granite cliffs of Pandim finally lay behind us, its north-west ridge falling steeply towards the Goecha La.

Above, in the middle of a field of glacial debris, lay another great glacial lake. On its north-western side rose a sheer cliff several hundred feet high, from which descended a hanging glacier, its enormous icy snout full of stone and gravel. We could hear this ice cracking and splitting in the early morning sun, melting the ice, which streamed down the cliff into the turquoise waters of the lake below; we were witnessing the origins of the Prek River.

We reached the Goecha La at 10.30 a.m. having crossed the last moraine. It was inevitably the most difficult, with enormous boulders and a deep layer of snow and ice, and a steep chasm below. Having trekked a gruelling eight hours, I wasn't prepared to give up at this stage, and four of us alongside our guides made the crossing.

Beyond there we crossed the dry bed of a substantial lake, the result of a catastrophic outburst in 1988. Such was the power of the escaping deluge that it shook the windows of houses at Tshoka, the Tibetan refugee village above Yuksam. At the time of our visit it looked like a desert plateau, as vegetation had only begun to recolonise the area.

The final ascent towards the Goecha La proved steep and challenging. At that altitude every step took enormous effort, and we were breathing in less than half the oxygen levels we were used to. Half the group had fallen back, content to have reached the head of this massive valley. The rest continued, some of us with pounding headaches, exhausted but determined to reach the pass. Our trail

Lamuney is named for the many mani stones in this part of the valley on which is inscribed the Buddhist mantra 'Om Mani Padme Hum', meaning 'Hail to the jewel in the lotus.'

Our next campsite was at Lamuney, and it acted as a base camp before we made our ascent towards Kangchenjunga and the Goecha La. Our tents may be seen on the far left, dwarfed by the great cliff face of Pandim.

Kangchenjunga loomed closer and closer as our days of trekking passed. Finally, we made the ultimate push. Shortly after we stopped to take in this view our group split; for some we had travelled far enough, the remainder of our party pressed on. It was tough going, and before long at this heady altitude we were breathing in less than half the oxygen we were used to in Ireland.

The pass made a spectacular sight and we rejoiced on reaching it. In front of us, just 5 km away, lay the stupendous eastern flank of Kangchenjunga, the mightiest mountain in all India, with the great Talung Glacier descending from its face. Our Tibetan guide, Thupden Tsering (a deeply devout Buddhist), with the aid of a Bhutia guide from Yuksam, tied prayer flags across the pass and beneath these, lit the boughs of *Juniperus indica* and *Rhododendron anthopogon* that he had carried up the mountain, as an offering to the deities of Kangchenjunga. We too were grateful to these deities for allowing us to reach the pass. It was a wonderful experience, and we stood there, high by Kangchenjunga, as prayer flags fluttered and incense rose into the clear blue skies.

J. Claude White crossed the Goecha La, in July 1890, and he was equally impressed by this mighty Himalayan pass and the surrounding ice peaks:

> The weather now cleared up, and I had one of the glorious breaks which occur at intervals during the rains, and crossed the Giucha-la [sic], 16,420 feet, in clear weather, with not a cloud in the sky. The view from the top is superb. Before one lies an amphitheatre of snow peaks, all over 21,000 feet, save in one gap, which is 19,300 feet. On the right hand Sim-vo-vonchin rises sharply over the 19,000-foot gap, then the splendid shoulder supporting the twin peaks of Kangchenjunga, which towers up to a height of over 28,000 feet, and with something like 11,000 feet of uninterrupted snow and ice falling in a sheer precipice on its south face to the great glacier at its foot, next the ridge connecting Kangchenjunga with Kabru, and on its immediate left a fine unnamed snow peak with hanging glaciers … On the south side of the Kangchen Glacier were some ancient moraines covered with exquisite green turf and masses of alpine flowers, whose simple beauty and vivid colouring stood out in sharp contrast to the grandeur of the surrounding snows, making a picture long remembered.[14]

Our group pictured here with Kangchenjunga, the guardian deity of Sikkim. Back row left to right: David Koning, Eileen Murphy, Philip Quested, Margie Phillips, Gail Gilliland, Debbie Bailey, Front row: Robert Wilson-Wright, Terry Smith, Billy Alexander, the author, Alan Ryder.

White camped on these grassy moraines and spent the following days exploring the area around the pass and the lower tongue of the Talung Glacier. He was in unexplored country, and his plan was to trace his way along the Talung River, to present-day Mangan and from there to his base at Gangtok. The following days were full of excitement:

> Next morning, I walked up the valley as far as I could go without crossing the glacier, and the scene, if possible, became still wilder and more magnificent. On the right was Kangchenjunga and to the left Kabru with its magnificent glacier, while joining the two mountains in front of me was a wall of snow and ice 21,000 feet high.....
>
> … Looking directly up the valley was the end of the glacier I had just descended, gloomy and forbidding, and on the right, to the north, was the limit of the glacier from the 19,000-foot-gap, adding to the scene of desolate grandeur; for I think there can be no more wild and desolate scene than these moraines, in which the large glaciers end in utter confusion......with the constant fall of stones as the ice melts, and the weird feeling that everything in addition is quietly though imperceptibly on the move.[15]

The final glacial lake we met before reaching the Goecha La. It was fed by the melting waters of the Onglathang Glacier.

We stayed for only half an hour on the pass. Time was pressing on and we had a testing return journey to Thangsing in front of us. We botanised on the same flowery moraines visited by Claude White 114 years before, and, despite the lateness of the season, there was plenty to admire.

Below the pass we found dozens of plants of the weird and wonderful *Saussurea gossypiphora*, a curious alpine that grows only by the highest mountain passes of the Himalaya. Aptly called the snowball plant, its rounded foliage is densely coated in a globular mass of fine white woolly hairs that protect it from the extreme climate it encounters in its native habitat. The specific epithet *gossypiphora* alludes to the plant's appearance; it literally means 'carrying cotton'.

Curiously, another plant with a similar specific epithet to the snowball plant grew nearby. Just metres away, we found another Hooker discovery, *Hippolytia gossypium* (syn. *Tanacetum gossypium*), growing in scree and rocky slopes. It reached no more than 7 cm high, bearing closely overlapping leaves, densely clothed on both sides with a layer of short cottony hairs. With it grew the curious *Tibetia himalaica* (*Gueldenstaedtia himalaica*), a tiny alpine, with short pinnate leaves and, in summer, the most beautiful purple-blue, pea-like flowers. Hooker found this near Talam, in the upper Lachen valley on 13 July 1849.

We also encountered another of Hooker's many discoveries there, including *Meconopsis horridula*, the wonderful prickly blue poppy, named for its the many bristle-like spines that cover its stems and protect it from grazing animals. In summer, during the height of the monsoon season, it carries masses of sumptuous blue blossoms, arising singly from the prickly leaf rosettes. Hooker found it growing on the Kongra La pass at 5,182 m in August 1849. This is, in fact, the highest-growing species of *Meconopsis* in the world.

The blue poppy, by chance, grew with another of Hooker's plants, the dainty little *Primula glabra*, a delightful species carrying masses of small mauve blossoms in a drumstick-like head, on slender 12 cm tall stems.

Below there we re-crossed the dry lake bed. By then a furious sandstorm raged, whipping dust high into the air and obscuring the view in front of us. Perhaps it was as well. Beyond it lay the steepest ridge we had to climb on our return journey, an unwanted obstacle for weary travellers. Once past this ridge it was all downhill on steeply descending lateral moraines to Samiti Lake past thickets of *Rosa sericea*, *Spiraea arcuata* and *Potentilla fruticosa* var. *arbuscula*.

The perils of alpine travel. A glacial sinkhole disguised by gravel.

The snout of the Onglathang Glacier, seen here hanging over a cliff face that fell steeply into a glacial lake below. We could hear its ice cracking in the early morning sunlight. Streams of water ran down the cliff.

By the edge of Samiti Lake, among *Juniperus indica* and *Rhododendron anthopogon*, the tiny *Lilium nanum* sheltered, its rounded, plump seed capsules ripening before the onset of winter. We also found several dozen wind-blown plants of the mint-relative, *Eriophyton wallichii*, one of my favourite Himalayan alpine plants, which in summer produces densely tufted 20 cm tall spikes of wine-red flowers, partially hidden by overlapping broadly ovate leaves, which are densely coated in fine, silken cottony hairs.

As we crossed the eastern shores of the lake we spotted a rare scene – blue sheep grazing on the pastures above us, and, since a broad stream divided us from them, they seemed not in the least frightened. Blue sheep, *Pseudois nayaur*, are a flagship species of high-altitude alpine ecosystems, and their best grazing in Sikkim is in the sedge meadows above Samiti Lake by the Onglathang Glacier; the area has been a blue sheep conservation zone since 1972. The surrounding meadows are among the best summer pastures of this highland region, with an abundance of highly nutritious fodder

species, among them *Kobresia capillifolia* and *Carex nivalis*, and showier flowering plants such as *Potentilla peduncularis* and various species of *Anemone* and *Lomatogonium*.[16] Native of the high Himalaya of Pakistan, India, Nepal, Bhutan and China, blue sheep are a major food source of the apex predator of the Himalaya – the snow leopard, of which just 6,000 remain in the wild.

J. Claude White, passing through here during his 1890 expedition to the Goecha La, was lucky to see this rare animal, though he lived in the age of the trophy hunter:

> While I was wandering some little way from camp I saw a snow leopard. He was on the other side of a glacial stream, so I could not get very close to him, and as besides I only had a shot gun with me, I contented myself with watching him, and a very pretty and most unusual sight it was. He was playing with a large raven, which kept swooping down just out of his reach, and to see him get on his hind legs like an enormous cat and jump at the bird was worth watching. Suddenly he saw me and went off up the hill at a pace that made me envious. He was a fine specimen, very large and with a beautiful coat, and I wish I had had the luck to bag him.[17]

Approaching the campsite at Lamuney the low shrubby thickets were smothered in a dense quilt of a cotton-like plant. On closer inspection this proved to be a dwarf willow, *Salix calyculata*, which, in common with other willow species, distributes its seeds through the air on cotton-like hairs, which often land far from the parent plant. It was so abundant on the landscape near Lamuney that in places it looked as though snow had fallen. Hooker's specimens at Kew record that he discovered it in the upper Lachen valley at 3,658 m in June 1849. After this we raced downhill to Thangsing, where we enjoyed a lazy evening following eleven hours of arduous trekking in the rarefied climes of Kangchenjunga.

On the pass at 16,207 feet. David Koning, Alan Ryder, Billy Alexander and the author.

Thangsing to Tshoka

We returned to the Tibetan village at Tshoka by a different route from the one we had taken on our ascent, this time trekking along a mountain ridge between the Prek valley and Phedang. From the expansive valley at Thangsing we retraced our route to the banks of the Prek River and followed a trail leading directly to Phedang. This trail is too narrow for yaks and other pack animals, and we were glad to get a break from the ill-tempered dzos.

We were back again in prime *Abies densa* forest and gradually began to ascend a steep ridge, smothered in primeval, species-rich forest. *Betula utilis* was abundant throughout the dark fir forests, forming enormous specimens, the best I have seen anywhere in Sikkim. Their lower trunks were smothered in *Hymenophyllum* and *Trichomanes*, common ferns in the damp, mist-laden Prek valley. In general, I haven't been impressed with the forms of *Betula utilis* elsewhere in the Sikkim Himalaya, believing them to be far inferior to those I've seen in SE Tibet, but the trees here were magnificent, both in stature and in terms of bark, which was here a warm cherry-brown, with prominent white lenticels.

Rhododendrons abounded, though the species that stole the show was *Rhododendron hodgsonii*, which formed dense thickets below the firs. I always think that large-leaved rhododendrons are best viewed from above, so in a garden situation, they should ideally be planted in deep dells, as seen, for example, in Derreen Gardens in Co. Kerry. Such was the situation we trekked through that day, our group on the upper ridge with tens of thousands of *R. hodgsonii* beneath us, its large, bold leaves adding an exotic element to the area.

Blue sheep, one of the inhabitants of this high-lying landscape. In north Sikkim they are preyed upon by snow leopards.

It was like walking through a wild, Robinsonian woodland garden, the sort of setting you'd expect to meet in Cornwall, SW Ireland, or the west coast of Scotland, and any of us travelling that day would have been happy to transplant this sylvan scene back to our homes in Ireland. My thoughts roamed back to Kilmacurragh and the single young specimen I grew near the pond. How lonely it seemed compared to this great gregarious mass and how worthwhile it is to travel the many thousands of miles from Western Europe to see this aristocratic species *en masse* in its native haunts.

Brian Hodgson's rhododendron dominated the scene for much of the day, contrasting beautifully with *Enkianthus deflexus*, which formed small trees with glorious amber-red autumn foliage. Autumn-flowered

Kangchenjunga from the Goecha La. We have just erected prayer flags across the pass, and incense from burning juniper and rhododendron give thanks to the guardian deity of the mountain for our safe ascent.

The dwarf alpine willow, *Salix calyculata*, seen here shedding its seeds above Lamuney where it formed extensive thickets alongside *Potentilla fruticosa* var. *arbuscula* (seen at bottom left), a common component of the surrounding morainic vegetation.

gentians carpeted the woodland floor, with a few late blue blossoms among Hooker's strawberry, *Fragaria daltoniana* (first found by him in the Lachen valley in 1849), and sheets of *Gaultheria pyroloides*. Midway through the *Rhododendron hodgsonii* woods, and again at Phedang, we found a few scattered plants of *R. × decipiens*, a natural hybrid between *R. hodgsonii* and *R. falconeri*. It was first discovered by the noted British botanist Charles Carmichael Lacaita (1853–1933) on the Singalila Range near Sandakphu (west of Darjeeling, on the Nepal frontier) in 1913, and described by him as a new species three years later. He chose to use the rather appropriate specific epithet *decipiens* meaning deceptive, and for many decades it remained a puzzling enigma.

Nowadays it is recognised as being a hybrid and occurs where the two species overlap, forming small trees to about 3 m tall, bearing terminal corymbs of 25–30 rose-pink, fading to almost white, ventricose-campanulate flowers, with dark crimson blotches at their throats.[18]

We reached the village of Tshoka late that evening, and over a hearty campsite dinner voted *Rhododendron × decipiens* plant of the day. The following evening, we finally reached civilisation, our hotel at Yuksam, clean clothes, hot running water and home comforts after an exciting, though long and sometimes difficult trek to Dzongri and the Goecha La.

Having a well-earned break between Lamuney and Dzongri. Dehydration and altitude sickness are the perils of alpine travel; we drank plenty of water. Here we stop for lemonade provided by Dawa, our Lepcha porter. Left to right: Debbie Bailey, Billy Alexander, the author, Thupden Tsering, Margie Phillips, Eileen Murphy and Gail Gilliland.

CHAPTER 10

EXPLORING THE LACHEN VALLEY

Following his exploration of the Terai, Hooker spent most of April 1849 in preparation for an expedition to the mountainous region of NE Sikkim, particularly the Lachen and Lachung valleys, which ultimately led to passes onto the Tibetan Plateau. From the outset Hooker was determined to cross the border into this mysterious, forbidden land, by any means possible, and, it seems, with, or without, the relevant permits.

The Admiralty, who were footing the bill for Hooker's travels, expected a little more than natural history collections in return. This was the era of the Great Game, the strategic rivalry between the Russian and British Empires for supremacy in central Asia. Britain had long worried, albeit unfoundedly, about Russia's influence in Tibet, and its expansion into central Asia posed a further threat to what was regarded as the jewel in the crown of the British Empire. The possibility of the Russians using either Afghanistan or Tibet as

a staging-post for the invasion of British India seemed very real, and so Hooker's travels in Sikkim, which acted as a perfect buffer between Tibet and India, were of enormous strategic value. Several of the valleys Hooker was about to explore and map led to passes that allowed access directly across the Tibetan frontier.

As we have seen already, Archibald Campbell had informed the Rajah of Hooker's intention to leave Darjeeling, and travel up to the frontier region north-east of Kangchenjunga. Permission was granted, though the Rajah's Dewan was to place innumerable obstacles in Hooker's way.

Since Campbell and Hooker's meeting with the Rajah the previous December at Samdong, no suitable Vakeel (agent or ambassador) had been appointed by the Sikkim court to Darjeeling. There was thus no direct means of communicating with the Rajah, and it was impossible to verify the many reports promulgated by the Dewan, that Hooker

Looking south down the Lachen valley below the village of Lachen. The Himalayan hemlock, *Tsuga dumosa*, bristles along a ridge. At this point the vegetation of the valley is cool-temperate.

'In the *Rhododendron* valley of the Himalaya mountains', mezzotint by William Tayler (1854) after Frank Stone. The scene depicts Hooker inspecting the botanical collections of his Lepcha men.

believed were intended to prevent him entering Sikkim. Indeed, news reached the exasperated Hooker that the Lasso headman was prepared to use strong measures if he crossed the frontier.

Towards the end of April he completed the task of sorting and packing his Nepalese collections and had them dispatched to Calcutta. During this time, he had an interesting visit from William Tayler (1808–1892), the Postmaster-General for India, who was also a skilled portrait painter. He and Hooker rapidly became friends, and Tayler produced a charming scene of Joseph Hooker and his collecting party at Darjeeling:

> He is pleased to desire my sitting in the foreground surrounded by my Lepchas and the romantic-looking Ghorka guard, inspecting the contents of a vasculum full of plants, which I have collected during the supposed day's march. My Lepcha Sirdar (which means great man's head man) is kneeling before me on the ground, taking the plants out of the box, that in his hand being a splendid bunch of *Dendrobium nobile*. He is picturesquely attired in costume, with a large pigtail. Another is behind me; the Ghorka Havildar and Lepchas, in their picturesque uniforms, are looking on, and my big Bhotea dog lies at my feet. On one side two Lepchas are making my blanket tent house, cutting bamboos, &c. I am in a forest, sitting on the stump of a tree, with the snowy mountains in the background; and a great mass of ferns and rhododendrons, brought in by another man, are on the ground close to me.

> My dress was the puzzle, but it was finally agreed I should be as I was when in my best, a Thibetan in the main, with just so much of English peeping out as should proclaim me no Bhotea, and as much of the latter as should vouchsafe my being a person

of rank in character. So I have on a large, loose, worsted Bhotea cloak, with very loose sleeves; it is all stripes of blue, green, white, and red, and lined with scarlet. Enough is thrown back to show English pantaloons, and my lower extremities cased in Bhotia boots. My shirt collar is romantically loose and open, with a blue neckerchief, which, and my projecting shirt wrists, shows the Englishman. My cap is also Thibetan, and only to be described thus: it is of pale gray felt, the upturned border stiff and bound with thin, black silk ribbon. On the top is a silver-mounted pebble, and a peacock's feather floats down my back. The latter are marks of rank.[1]

Hooker had the sketch sent home to Kew so Walter Hood Fitch could make a copy for his father. Fitch made some improvements, though Hooker junior, always a stickler for accuracy, bemoaned the fact that he had included the autumn fruits of Hodgsonia, with spring-flowered rhododendrons. Perhaps the most pleasing of all these scenes was the re-working of Fitch's scene by Frank Stone (1800–1859) in black and white.

Hooker left Darjeeling on 3 May 1849, and Archibald Campbell rode with him, and his sizeable party, as far as the Great Rangit River. Hooker's Luso-Indian servant, John Hoffman, still acted as his cook and helped with many other chores.

He had hoped that his school friend from Glasgow, and fellow botanist, Thomas Thomson, could join him, though sadly this proved impossible.

Hooker's route took him from Darjeeling, north past Tendong, a mountain of some 2,660 m. Tendong in Lepcha dialect means 'hill of the raised horn'. Legend has it that when Sikkim was submerged beneath a great flood, the horn-like peak of Tendong rose to save the surrounding inhabitants. Tendong was clearly visible from Darjeeling, where, curiously, it was known by the Lepchas as Ararat, and this name, and the story of the great flood, was used by the Lepchas long before European settlement of the area.

The ascent from the Great Rangit River valley (249 m) passed through woods of *Shorea robusta,* and *Pinus roxburghii,* to the campsite at the village of Mikola (Mikk) at 1,189 m. From there Hooker trekked to Namchi (Namtchi), once an important town, with a monastery. From the latter, a nervous lama came to visit Hooker as soon as he had set up camp, bringing no presents, as was normally the custom, and the reason for his appearance soon became apparent; the Lasso Kaji had ordered him to stop Hooker's progress. Hooker stubbornly refused to listen, and told the lama he had every intention of continuing on his way the next morning. Shortly after the lama left, he sat peering out of his tent onto nearby mountain ridges and spotted a party of men moving down the mountainside:

> Soon afterwards, as I sat at my tent-door, looking along the narrow bushy ridge that winds up the mountain, I saw twenty or thirty men rapidly descending the rocky path: they were Lepchas, with blue and white striped garments, bows and quivers, and with long knives gleaming in the sun: they seemed to be following a figure in red Lama costume, with a scarlet silk handkerchief wound round his head, its ends streaming behind him. Though expecting this apparition to prove the renowned Kajee and his myrmidons, coming to put a sudden termination to my progress, I could not help admiring the exceeding picturesqueness of the scenery and party. My fears were soon dissipated by my men joyfully shouting, "The Tchebu Lama! The Tchebu Lama!" and I soon recognised the rosy face and twinkling eyes of my friend of Bhomsong, the only man of intelligence about the Rajah's court, and the one whose services as Vakeel were particularly wanted at Dorjiling.

The Chebu Lama, who acted as interpreter to the Sikkim Rajah. Like several other members of the Sikkim Court, he detested the Dewan (Prime Minister). Watercolour by Lieutenant H. M. Maxwell.

> He told me that the Lassoo Kajee had orders (from whom, he would not say), to stop my progress, but that I should proceed nevertheless, and that there was no objection to my doing so; and he despatched a messenger to the Rajah, announcing my progress, and requesting him to send me a guide, and to grant me every facility, asserting that he had all along fully intended doing so.[2]

The following morning the Chebu Lama headed south to Darjeeling, while Hooker began his ascent of Tendong, sending his men on to the village of Temi in the Teesta valley, where they were due to camp that night. A narrow, winding path brought Hooker along the mountain's steep slopes, covered with a dense forest of *Lithocarpus*, *Quercus*, *Rhododendron* and various showy shrubs, of which Hooker singled out the evergreen laburnum, *Piptanthus nepalensis*, for mention. Orchids abounded on the lower warm, humid slopes, and Tendong was to offer up several new species, including

Rhododendron grande, a Hooker original seen here in bloom at Kilmacurragh where it forms an enormous multi-stemmed tree. At Kilmacurragh it begins to bloom in early February, thus kick-starting the flowering season.

Anoectochilus grandiflorus, a showy white-flowered terrestrial species found on the forest floor, and a little higher still, growing as an epiphyte in dense forest, he was to find the stunning yellow and purple flowered *Sunipia cirrhata*.

The views from the summit were obstructed by a vast forest of *Rhododendron grande*, then smothered in white blossoms. A small Buddhist temple and some stupas crowned the top, and the path descending along the northern slopes of the mountain presented a vegetation type very different from that on the wetter south-facing side; the oaks were particularly magnificent, one tree having a girth of 15 m at 1.5 m above ground level.

Alpines grew on the summit; Hooker's best find here was the lovely *Androsace selago*. On the highest peaks it formed dense hummocks with stems radiating out from a central rootstock, and with foliage covered in silvery hairs. The flowers, which appear in late spring, are purple, and smother the plant. It is one of the loveliest of all Himalayan alpines, though probably not in cultivation at the present time. Hooker found it again later that year in the mountains above Thangu and across the Sikkim frontier in Tibet.

Having completed his exploration of Tendong he and his men moved on, and further north, near the village of Temi 1,454 m, he gathered the globular fruiting clusters of *Kadsura heteroclita*, a handsome evergreen climber closely allied to the magnolias. The fleshy red aromatic fruits were a popular food with inhabitants of the surrounding forests according to Hooker, as they still are today.

Hodgsonia heteroclita, a rampant vine related to the gourds and cucumbers, that commemorates Hooker's friend at Darjeeling, Brian Hodgson. The female flowers are illustrated in the first image, followed in the second by the male pollen-bearing inflorescence. The seeds were roasted and eaten and considered a delicacy by the Lepchas. Illustrations by Walter Hood Fitch based on drawings made by John Ferguson Cathcart's Indian artists.

The tropical Teesta valley

From Temi the road dipped into the Teesta valley and followed its course north. It was early summer and the valley was as hot as hell. The curious Dutchman's pipe, *Aristolochia saccata*, formed gigantic vines in the riverside jungles, covering the loftiest trees and bearing purple, pitcher-shaped blossoms. The damp, gravel flats above the river made for especially good botanising, and the herbaceous *Houttuynia cordata* was everywhere abundant. Like *Kadsura* and *Aristolochia*, the *Houttuynia* is part of a large, ancient, and very primitive relict flora that survived in Sikkim, while wiped out elsewhere through glaciation and other natural cataclysms.

Of the many vines and lianas in the tropical jungles of the Teesta valley, Hooker seemed most impressed with the gourd-relative, *Hodgsonia*, a rampant woody climber up to 30 m long, with immense, creamy-white pendulous blossoms, whose petals were terminated by a fringe of long, buff-coloured thread-like appendages. The fruits, resembling small brown melons, contain up to six large kernels, which were roasted and eaten, and were said to taste like pork scraps.

Tropical oaks and terminalias were the giants of this hot valley, particularly a species of the latter commonly known by the Lepchas as *sung-lok* or *sungloch*. Hooker's *sungloch* collections from the Teesta valley were later identified as *Terminalia myriocarpa*, and one tree measured by him had gained a girth of a whopping 14 m at 1.5 m and was fully 61 m high. Around Darjeeling its timber was popularly used for house building and tea chests.[3]

In these boiling, humid, malarious forests Hooker found several new trees including *Ficus hookeriana*, a large fig to 25 m. It is native to the hot valleys of north-east India, Nepal, Bhutan and western China, where it is often planted around temples. These trees act as hosts to numerous epiphytes, including *Agapetes sikkimensis*, a shrubby species bearing tightly packed fascicles of orange tubular blossoms directly from older stems. Here it shared this habitat with several

orchids, including the lovely yellow and rose-purple *Phalaenopsis deliciosa* ssp. *hookeriana*, a far more refined plant than the mule *Phalaenopsis* hybrids that have become so popular as houseplants in recent years. Beneath, on the densely shaded forest floor, Hooker was to discover the crimson and purple flowered *Calanthe fulgens*, one of a large group of orchid species endemic to Sikkim.

The Teesta valley is also the best place in Sikkim to see the yellow-flowered orchid, *Crepidium josephianum*. This terrestrial species was discovered in 1863 in the tropical valleys of Sikkim by Thomas Anderson, Superintendent of the Royal Botanic Garden, Calcutta. He brought back plants to his base in Calcutta where it flowered in April 1867. It was introduced to cultivation by James Alexander Gammie, Superintendent of the Sikkim cinchona plantations, who sent plants to Kew, where it flowered for the first time in May 1877. When featured and first named by the German orchidologist, Heinrich Gustav Reichenbach (1824–1889) in *Curtis's Botanical Magazine* (as *Microstylis josephianum* (t. 6325) in 1877); he dedicated the species to Joseph Hooker, 'in recognition of his services to orchidology, when exploring for the first time by any botanist, the primeval forests of the Sikkim Himalaya'.

By May 10th, the pre-monsoon rains had commenced, drenching Hooker and his men, but moderating the stifling heat of the valley. The following day they reached Samdong, where, the previous Christmas, Hooker and Campbell had met the Rajah. It looked a little different, surrounded by verdant tropical foliage and with a bright green crop of rice where he had pitched his tent.

A little further north, the banks of the river became steep, and the road disappeared, so they were forced to climb a path above the river, to the village of Lingkiang (Lingtuim or Lathiang). From here, a tributary running into the Teesta was bounded by cliffs covered with wild bananas and screw-pines up to 15 m high, clinging to rocks with cable-like roots. Two palms, the climbing rattan, *Calamus acanthospathus*, and the slender *Pinanga gracilis*, ascended thus far up the Teesta valley.

The view from the village was superb: the tropical gulley, the great Teesta valley, Samdong with its rice fields, the fir-covered summit of Mount Maenam and the cone-like cap of Tendong to the south, while to the east the Chola range broke the horizon. Beyond there his route followed the lower eastern slopes of Maenam, to the village of Gar (Gorh), where the local Lama informed him the roads ahead were impassable and he had no authority to allow Hooker to go any further.

Hooker made it clear he had every intention of travelling on, at which the Lama promptly left, as Hooker suspected, to damage the paths and bridges along the route. The next morning he returned, volunteering to guide Hooker along the way. As he had guessed, Hooker found every bridge torn apart, but their components hidden in bushes nearby. Each time Hooker made the Lama repair them, which he did with rather indifferent grace.

As the Lama was busily repairing yet another bridge, Meepo, the Sikkim court messenger who had met Hooker in East Nepal, arrived with the news that the Rajah had granted him permission to travel in Sikkim. He also bore the instructions that he was to proceed with

Hooker, though not to the Tibetan frontier. The Rajah asked Hooker only to visit the west branch of the Teesta (the Lachen River), and he was threatened with Chinese interference on the border. Hooker overruled every argument on the basis that no instructions had been communicated to the Superintendent of Darjeeling, Archibald Campbell. Like it or not, he was determined that he would make it to the Tibetan frontier. At this point the Lama from Gar took his leave, surprised at the Rajah's instructions and pretending to be a friend. He pompously charged Meepo with Hooker's care, bidding him a polite farewell. Joseph Hooker relates in *Himalayan Journals*, 'I could not help telling him civilly, but plainly, what I thought of him; and so we parted.'

Crepidium josephianum, a Teesta valley orchid that commemorates Joseph Dalton Hooker. It flowered for the first time in cultivation at Kew in 1877.

W.Fitch, Del et Lith.

From the village of Lingdong (Lingo), Hooker, with his new guide Meepo, and his party of porters, sepoys and collectors, crossed a cane bridge to the east bank of the Teesta, and from there proceeded over a lofty spur around which the Teesta made a great arching sweep. Beyond there they climbed a steep ridge rising from the Teesta valley, and just before the village of Sentam (Singtam), at a place called Singhik, he sketched one of his most famous views of Sikkim, that of Kangchenjunga rising above the Talung River valley, with his tent pitched in the foreground.

It was at Sentam that Hooker met the Subah (chief officer) of the district, a short, portly Bhutia, who was to become a very active enemy to his travels and exploration. This man governed a large tract of country between Gar and the Tibet border, for the Rajah's wife. These lands were her dowry, and since she was a relation of the Dewan, Hooker expected little from her agent. Though polite, he delayed Hooker by a day, pretending to organise food, and so Hooker was forced to move on, hoping to obtain supplies in the next village, or from Darjeeling. However, owing to the increasing distance and the destruction of the roads by the heavy rains, his supplies from Darjeeling were becoming very irregular so Hooker reduced his team, by sending back his guard of five sepoys.

Beyond Sentam, the bed of the Teesta narrowed and it became an angry torrent, contained by a steep, precipitous valley. The cliffs there were scaled by honey collectors, who climbed the rock faces on bamboo cane ladders to reach pendulous beehives that were so enormous that they could be spotted for over a mile away. This was an extremely dangerous operation. The flimsy cane ladders were the only footing across cliffs that dropped sheer into the riverbed, but collecting honey was the only way many of the poor hill people could raise the rent payable to the Rajah.

Because of the steepness of these upper valleys, enormous landslides were common, particularly during the monsoon or when the snows had melted, and the previous year a local village, with 12 of its inhabitants and all the cattle, had been swept away. Hooker came across one such landslip just beyond Sentam, where he had to cross a field of debris after the entire face of a mountain fully a mile across was torn away, presenting a face of white micaceous clay, full of angular masses of rock. The path across here was extremely dangerous, and Hooker had the misfortune to lose one of his sheep, which was swept away by the raging currents of the Teesta. They were still at low altitude and leeches, the dread of every explorer, appeared in tremendous numbers:

> The weather continued very hot for the elevation (4000 to 5000 feet), the rain brought no coolness, and for the greater part of the three marches between Singtam and Chakoong, we were either wading through deep mud, or climbing over rocks. Leeches swarmed in incredible profusion in the streams and damp grass, and among the bushes: they got into my hair, hung on my eyelids, and crawled up my legs and down my back. I repeatedly took upwards of a hundred from my legs, where the small ones used to collect in clusters on the instep: the sores which they produced were not healed for five months afterwards, and I retain the scars to the present day. Snuff and tobacco leaves are the best antidote, but when marching in the rain, it is impossible to apply this simple remedy to any advantage. The best plan I found to be rolling the leaves over the feet, inside the stockings, and powdering the legs with snuff.[4]

Pandanus furcatus, the Himalayan screw pine. One of Hooker's sketches that was later engraved by Josiah Wood Whymper (1813–1903) and published in *Himalayan Journals*.

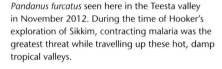

Pandanus furcatus seen here in the Teesta valley in November 2012. During the time of Hooker's exploration of Sikkim, contracting malaria was the greatest threat while travelling up these hot, damp tropical valleys.

Three views of the Talung River valley from Singhik with Kangchenjunga soaring skywards. The first image is Hooker's field sketch, followed by W. H. Fitch's re-worked watercolour. Hooker employed some artistic license: it is not possible to see the flat on which he pitched his tent from his vantage point, and for some reason he chose not to include the Talung River (pictured right).

The forests along this stretch of river combined temperate and subtropical floral elements with birch, willow, *Alnus nepalensis* and walnut, growing alongside gigantic bamboos, *Erythrina arborescens*, *Wallichia oblongifolia*, *Toona ciliata*, various species of *Ficus*, *Paris polyphylla* and *Rubus ellipticus*. Several epiphytic orchids, particularly the lovely yellow-flowered *Dendrobium densiflorum* and the violet-coloured *D. nobile*, were especially abundant, while on rocks Hooker collected *Lyellia crispa*, a superb moss he had last seen on rocks in east Nepal. He was to make several important discoveries in the area, the most beautiful of these perhaps being the lovely *Codonopsis gracilis*, a wiry herbaceous climber bearing exquisite little blue tubular-shaped blossoms.

Paris polyphylla. Hooker collected it several times during his travels in Sikkim, including the upper Teesta valley approaching Chungthang.

A cane bridge across the upper Teesta River with the peak of Lamo Angdang in the far distance. Hooker's Lepcha porters carry provisions in wicker bamboo baskets as they make their way to the nearby village of Chungthang. Drawing by Joseph Dalton Hooker.

Trouble near Chungthang

As he approached the village of Chungthang (Choongtam) and the upper course of the Teesta River (Lachen-Lachoong), the valley became higher and steeper, resulting in more and more immense landslides.

Chungthang was then a tiny village of 20 houses and a small wooden monastery, which, much to Hooker's interest, contained several drawings of Lhasa, depicting the city's enormous monasteries and temples, with their gilded roofs and white-washed walls. To the rear of the village, a steep rocky mountain rose to over 3,048 m (10,000f t), there dividing the Teesta into its two primary tributaries, the Lachen and the Lachung (Lachoong) Rivers. Both rivers had cut deep valleys, separated from one another by a long mountain ridge beginning immediately behind Chungthang Peak (or Chameringu as it is also known) and running for 35 km (22 miles) north to meet the summit of Kangchengyao (Kinchinjhow).

The southern side of Chungthang peak was almost entirely bare of trees, and except for occasional clumps of fir and hemlock near the top, its slopes were steep, grassy and dry. It therefore made a stark contrast with the heavily forested slopes further south and immediately indicated a drier, sunnier climate. The scenery reminded Joseph Hooker of Switzerland.

At the foot of Chungthang Peak lay a broad triangular flat area, 1,606 m above sea level, and 91 m above the river, to which it descended in a succession of three cultivated terraces. Hooker camped here, surrounded by scarlet-flowered forms of *Rhododendron arboreum*, walnuts, *Pieris formosa, Elaeagnus parvifolia* (then bearing edible fruits) and *Photinia argyrophylla*, a small evergreen tree from which the local inhabitants made tea.

Rhododendron arboreum, a columnar form bearing red-scarlet blossoms, seen here at Kilmacurragh. Hooker found it in flower at Chungthang. Rare in cultivation, these red-flowered variants are tender and suited only to mild coastal gardens in Great Britain and Ireland.

The modern-day village of Chungthang. The peak above it divides two major valleys. To the left can be seen the terminus of the Lachen valley, to the right is the Lachung valley. At this point the two major rivers unite to form the Teesta River.

The people of this village were of Tibetan origin or Bhutia, an intermixing of Tibetan migrants with Lepcha people. He was greeted by them in the traditional Tibetan way of sticking out their tongues, and scratching their ears, and was accorded the same welcome by the local Lama and the Phipun, or chief officer, of the nearby Lachen valley. Hooker was still travelling through an area controlled by the Sentam Subah, and continued to experience opposition, particularly from the Lama from the Chungthang monastery, who threatened the wrath of the gods against any further travel. Luckily, Hooker spotted a monk from the monastery at Sanga Choeling, who spoke of his kindness to the monks resident both there and at Pemayangtse, as did Hooker's guide, Meepo.

Once again provisions had run low, and the roads behind had been destroyed by landslides, preventing the delivery of food, so the only way to survive was to replenish stores from the village itself. While the villagers certainly supplied food to Hooker directly (since the Rajah had ordered that he be cared for), it was impossible to buy anything for his men thanks to the Dewan, who had issued strict orders that no food be supplied to Hooker's porters, nor should the roads be repaired while he and his team explored Sikkim.

Rhododendron maddenii, W. H. Fitch's superb plate from *Rhododendrons of the Sikkim Himalaya*. It is one of the rarest *Rhododendron* species in Sikkim.

Rhododendron maddenii, a cultivated specimen seen here at Kilmacurragh, Co. Wicklow.

At Chungthang, Meepo received a letter from the Sikkim court, purporting to be from the Rajah, commanding Hooker to return at once, claiming that he had spent long enough in the kingdom for the object of his studies. Hooker refused to receive it as an official communication, since it wasn't addressed to him, and unless directed by Campbell, he intended to continue; the long-expected provisions arrived from Darjeeling soon afterwards.

Though at similar altitudes, the vegetation on the hills above Chungthang was quite different from that at Darjeeling, owing to the region's drier climate. Another strange anomaly was that Darjeeling plants grew at higher altitudes than their counterparts at Chungthang. For example, on the hill above the village, Hooker collected *Rhododendron arboreum* and *R. dalhousieae* at 1,524–1,829 m, whereas at Darjeeling neither of these plants ever descended below 2,286 m. In this district, *Taxus wallichiana*, the Himalayan yew, appeared at 2,134 m; on Tonglu it grew near the summit of the mountain between 2,896 and 3,048 m. At 2,438 m *Tsuga dumosa* appeared on the scene, a tree normally associated with far higher altitudes.

The peeling bark on older plants of *Rhododendron maddenii* is particularly attractive. Seen here is a venerable specimen in the island garden on Ilnacullin (Garinish Island), Co. Cork, Ireland.

Rhododendron edgeworthii, one of Hooker's Chungthang discoveries, which was named for M. Pakenham Edgeworth. An epiphytic species, it is still common in the valley area between Chungthang and Lachen.

Hooker collected ten species of *Rhododendron* on the hills surrounding Chungthang, of which two were new to science, and both of which he chose to name after notable Irish amateur botanists. *Rhododendron maddenii* and *Rhododendron edgeworthii*. *R. maddenii* bears the name of a Kilkenny man, Edward Madden, and in his description of the species in *Rhododendrons of the Sikkim Himalaya*, Hooker completed his text with the following compliment:

> I do myself the pleasure to name this truly superb plant in compliment to Major Madden of the Bengal Civil Service, a good and accomplished botanist, to whose learned memoirs on the plants of the temperate and tropical zones of [the] north-west Himalaya, the reader may be referred for an excellent account of the vegetation of those regions. The same gentleman's paper on the Coniferae of the north of India may be quoted as a model of its kind.[5]

It is a lovely species, one of my favourites, and is especially at home in the milder, coastal gardens of Ireland. It is quite rare in Sikkim, and in his original description Hooker stated that it was a very rare plant. It is worrying therefore that the type locality at Chungthang has been affected by a massive new dam being built in view of the tiny population to the south of the village. It is a popular garden plant even in Sikkim, and in Gangtok the Sikkim royal family grew good forms in the palace gardens.

In gardens it forms a multi-stemmed shrub of about 3.6 m, with stout stems, and smooth brown flaking bark. It bears lax trusses of white, delicately flushed pink, funnel-shaped very fragrant flowers in June. Hooker introduced it to cultivation in 1849, and it was reintroduced from Bhutan by Roland Cooper, and later by Ludlow and Sherriff.

In gardens it has been put to good use by the hybridisers, and following our visit to Chungthang, one of my travelling companions, Neil Porteous, Head of Gardens for the National Trust in Northern Ireland, and Head Gardener at Mount Stewart in Co. Down, has planted a wide range of *Rhododendron maddenii* forms and cultivars at Mount Stewart.

Mount Stewart, arguably the greatest garden in Ireland, has a long-established *Rhododendron* collection, planted by the Marchioness of Londonderry, Edith Vane-Tempest-Stewart, who sponsored the great plant hunters including George Forrest and Frank Kingdon Ward. At Mount Stewart, the new *Rhododendron maddenii* collection, purchased from various nurseries in England and Scotland, is grown in semi-decayed tree trunks in a woodland setting, thus re-creating a Sikkim-like setting, as seen on our travels, but crucially, giving young plants exceptionally good drainage, which this species demands in cultivation. This was made absolutely clear to us from our Chungthang visits. Plants growing there were all on steep, sharply drained, drier than usual riverside cliff faces above the upper Teesta River. The fastest way to kill *R. maddenii* in cultivation is to plant it in a wet, poorly drained position. The same goes for *R. edgeworthii,* and the other epiphytic species.

The surrounding forests supported many epiphytes, and here Hooker found *Agapetes hookeri,* an ericaceous shrub to about a metre tall, with greenish-yellow bell-shaped flowers hanging from the undersides of arching branches. Like many other epiphytic Asian shrubs, this species has a swollen turnip-like stem base, which stores water and sustains the plant during the dry, late autumn period.

Chungthang abounded (as it still does) with orchids, and Hooker made a rich harvest. Several new species were described from his collections here, including the sweetly scented, green-flowered *Habenaria pantlingiana,* named in honour of Robert Pantling (1856-1910), Curator of the Royal Botanic Garden, Calcutta.

He also collected several snakes, skinks (lizards) and other reptiles for the British Museum, and a number of these proved to be new to science. Snakes, though relatively uncommon and shy in the Himalaya, occurred at Chungthang, including an extremely venomous black viper. Of the 12 species collected by him in Sikkim, seven were venomous, and all were dreaded by the Lepchas; one, a venomous mountain pitviper, *Ovophis monticola,* proved to be new to science.

A scene in the lower Lachen valley above Chungthang. At this point the valley is precipitous and cliffs rise sharply above the river. Giant trees of *Acer campbellii* frame the bridge and waterfall.

A route to Tibet through the Lachen valley

From Chungthang there were two routes to the forbidden land of Tibet, each six days' journey away. The first, and more difficult option, was up the Lachen valley to the Kongra La (Kongra Lama pass), which, at a lofty 5,133 m, led directly onto the Tibetan Plateau. The second option was to trek up the more easterly Lachung (Lachoong) valley to the Dongkya La (Donkia pass). The source of the Lachung River originated in a series of small lakes, west of Pauhunri (Dongkya mountain) in Sikkim, while that of the Lachen (the headwaters of the Teesta) rose just 4 km (2.5 miles) from the Tibetan border in the Cholamu or Tso Lhamo (Cholamoo) Lake, one of the highest lakes in the world at 5,330 m.

Hooker was absolutely determined to reach both the Dongkya and the Kongra La passes, despite Meepo's ignorance of their location, and his guide's persistence in trying to frighten his porters away from pressing further ahead. Hooker claimed his permits gave him every right to travel there, though they certainly didn't authorise him to cross the passes into Tibet.

Juniperus recurva, a large tree on the peak above Chungthang. At Chungthang it grew at a far lower altitude than it would generally be found elsewhere in Sikkim.

Since the impending monsoon rains made travel in the Lachen valley impossible in June, Hooker decided to attempt exploring it first. It was an exciting prospect. The frontier lay deep in the Himalaya, on a high plateau, and the plants and animals found there were wholly different from anything he had encountered in his travels thus far.

The May rains continued, destroying the roads, and as a consequence Hooker's porters arrived a week late, having eaten a great deal of the food they had carried from Darjeeling, leaving just an eight-day supply. As a result, he split the team, leaving a large number at Chungthang, with instructions to forward any food to him as it arrived. On 25 May 1849 he set out for the village of Lachen (Lamteng), a three-day trek from Chungthang.

His route brought him along the course of the Lachen River, a torrent of turquoise waters racing through a narrow, luxuriantly forested valley, with mountains rising on either side from 3,048 to 4,572 m. The road disappeared in places, forcing Hooker and his men to scale rocks 305 m above the river, or to descend into the gorge, where they waded through the torrential waters of tributaries. On several occasions they were forced into the river, fording rocky promontories, where they were dragged through chest-high foaming torrents by the strong Lepchas.

In the lowest part of the valley, just above Chungthang, he found *Saccharum sikkimensis*, a wild relation of sugar cane, though in this case a dwarf species reaching little more than 60 cm. The surrounding forests still contained subtropical elements, including several new broad-leaved evergreen trees such as *Elaeocarpus sikkimensis*. While the latter are probably of botanical interest, another of his finds, *Begonia josephi*, recently introduced to cultivation, is a horticultural sensation, a plantsman's dream come true. Here it grew on wet, shaded banks and bore large peltate leaves to 15 cm (6 in) across, which were beautifully marbled with chocolate-coloured bands and varying intensities of green. If that wasn't enough, it also carried delightful lemon-yellow blossoms on long stems that contrasted beautifully with its dark foliage. It has huge potential for British and Irish gardens, since its tubers are exceptionally hardy.

In this valley the flora changed constantly, particularly with altitude, and Hooker was later to describe Sikkim as 'the perfect microcosm of the Himalaya'. First-time visitors to this area are always struck by the richness and diversity of the flora and fauna. New plants appeared all the time, particularly rhododendrons, and in this warmer part of the valley he found *Rhododendron vaccinioides*, a small, slender, straggling species growing epiphytically on the mossy branches of trees and damp cliff faces overhead. Hooker collected it in the Lachen valley, and again in the valleys below Darjeeling, though he never saw it in flower. He did however send seeds to Kew, where seedlings soon died off.

The village of Lachen, in Hooker's sketch that appeared as a wood engraving in *Himalayan Journals*. It sits on a small flat area to the west of the Lachen River. The mountains behind are thickly forested by *Abies densa* and *Tsuga dumosa*.

J. D. H. delt.

J.W. WHYMPER.

Picea spinulosa, a common conifer in the higher valleys of north Sikkim. Hooker first saw it below the village of Lachen.

This species has limited value horticulturally. Its tiny, white, lilac-tinged blossoms are pretty, though it is not likely to appeal to anyone outside specialist growers. It is, however, botanically interesting, being a member of the section vireya, characterised by long-tailed seeds, which has its main distribution in Malaysia, New Guinea and the Philippines.

On the forest floor Hooker also discovered a new aroid, *Remustia hookeriana*, commonly known as 'Hooker's hitcherhiker elephant ear'. The bold foliage of this plant is its most striking feature, the peltate, cordate olive-green leaves being beautifully splashed purple beneath. Added to that are golden-yellow flowers that appear before the emergence of foliage in late spring. In the lower Lachen valley it grew on mossy rocks, on tree stumps and occasionally as an epiphyte on trees.

Lachen and other places. One of Joseph Hooker's 1849 notebooks, now preserved in the archives at the Royal Botanic Gardens, Kew. In it he recorded details of his travels.

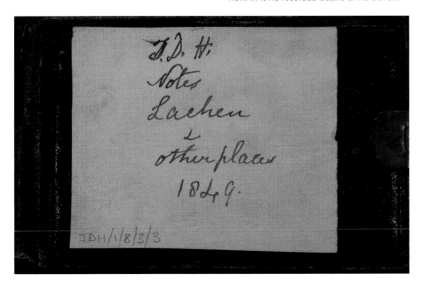

The valley was virtually uninhabited between Chungthang and Lachen, and no one attempted any form of travel during the rains. The roads for the most part were badly damaged throughout their route. Landslides were everywhere, and reaching the junction of the Lachen and Tarum (Taktoon) River, they were almost turned back. This was due to the actions of Hooker's rather reluctant guide, the Sentam Subah, who continued to place obstacle after obstacle in front of his party:

> My unwilling guide had gone ahead with some of the coolies: I had suspected him all along (perhaps unjustly) of avoiding the most practicable routes; but when I found him waiting for me at this bridge, to which he sarcastically pointed with his bow, I felt that had he known of it, to have made difficulties of it before would have been the work of supererogation. He seemed to think I should certainly turn back, and assured me there was no other crossing (a statement I afterwards found to be untrue); so, comforting myself with the hope that if the danger were imminent, Meepo would forcibly stop me, I took off my shoes, and walked steadily over: the tremor of the planks was like that felt when standing on the paddle-box of a steamer, and I was jerked up and down, as my weight pressed them into the boiling flood, which shrouded me with spray. I looked neither to the right nor to the left, lest the motion of the swift waters should turn my head, but kept my head on the white jets d'eau springing up between the woodwork, and felt thankful when fairly on the opposite bank: my loaded coolies followed, crossing one by one without fear or hesitation. The bridge was swept into the Lachen very shortly afterwards.[6]

One of the most beautiful shrubs in the lower Lachen valley was *Deutzia hookeriana*, which bore wide corymbs of white, vanilla-scented blossoms, that faded lilac-pink as they aged. Approaching the village, he and his party of men made their way through a forest of the pendulous branched *Picea spinulosa,* and at 2,743m (9,000ft) *Larix griffithii* grew alongside fine trees of the Himalayan hemlock, *Tsuga dumosa*. It was here, on 26 May 1849, that he discovered *Malus sikkimensis*, which he mentions briefly as an austere crab. It is, in fact, one of the finest of the crab apples, forming a small elegant tree, bearing dense corymbs of snow-white blossoms in May, and masses of red, pear-shaped fruits in autumn. Hooker was obviously impressed by this, and sent seeds to Kew, where the original tree lived for well over a century.

It was also near Lachen that Hooker found *Fraxinus suaveolens*, a handsome flowering ash whose description was based on material gathered by George Forrest on the Jade Dragon Snow Mountain in NW Yunnan, in western China in May 1917. Hooker described both the crab and the ash as being rare in other parts of Sikkim, though they were abundant in the Lachen valley between 2,134 and 2,438 m, as was the Himalayan ivy, *Hedera nepalensis*, then covered in colourful umbels of reddish berries.

On damp rocks Hooker found a new species of bladderwort, *Utricularia furcellata*. Bladderworts are carnivorous, capturing small organisms by means of small, though sophisticated, bladder-like traps. The genus is of exceptional botanical interest and species are either terrestrial or aquatic. Hooker's bladderwort is a lithophytic terrestrial and is distributed through the eastern Himalaya and northern Thailand, though it is rare throughout.

The Himalayan ivy, *Hedera nepalensis*. Hooker found it in fruit as he travelled up the Lachen valley. Seen here in the lower Lachung valley in May 2015.

One of Hooker's Lachen discoveries, *Vaccinium glaucoalbum*, a watercolour of a Kilmacurragh plant by the Irish botanical artist Lydia Shackleton (1828–1914).

Perhaps of more horticultural interest was his discovery of *Chirita lachenensis*, a superb gesneriad, bearing cymes of showy white, blue-tinged blossoms. The chiritas are one of the loveliest inhabitants of moist, shaded rocky forested Himalayan slopes. He also found a second species in the valley, the purple-blue flowered *C. calva*.

The Irish army officer Edward Madden was obviously in favour with Hooker, as another plant discovered here was named for him. *Maddenia himalaica*, a small bushy tree to about 10 m, has wonderful copper-suffused foliage in spring followed by short, densely packed racemes of white flowers. It is a close ally of the cherries, and, like them, has wonderfully polished brown bark; I still remember finding it as we re-traced Hooker's route up the Lachen valley in May 2015.

I have to admit I'm fond of the hypericums, not always a fashionable group. Hooker found several new species in the lower reaches of the Lachen valley, including the elegant and quite showy *Hypericum tenuicaule*, which bears relatively large yellow, pendent, cup-shaped flowers on long, thin red pedicels.

A few miles south of Lachen the road crossed over Chateng, a steep spur extending from Lamo Angdang (Tukcham), a high peak rising above the west bank of the Lachen River. The grassy, boulder-strewn landscape gave way to the most spectacular alpine scenery, with ridges bristling with large trees of Himalayan hemlock, while on the summit were small pools fringed with bushy trees of *Rosa sericea* laden with white blossoms.

Vaccineum albo-glauca.
Kilmacurragh
Sept.
1899

July, 1902

Inula hookeri, one of Joseph Hooker's best-known Lachen valley discoveries. Seen here by the pond at Kilmacurragh, Co. Wicklow.

Red-flowered trees of *Rhododendron arboreum* and scarlet *R. barbatum* were ablaze with colour, and it was in these forests that Hooker was to find the rhododendron look-alike *Daphniphyllum himalayense*. Though it was late in the season, the rose-purple flowered form of *Magnolia campbellii* still created a fine show, and in his account of his travels through this remote valley he singled out *Schisandra grandiflora*, with beautifully scented, pale pink flowers, for mention.

One of the prettiest new shrubs from this area was *Hydrangea stylosa*, a deciduous species forming mounds to around 1.5 m. This bears (in autumn) masses of striking bi-coloured lace-cap flowers, with central fertile flowers of a striking mid-blue, surrounded by pure white, frilly edged ray florets. It was introduced to cultivation from Bhutan during the early 1850s by Thomas Jonas Booth, flowering at Kew in 1858, but has never really been widely available, which is a great shame.

The honeysuckle, *Lonicera tomentella*, a wiry shrub of about 3–4 m tall, was also common here. Hooker introduced this new species to cultivation, sending seeds to Kew where it bloomed every July. It has a quiet charm, and though rare in cultivation, there is a good plant at Cambridge Botanic Garden.

On the forest floor he also discovered *Corydalis ophiocarpa*, a short-lived perennial bearing spikes of cream-coloured flowers with purple tips. The flowers are not particularly exciting, though its blue-green ferny foliage is outstanding, particularly when it turns bronze in autumn and winter. Hooker referred to the hemlock and fir forests as 'pine woods' and beneath them, on the moist forest floor he found *Neottia pinetorum*, a tiny green-flowered orchid, whose name stems from the Latin *pinetorum*, meaning 'of pine forests', though there is not a single true pine in the Lachen valley. Another of Hooker's orchid

discoveries from the Lachen valley is the dainty *Tipularia josephi*, a species still relatively common in the Lachen and Lachung valleys. Hooker was to have many orchids named for him, including *Josephia*, a genus of Indian orchids named in recognition of his contribution to Antarctic and Indian floras. He is also commemorated in *Sirhookeria lanceolata*, from peninsular India.

At this point, the river ran through an immense valley, so steep that it ran for several miles through a continuous landslide over 610 m high. On 27 May 1849, Hooker reached the village of Lachen, which consisted of little more than a cluster of 40 houses perched at an altitude of 2,750 m, and 305 m above the river. All around it to the north and to west, lay an amphitheatre of steep, fir-clad mountains, with snowy peaks and dark soaring cliffs. Lachen means 'big pass', a major incentive to Hooker in his determination to reach the Tibetan passes. The village was the winter quarters of the Bhutia people who inhabited the valley and who moved their livestock to graze the alpine pastures on the Tibetan frontier during the summer months. Their houses reminded Hooker of those at Walungchung Gola in Nepal, though due to the dampness of the valley they were raised higher off the ground, and the roofs were tiled with shingles made from *Abies densa*.

Though subjects of the Sikkim Rajah, the people of this valley were Tibetan, rarely descending to Chungthang, and travelling instead to trade with the Tibetan towns of Shigatse and Lhasa. Altitude reduced the ability of the inhabitants to cultivate a wide range of crops, though Hooker found the small gardens surrounding the houses full of buckwheat, radishes, turnips and mustard. By then the snows had disappeared and the villagers were preparing to send their yaks, sheep and ponies to the alpine slopes higher up the valley to Talam (Tallum Sandong) and Thangu (Tungu).

On his arrival, Hooker spotted a large troop of the western Assamese macaque, *Macaca assamensis* subsp. *pelops*, swinging their way through a wood of *Tsuga dumosa*. Described as a new subspecies in 1840 by Brian Hodgson, habitat destruction, hunting and trapping has brought about a massive decline in the numbers of this primate in recent years, and there may be less than a thousand individuals remaining today.

The same hemlock and fir woods were to yield 18 new fungi, including *Boletus gigas*, a magnificent species that grew on the edge of the Lachen River, among copses of *Pieris formosa* and *Betula utilis*. Like Hooker's Darjeeling fungi, these were named by the English cryptogamist, Miles Joseph Berkeley, and for this particularly impressive species he chose the specific epithet *gigas*, alluding to the Greek mythological Gigantes, a race of huge men of incredible strength who stormed the heavens and battled with the Olympian gods.

Hooker was to make an incredible number of new discoveries from this upper part of the valley, and among the best of these was *Vaccinium glaucoalbum*, a low-growing shrub with vivid blue-white under surfaces to its leaves.

Several of his discoveries bear the name of the Lachen valley, including *Polystichum lachenense*, a dainty fern, now known to be widespread across alpine habitats in the Himalaya and China. Another, *Strobilanthes lachenensis*, a blue-flowered perennial species to about a metre tall, was collected on the return journey on 31 July 1849, when Hooker took a single specimen from meadows just above Lachen.

This *Strobilanthes* has never been found in Sikkim since, and is therefore presumed locally extinct, althought is widespread elsewhere. George Forrest rediscovered it in Yunnan in 1906, Frank Kingdon Ward found it in Arunachal Pradesh, India, in 1938 and Ludlow and Sherriff in Bhutan in 1949. It was also collected in Tibet by Richard Hingston during the 1924 British Mount Everest Expedition, on which Mallory and Irvine vanished close to the summit.

Another of Hooker's discoveries from this region, *Ceropegia hookeri*, is now critically endangered, with a handful of plants still found in the Zemu valley and a single population, comprising just eight to ten plants, clinging to existence in an area of one square kilometre near Lachen village.

Perhaps Hooker's best-known Lachen discovery was *Inula hookeri*, a showy perennial to about 60 cm with soft, downy stems that carry large, bright yellow daisy-like blossoms in profusion in late summer. Lift the flower and look beneath and you'll find another lovely feature, the wonderfully decorative hairy calyx. It is a popular garden plant, thriving on neglect and spreading rapidly in rich ground. Hooker sent seeds to Kew in the autumn of 1849 and plants flowered there, for the first time in cultivation, in the summer of 1851. It is a charming perennial, bright and cheery, and beloved by bees and butterflies.

Wildflowers teemed in open meadows above the village, and as the summer progressed they would become a riot of colour. One of the most charming inhabitants of those grassy habitats was the lovely gentian-relative *Swertia hookeri*, a gorgeous perennial, producing 60 cm tall candelabra-like spikes of apricot-purple nodding blossoms. Its bedfellows included further new discoveries like *Anaphalis hookeri*, *Anchusa sikkimensis*, *Geranium polyanthes* (with lovely rose-pink flowers), the perennial teasel *Dipsacus atratus*, and great drifts of the dramatic purple lousewort *Pedicularis pantlingii*. The grasses also proved new; one of them, *Agrostis hookeriana*, bears his name.

In thickets, on the edge of fir and hemlock forest, he found several new climbers including the delightful herbaceous, climbing *Codonopsis benthamii*, with yellow-green, bell-like blossoms on wiry, scrambling stems. He later named it for the great 19th-century systematic botanist, George Bentham (1800–1884).

Hooker discovered several species of *Codonopsis* during his travels in north Sikkim. At Lachen he found *C. benthamii*, though far more beautiful was the wiry climber, *C. gracilis*, which he found in the upper Teesta valley below Chungthang. This image appeared in *Illustrations of Himalayan Plants*.

Enjoying the same habitat, though far more robust, was *Celastrus hookeri,* a deciduous woody climber to 6 m tall. It was later rediscovered in China, at Wushan, in Sichuan (near the famous Three Gorges) by Augustine Henry in 1888, and E. H. Wilson, following in Henry's footsteps, introduced it from Wa-shan, Sichuan in October 1908, while collecting for the Arnold Arboretum. His seed collection (W. 1184) was further distributed from Boston and a consignment was received at the National Botanic Gardens, Glasnevin, Dublin, shortly thereafter, thus reaching Irish gardens. It is a handsome climber bearing masses of orange-coloured seeds that persist well into the new year.

The mean annual temperature of Lachen (10ºC), was that of the isothermal which passed through Britain and Ireland at 52 degrees north, indicating that many of the plants growing on the surrounding slopes would be perfectly hardy in sheltered gardens throughout both islands. Hooker was amazed to find that such a small area, buried in the depths of the Himalaya, should harbour almost all the vegetation types of the north temperate zone, and also exhibit elements from the Malaysian, Tibetan and Chinese floras.

A few days after his arrival, the villagers asked Hooker to desist from shooting local birds and mammals, since they believed it brought about violent rains that damaged their crops. He at first ignored their calls, dismissing their superstitions and claiming he could procure food by no other means. The villagers, however, became even more vocal and gained the support of the Chungthang Lama, so Hooker submitted, knowing full well that the Lama had supported the claims only to force him to leave. He finally yielded, on the condition that he be provided with supplies, which they agreed to, giving him and his men limited rations until provisions finally arrived from Darjeeling.

The local people, though superstitious, regarded a doctor with veneration. When they saw Hooker writing, drawing and painting, they flocked around his tent for hours gazing in admiration, though they were soon driven away by Hooker's dog. The village men spent most of their day loitering about, smoking, spinning wool or occasionally carving yak-saddles and spoons from *Rhododendron* timber, while the women were busily occupied in drying the leaves of *Symplocos* for the markets in Tibet, where they were used as a yellow dye.

Though the first day of June was ushered in with brilliant sunshine, the weather soon deteriorated. Cloud, mist and drizzling rain were the norm by day, while at night the rain was torrential, and beyond Hooker's tent the roar of landslides and avalanches continued for hours, followed by the crashing of immense timber trees on the mountain slopes surrounding him.

In the subalpine forests above the village Hooker made a rich harvest of exciting new plants. One of the prettiest of these was *Corydalis flaccida*, a perennial species bearing dense racemes of dark-purple blossoms with pink spurs and blue margins. Where the forests petered out, the first of the alpines encroached, including *Androsace hookeriana*, a fragile carpeter forming wide spreading hummocks, smothered with small rose-coloured blossoms.

Higher still, he found the tiny tufted *Primula dickieana*, a variable species in the Himalaya where it ranges from Mount Everest in E Nepal

Salix daltoniana, named for Joseph Dalton Hooker, pictured here covered in caterpillar-like catkins in the Lachen valley in May 2015.

to Yunnan in western China. Over that area the colour of its blossoms varies from yellow to white and purple, and while all these shades are found in Sikkim, yellow appears to be the predominant colour.

In gullies and ravines, he was to find *Salix daltoniana*, a small bushy species, which in the Lachen and Lachung valleys ascends to high altitudes where it grows with junipers and dwarf rhododendrons. With it grew dense, rounded bushes of *Rhododendron campanulatum.* This was not a new discovery. It had been previously found by Nathaniel Wallich in Nepal, and Wallich was also responsible for its introduction to cultivation in 1825. This species is variable in flower colour, and whereas Wallich's original introduction bore white, flushed rose-coloured flowers, Hooker's seeds were of forms with lavender-blue and rose-purple blossoms; these superior forms were put to immediate use by keen *Rhododendron* hybridisers.

Lachen, Lamo Angdang and the Zemu River

Occasionally conditions were perfectly clear at dawn, and Hooker climbed Lamo Angdang several times, hoping for a view of the mountains towards the Tibetan passes, though he only once saw Kangchengyao, one of the most remarkable peaks of NE Sikkim.

On the second day of June he received the disastrous news that the large team of porters travelling from Darjeeling with rice had returned, having met with a series of enormous landslides, leaving Hooker with only meagre rations. Added to that, it was impossible to get accurate information of where the frontier with Tibet actually lay. Archibald Campbell had provided Hooker with details of how he would recognise the border once he reached it, though to add to the confusion, the Sikkim–Tibet border had moved over time from Chungthang. Over the decades the Lepchas had pushed the frontier further north having gradually beaten the Tibetans back to Zema (Zemu Samdong), then again to Talam, and finally to the present border area at the Kongra La.

Rhododendron campanulatum is variable in terms of flower colour across its range of distribution. The form above, photographed at the Royal Botanic Garden Edinburgh, is of Nepalese origin and is similar to Nathaniel Wallich's early introduction. Several colour forms ranging from rose to lavender-blue grow in the Lachen valley.

Eventually the chief officer (Phipun) of the Lachen valley offered to bring Hooker to the borderlands, lying about its location and claiming it was only two hours away at Zema, a remote spot, just north of Lachen, where the Lachen River meets its major tributary, the Zemu. Their route took them through fine forests of *Picea spinulosa* and *Tsuga dumosa*, and in open glades *Rosa macrophylla* was common, bearing large clusters of enormous red blossoms. Hooker was enthralled by it, claiming it to be one of the most beautiful of all Himalayan plants.

Just above the fork of the valley, where the two rivers met, a wooden bridge (samdong) crossed the Zemu River, and this was pointed out to Hooker as the frontier. Knowing better, he crossed the bridge, throwing to one side some sticks and a woollen rope stretched across it, which the Phipun rather ridiculously expected him to believe was the frontier. He camped on the banks of the Lachen River, and at this altitude, 2,734 m, the Zema area proved to be good ground for botanising, so the Lachen villagers returned home, while he set his men to the task of plant collecting in the vicinity for the next three days.

The few scattered inhabitants of these mountains made a living from exporting timber to the tree-less Tibetan Plateau. While exploring the hills above the two rivers, Hooker came across great stacks of prepared planks, mainly harvested from trees of *Tsuga dumosa, Larix griffithii* and *Abies densa*. The latter was easily split and very durable, as was the larch, and all the planks were thatched with the bark of *Tsuga dumosa*. *Picea spinulosa* was also planked for timber and mainly used for posts and beams.

These were the principal timber-producing conifers of N Sikkim. In the hot, low altitude valleys *Pinus roxburghii* was commonly felled for timber, and at that time it was a common tree in the warm, drier parts of Sikkim. Both *Juniperus recurva* and *Juniperus indica* (in its tree form) formed small trees, and while their wood was highly prized, their timber was scarce, as was that of the Himalayan yew, *Taxus wallichiana*.

It was on the river banks near Zema that Hooker made one of his most beautiful discoveries, *Magnolia globosa*, there forming small trees and bearing masses of pendent, globular snow-white flowers. Balsams, a group that later became one of his great specialities, abounded, as did several species of showy *Maianthemum, Cotoneaster, Gentiana, Spiraea, Euphorbia, Lonicera* and *Pedicularis*. On the hillsides primulas abounded, growing literally in millions, and in shades of yellow, purple, pink and white. They grew alongside white-flowered thalictrums and anemones, *Berberis, Podophyllum, Fritillaria* and *Lloydia*.

Hooker also re-ascended Lamo Angdang, climbing the northern flank of the mountain that fell steeply into the Zemu River. There, strangely, alpine plants descended to the relatively low altitude of 3,048 m, where normally the mountains were heavily forested with conifers and cold-temperate trees. The forests were replaced with diminutive alpines such as dwarf willows, and on its upper slopes he discovered *Androsace geraniifolia, Tofieldia himalaica* and a new Himalayan heather, *Cassiope selaginoides*, a dwarf shrub of dense habit bearing tiny white, nodding bell-shaped flowers in summer.

Though not new to science, the alpine butterwort, *Pinguicula alpina*, had never been recorded from the Himalaya before Hooker's collection. This curious carnivorous plant is native to the high latitudes and altitudes of Eurasia and is distributed widely from Iceland, the Alps, Siberia, Mongolia and the Himalaya, surviving bitterly cold winters in those regions by means of stout hemicryptophytic winter buds.

Returning to the same area a month later, Hooker found the woods ablaze with the flowers of *Buddleja colvilei, Cardiocrinum giganteum*, with their 4 m tall spikes of white, trumpet-shaped scented flowers,

Logs of *Betula utilis* stacked for winter burning. A scene from the village of Lachen in November 2012.

The Lachen valley above the village of Lachen, seen here in late autumn. In the foreground are trees of *Hippophae salicifolia* laden with fruit. At this level *Abies densa* and *Picea spinulosa* dominate the forest cover.

sheets of purple-flowered *Roscoea auriculata, Morina longifolia, Neillia rubiflora,* and several terrestrial orchids. The giant lily is a relatively common sight in the higher parts of Lachen and adjoining valleys. It was first discovered by Nathaniel Wallich in 1825 in Nepal, while he was based at Kathmandu for the British East India Company. Sir Henry Elwes, following Hooker's route up the Lachen valley, found it again, describing the scene as follows:

> In 1870, during my journey in the almost unknown mountains of Sikkim to the frontiers of Tibet, I saw this noble plant growing abundantly in the Lachen valley, at about 8000–9000 feet elevation. Here it flourishes, in a climate which may best be described by saying that for many weeks it was impossible to dry thoroughly our tents, clothes, or bedding ..... In this moist though not ungenial climate the vegetation is of the richest and most varied description ...
>
> The gigantic lily towers in all its glory above most of the other herbaceous plants, scenting the air for yards around with its sweet perfume, in the month of July, and lying buried under the snow from December to March, when it again begins to push forth leaves.
>
> Though long known to botanists, this plant was not introduced to Europe till 1847, when Col. Madden sent seeds to the Botanic Gardens, Glasnevin, Dublin, which were successfully raised and liberally distributed. It had previously been sent over several times, but, owing to the long time it takes to germinate, had probably been thrown away as bad.[7]

Magnolia globosa, photographed in the wilds of west Sikkim by our Tibetan guide, Thupden Tsering. It is also native to western China from where it was introduced by George Forrest.

Hooker's guide, the Phipun, had remained behind at Lachen, apparently because he had multiple sores on his legs due to leech bites, though his real object, apparently, was to stop a party of traders on their way to Tibet with madder and canes, who, had they proceeded, would have guided Hooker along the correct road to the Tibetan passes.

The Lachen villagers wanted to herd their animals to higher pastures on the frontier, so the Phipun sent word that Hooker could travel with them as far as he liked along the east bank of the Zemu River. A meagre supply of rice reached his camp on the fifth of June and so, he and eight of his men, subsisted on this and some Tibetan meal or Tsampa, which he had bought by stealth from the villagers. They supplemented their diet with wild food such as nettles, wild leeks and *Fagopyrum acutatum*, whose leaves are highly nutritious. His greatest concern however, was not lack of food, but the worsening weather. The monsoon approached and he was losing pressed specimens owing to the damp, which over time reduced his collections to a fraction of what they might have been.

Cardiocrinum giganteum, a plate from Henry Elwes's *A Monograph of the Genus Lilium* (1880). The plate is by the Kew botanical artist Walter Hood Fitch, who cleverly worked the vignette into the background. Elwes saw flowering plants of the giant Himalayan lily in the Lachen valley.

The giant Himalayan lily, *Cardiocrinum giganteum*, a group in bloom at the National Botanic Gardens, Kilmacurragh. In July 2014 over one hundred plants blossomed in the garden's herbaceous border. For ten days the scent was intoxicating.

CHAPTER 11

CONFUSION IN THE ZEMU VALLEY

From Zema and the mouth of the Zemu River the valley ran north-west for the equivalent of a two-day trek. Joseph Hooker claimed to have explored both the Zemu and the Lhonak (Thlonok) Rivers, though later explorers dispute his accuracy. The English mountaineer, Douglas Freshfield (1845–1934), who visited the region half a century later, claimed that both rivers were often misnamed and that Hooker called the Zemu the Lhonak, and the Lhonak the Zemu.[1] This certainly appears to be the case when one compares Hooker's map in *Himalayan Journals* to contemporary maps of Sikkim, and so I have

adjusted his accounts of travels in these valleys accordingly. Hooker later attributed this blunder to the Lachen Phipun and his guides, who, he believed, deliberately misinformed him, as they believed that by leading him to think that the Zemu valley led into Tibet, they hoped the rhododendron thickets, snowfields and glaciers would turn him back.

At Zema, near the confluence of the Zemu and Lachen Rivers, he camped for three days while waiting for provisions to arrive. While trekking along the Zemu through a forest of *Betula utilis*, *Alnus*

The mouth of the Zemu River near its confluence with the Lachen River. The lower slopes of Lamo Angdang may be seen to the left. The track above the river bank to the right was used by Hooker when he explored the valley in 1849.

Betula utilis. This species is variable in the eastern Himalaya and seen here is the shaggy brown-barked form typical of Sikkim trees. Hooker trekked through forests of it in the Zemu River valley. He also introduced this species to cultivation.

made it into the campsite cooking pot, including a large mushroom that proved to be a new species, *Cortinarius emodensis*. Named by Hooker's friend, the Rev. M. J. Berkeley, the specific epithet alludes to the plant's area of distribution, the Himalaya or the *Emodi Montes* of classical geography. Hooker discovered it on the last day of May 1849, growing in *Abies densa* forest.

Cortinarius is the largest genus of mushrooms in the world, with well over 1000 species. Many are colourful and used in dyeing. Most are poisonous, and some are lethal, so as a rule none should be eaten unless the collector is familiar with the group.

The most important wild herb, of which they ate large quantities, was *Maianthemum oleraceum*. This is a pretty perennial growing from

Snowbeds in the Zemu valley at 3,962 m (13,000 ft). Scarlet-red flowered *Rhododendron arboreum* ssp. *cinnamomeum* is at its spectacular best. On the steep slopes are pendulous branched trees of *Picea spinulosa* and dark firs, *Abies densa*. The river surges through a natural ice bridge and Kangchenjunga dominates the scene behind. Hooker mistakenly thought this was the Lhonak River valley.

nepalensis and *Acer sikkimense*, he found the unusual *Balanophora*, a curious parasite, in this case growing on the Sikkim maple. These parasites produced enormous knots on maple roots, from which the Tibetans carved cups, later selling them for large sums. Hooker was to find a store of these knots, cleaned and cut for the turner, and hidden behind a stone by some poor Tibetan, who never returned for them. Today, the same knots are part of the economic botany collections at the Royal Botanic Gardens, Kew. Hooker retained a special interest in this parasitical group, discovering *Balanophora involucrata* on the Cho La, in east Sikkim, in November 1849.

Though it was June, summer was late reaching these high valleys, and the surrounding ravines were still choked with enormous accumulations of ice. Following several avalanches and landslides, in one place a natural bridge 9 m thick and 91 m broad was formed, flanked by heaps of boulders. All this occurred at a relatively low altitude of just 2,987 m, though such features were generally more commonplace in alpine zones well above the tree-line.

The stony slopes there were covered in a good range of herbaceous plants including *Primula*, *Sedum*, *Polygonum* and the purple-flowered *Cardamine macrophylla*, known to the Lepchas as *kenroop-bi*; this last was cooked and eaten on the mountainside. Other wild plants

Arisaema utile, so named by Hooker because its tubers were eaten by the poorest inhabitants of the interior valleys of north Sikkim. It is still abundant there; this plant was photographed in the Yumthang valley in May 2015.

plunging into the main Zemu valley as a 305 m high cataract. At that point the Zemu was the smaller of the two, and wound its way from the snows of Kangchenjunga, which could be clearly seen at the head of the valley, just 32 km away. All around were soaring mountain peaks, sparsely clothed with fir, hemlock and larch, mainly on the southern flanks, which received the warm, moisture-laden winds, while those with northerly exposures were not at all as well vegetated.

Hooker pitched his tent at the junction of the Zemu and Lhonak, at 3,307 m, surrounded by extensive riverside thickets of spectacular scarlet-flowered forms of *Rhododendron arboreum* ssp. *cinnamomeum*. This place provided rich pickings for an enthusiastic botanist; his camp lay among willows, white-flowered cherries, thorn, maple and birch. Cobra lilies were abundant, especially the giant *Arisaema utile*, and the surrounding ground was covered in small pits, with large wooden pestles, used in the preparation of food from the tuberous roots of the *Arisaema*. The people there were poverty-stricken, eking out a miserable existence at that time of year, having brought their yaks up to graze and calve.

The scene on the page opposite appeared as a lithograph in *Himalayan Journals*. The image below is Hooker's field sketch, hastily made as he trekked deeper into this unexplored valley.

0.6 to 1.5 m tall, with handsome ribbed foliage and densely packed panicles of scented, pink-purple bell-shaped blossoms. A favourite potherb of the Lepchas even today, it was described as a new species from Hooker's Lachen collections, by the Kew taxonomist John Gilbert Baker. In his original description, Baker also cited Griffith's earlier collections from Bhutan, and a drawing of the plant made by Cathcart's Darjeeling-based artists. Inevitably bamboo shoots also became part of their everyday diet, eaten either raw or boiled.

On rocky slopes Hooker discovered *Rheum acuminatum*, the highly ornamental Himalayan rhubarb. He must have been enormously impressed by it, since he later sent seeds to his father at Kew where it was cultivated and later featured in *Curtis's Botanical Magazine* (t.4877) in 1855.

It was while exploring this valley that he discovered the pretty woodlander, *Asarum himalaicum*. Not only was this a new species, but prior to this collection the genus was wholly unknown in the Himalaya. Rhododendrons formed dense, tangled thickets and Hooker gathered 13 species while exploring the Zemu valley.

The Zemu River raced by at a relentless pace, with enormous boulders dragged along its bed by cold, glacial waters. The Lhonak River hurtled down a steep gulley from the north, before finally

William Hooker, a portrait dated to 1834 and attributed to Thomas Phillips (1770–1845). Joseph, though a fine botanist in his own right, was undoubtedly helped by his father's influential contacts, and sent him seeds and plants for Kew.

mass of glacier and ice, fully 14,000 feet high, to the head of the Thlonok River, whose upper valley appeared in a broad bay of ice; doubtless forming one of the greatest glaciers in the Himalaya, and increased by lateral feeders that flow into it from either flank of the valley. The south side of this [the Thlonok] valley is formed by a range from Kinchinjunga, running east to Tukcham, where it terminates: from it rises the beautiful mountain Liklo, 22,582 feet high, which, from Dorjiling, appears as a sharp peak, but is here seen to be a jagged crest running north and south.[2]

Again, Hooker had confused the Lhonak River with the Zemu, and the glacier he saw at the head of the valley was in fact the gigantic Zemu Glacier. He made two failed attempts to trek up to the glacier, from where he had hoped to continue north into Tibet. Thirteen kilometres was as far as he managed to travel from his campsite before he was defeated by a valley full of dense, impenetrable rhododendron thickets, while the sides of the mountains boxing the valley in were completely inaccessible. At the highest point of the valley, water boiled at 88.5°C, indicating he had reached an altitude of 3,628 m. His father constantly reminded him to collect seeds, particularly rhododendrons, though for Joseph it was tough going and the persistent rains made the collection of specimens and seeds a very challenging operation at times:

Maria, Lady Hooker, Joseph's mother. Born Maria Sarah Turner, she was the eldest daughter of the banker and keen botanist, Dawson Turner. Joseph Hooker enjoyed a very close relationship with his parents, particularly with his father.

The tubers were bruised with the pestles and immersed into the pits with water. After about a week acetous fermentation began, a sign that many of the poisonous compounds had dissipated, and the sour, fibrous mass was boiled and eaten, its nutritional value being the small quantities of starch. The fermented roots kept for only a few days and in most cases caused bowel complaints, and loss of skin and hair, especially when the fermentation process was not fully complete.

Sir Henry Elwes was very impressed by the cobra lilies of Sikkim, and during his visits there in the 1870s he collected the tubers of several species, introducing *Arisaema nepenthoides* and *A. utile*.

Hooker was also mindful of the fact that he was supposed to be mapping the area, as well as collecting plants:

I repeatedly ascended the north flank of Tukcham along a watercourse, by the side of which were immense slips of rocks and snow-beds; the mountain-side being excessively steep. Some of the masses of gneiss thus brought down were dangerously poised on slopes of soft shingle, and daily moved a little downwards. All the rocks were gneiss and granite, with radiating crystals of tourmaline as thick as the thumb. Below 12,000 to 13,000 feet, the mountains were covered with a dense scrub of rhododendron bushes, except where broken by rocks, landslips, and torrents: above this the winter's snow lay deep, and black rocks and small glaciers, over which avalanches were constantly falling with a sullen roar, forbade all attempts to proceed. My object in ascending was chiefly to obtain views and compass-bearings, in which I was generally disappointed: once only I had a magnificent prospect of Kinchinjunga, sweeping down in one unbroken

Alas, one of my finest collections of rhododendrons sent to Darjeeling got ruined by the coolies falling ill and being detained on the road, so I have to collect the troublesome things afresh. If your shins were as bruised as mine tearing through the interminable *Rhododendron* scrub at 10–13,000 feet you would be as sick of the sight of these glories as I am … a hole in a rock or a shed of leaves is very often my residence for days, and my fare is just rice and a fowl, or kid, eggs, or what I can lay my hands on – no beer or luxuries.[3]

Himalayan poppies and giant rhubarb

It was on the cold, icy slopes of the Zemu valley that Hooker discovered the diminutive *Rhododendron pumilum*, one of the tiniest of all Himalayan *Rhododendron* species. In Sikkim it is quite rare, found occasionally in the Zemu and Lhonak valleys and near the Yumthang valley. There it is happiest creeping though gravel in alpine and avalanche slopes, where its stems and roots are embedded in moss, forming shrubs of no more than 15–30 cm tall. Being a high-altitude species it carries its blossoms late; usually in June and July, and though small, the bell-shaped flowers are pretty, appearing in shades of pink, rose and purple. Hooker introduced it to cultivation, and seeds from his collections were raised at Kew in the spring of 1850. He described it as the smallest of the Sikkim rhododendrons, an extremely elegant species, its flowers appearing soon after the snows had melted, peeping above the surrounding short heath-like vegetation, reminding him of *Linnaea borealis*.[4]

Hooker discovered *Rhododendron pumilum* on 12 June 1849, the same day he first encountered the lovely *Meconopsis simplicifolia*. This blue Himalayan poppy had been discovered during the early 1820s by Nathaniel Wallich on Gosainkund in central Nepal. In Sikkim the flower colour varies from reddish-purple to varying shades of blue, and Hooker introduced this to cultivation also.

It was also in the same area of dwarf *Rhododendron* scrub that Hooker found the lovely sulphur-yellow flowered *Primula elongata*, bearing dense umbels of flowers on short scapes held above beautifully mealy, newly emerged foliage.

Rhododendron pumilum, the smallest of the Sikkim species. It is found across much of the eastern Himalaya, Myanmar and western China.

The spectacular blue poppy, *Meconopsis simplicifolia*, introduced to cultivation by Joseph Hooker. This superb plate was prepared for *Illustrations of Himalayan Plants*.

Meconopsis simplicifolia; the solitary flowers are carried on leafless stems, and vary in colour from the deepest ink blues to rose-purple.

Hooker was undoubtedly one of the most determined Victorian explorers, thinking nothing of crossing massive snow-beds by cutting holes in their steep faces, or being dragged through freezing water in a torrential river. At times like these, he was greatly impressed by the strength and energy of his chief porter, his Lepcha guide from the Sikkim court, and of his team of Lepcha boys. This he put down to the fact that they were as keen to reach the holy lands of Tibet as he was. Despite long, exhausting days and a miserable diet, they never complained, and spent the mornings and evenings changing the papers in Hooker's plant presses, drying the damp papers over an open fire by his tent door, sleeping at night wrapped in blankets and huddled under a rock.

Hooker was extremely fond of his Lepcha porters and treated them kindly, especially when after taking altitude readings in elevated places, there was a good supply of boiling water:

> After boiling my thermometer on these occasions, I generally made a little tea for the party; a refreshment to which they looked forward with child-like eagerness. The fairness with which these good-hearted people used to divide the scanty allowance, and afterwards the leaves, which are greatly relished, was an engaging trait in their simple character: I have still vividly before me their sleek swarthy faces and twinkling Tartar eyes, as they lay stretched on the ground in the sun, or crouched in the sleet and snow beneath some sheltering rock; each with his little polished wooden cup of tea, watching my notes and instruments with curious wonder, asking, "How high are we?" "How cold is it?" and comparing the results with those of other stations, with much interest and intelligence.[5]

On 11 June, Hooker's men finished building a bridge constructed from the branches of trees and bound by willow withies, which allowed him to cross to the northern bank of the Zemu River. There he camped on a gravel area at the junction of the Zemu and the Lhonak (though to confuse matters, another later explorer, Lt Col. H. W. Tobin, claimed this to be the Tumrachen Chu). The northern river bank rose in three great alluvial terraces, the uppermost of which proved a rich field for botanising, being covered with dense thickets of dwarf, small leaved holly and various rhododendrons. The soil was rich in humus, and where not forested, supported a rich herbaceous flora including the superb *Meconopsis paniculata* (mistaken by Hooker to be the closely allied *M. napaulensis*), a monocarpic species, then providing a spectacular display of superb yellow flowers on broad 0.9 to 1.8 m tall panicles.

Towering over the poppies were stately plants of *Heracleum nepalense*, an annual species reaching 3 m tall. Great trees of *Abies densa*, *Juniperus recurva* and *Larix griffithii* grew nearby, and on the edge of the gravel grew a colourful medley of rhododendron, scarlet *Spiraea*, several honeysuckles, white *Clematis* and *Viburnum*.

Hooker's first discovery after the crossing was a good existing bridge across the Lhonak River just above its confluence with the Zemu, and some overgrown tracks leading to stone huts, which his men soon occupied. A few days later he returned from botanising to find his men in abject terror, since the Lachen Phipun had announced that they were on the Tibetan side of the rivers, and that troops from there were marching towards the area to plunder his goods and to carry his men off into slavery.

Hooker managed to persuade the nervous Lepchas that it was all a ruse, and, taking a tent, began the ascent of the Lhonak valley with some of his men, while the others remained in the stone huts, where they continued to dry specimens with the aid of a wood fire. Once again, rhododendron thickets and the cliff-like walls of the valley defeated his endeavours, just as had happened in the Zemu

Meconopsis paniculata seen here flowering in the bog garden at Kilmacurragh. A monocarpic species, it is relatively easy to grow.

Meconopsis paniculata, a common perennial in the alpine valleys of north Sikkim. Hooker compared its lovely orb-like golden blossoms to hollyhocks.

sparkle like sapphires on the turf. Gentians begin to unfold their deep azure bells, aconites to rear their tall blue spikes, and fritillaries and *Meconopsis* burst into flower. On the black rocks the gigantic rhubarb forms pale pyramidal towers a yard high, of inflated reflexed bracts, that conceal the flowers, and overlapping one another like tiles, protecting them from the wind and rain: a whorl of broad green leaves edged with red spreads on the ground at the base of the plant, contrasting in colour with the transparent bracts, which are yellow, margined with pink. This is the handsomest herbaceous plant in Sikkim: it is called "Tchuka," and the acid stems are eaten both raw and boiled; they are hollow and full of pure water: the root resembles that of the medicinal rhubarb, but is spongy and inert; it attains a length of four feet, and grows as thick as the arm. The dried leaves afford a substitute for tobacco; a smaller kind of rhubarb is however more commonly used in Tibet for this purpose; it is called "Chula."[6]

The Sikkim rhubarb, *Rheum nobile*, one of the most astonishing plants from the eastern Himalaya. The outer translucent cream-coloured bracts create a glasshouse-like effect for the delicate flowers beneath, allowing light through, while blocking ultraviolet radiation.

valley, and frustratingly, from that point in the upper Lhonak valley, he could clearly see the arch of a brilliant blue sky that spanned the horizon to the north, indicating the drier climate of the Tibetan Plateau. For the eight days Hooker spent in the Zemu and Lhonak valleys, a black canopy of clouds prevailed and the rain poured every single day.

Hooker retraced his route down the steep, rocky valley, following the course of the Lhonak River, which in some places was a raging torrent. Enormous rocks and boulders constantly fell into water as a result of the heavy rain, and at night the sound of the rock falls was terrifying.

Hooker described June as the most glorious month for flowers in Sikkim, and for such a dedicated, pioneering botanist the Lhonak valley provided many rewards:

> Rhododendrons occupy the most prominent place, clothing the mountain slopes with a deep green mantle glowing with bells of brilliant colours; of eight or ten species growing here, every bush was loaded with as great a profusion of blossoms as are their northern congeners in our English gardens. Primroses are next, both in beauty and abundance; and they are accompanied by yellow cowslips, three feet high, purple polyanthus, and pink large-flowered dwarf kinds nestling in the rocks, and an exquisitely beautiful blue miniature species, whose blossoms

The author, in south-east Tibet in July 2001 with an enormous flowering spike of *Rheum nobile*.

This gigantic rhubarb was a spectacular plant, and Hooker aptly named it *Rheum nobile*, the noble rhubarb. One of the highest-growing of all the Himalayan alpines, Hooker would also later find it in the higher reaches of the Lachen valley. It is a denizen of boulder-strewn alpines slopes and high mountain passes, and instead of cowering in cushion form like many other high-altitude alpine plants, to escape icy buffeting winds, it instead throws up an immense towering spike covered in enormous semi-transparent, yellow, pink-tinged, scallop-shaped bracts. These on average reach well over a metre tall, though many years ago, while trekking in south-east Tibet during the July monsoon rains, I stumbled across an entire mountainside lit up by its bizarre triffid-like spikes, and some of these were more than twice that height.

In Sikkim the plant was colloquially known as 'chuka', and in her autobiography, the last queen of Sikkim, Hope Cooke, mentions that people would bring stems of this wonderful rhubarb to the royal palace at Gangtok, where they were stewed and eaten. Hooker introduced it to cultivation, sending seeds to Kew, though he also stated that plants lived for just two years.

At 3,862 m, Hooker was well above the tree-line, and the vegetation consisted of dwarf shrubs including new discoveries like *Lonicera myrtillus* and *Berberis concinna*, two new species of *Ribes*, including the highly attractive frilly-leaved currant, *R. laciniatum*, and *Bryocarpum himalaicum*, a new genus in the Primulaceae bearing wonderful golden-yellow *Soldanella*-like blooms.

Trouble near the frontier

The Lachen Phipun continued to threaten Hooker, and on 28 June a party of 50 men arrived on the banks of the Zemu River and evicted his men from the stone huts they had occupied, together with the plants they were drying, the plant presses, papers and other equipment. The Lepcha boys raced to Hooker, saying there were soldiers among the rabble, who had declared that Hooker was in Tibet, and that they planned to return the next morning to take his goods and drive him back to Darjeeling.

Hooker's greatest fear was for his collections. Meepo, his Sikkim court guide, was terrified, but Nimbo, Hooker's chief porter, a hardened, tough Bhutia, volunteered with two others to take away his collections and drying paper, following a parley with the enemy camp. He brought back the news that though half-clad and without food, they swaggered and bullied, though he gave as good as he got, having a good vocabulary of Tibetan slang. The following morning 20 or 30 ragged Bhutias, soaked by the rain, streamed up the valley and Hooker sat by his tent door, with a gun beside him.

They were rather taken aback to find Hooker alone, ignoring them, while his dog bayed violently and kept them well back from the tent. He succeeded in sending most of the party back down the mountain, and finally met only with the Phipun and his brother, whom he angrily rebuked, and the Phipun finally admitted defeat.

The last day of June marked the arrival of the monsoon. Violent storms crossed the mountains, and for the next two days the tent leaked as never before. Humidity reached saturation point and a raw, cold southerly wind with drizzling rain became a constant feature.

Hooker recorded the temperatures of Sikkim's rivers throughout his travels and observed that, along its course below 610 m, the Teesta was actually colder in summer than in mid-winter, when there was far less cloud and the snows were not melting. A curious fact was that the temperature of the northern headwaters of the Teesta actually rose with increasing altitude, confirming Hooker's assumption that all these rivers rose in a drier, and comparatively sunny, climate, and flowed through a less snow-clad landscape than the icy valleys he was then exploring.

Returning down the Zemu valley, he found that the herbaceous plants on the broad gravel flat had grown a further 60 cm during the fortnight he explored the upper valleys. From there he reached Zema in a single day, finding the woods and glades even more colourful and beautiful, with *Cardiocrinum giganteum* in full bloom, sweetly scenting the air all around, alongside the lovely *Rosa macrophylla* and *Neillia rubiflora*.

At Zema Hooker received letters from Archibald Campbell, who had written in protest to the Rajah condemning the harsh treatment his colleague was receiving. The Rajah replied that his strict command had been for Hooker to be well provisioned and led safely to the border. Letters also arrived on the Rajah's behalf from the Chebu Lama, ordering the Lachen Phipun to take Hooker to the pass. Victory was at hand and so preparations began for travel towards the Kongra La.

On 5 July, the Sentam Subah arrived from Cho La, the Rajah's summer residence, carrying presents from the Rajah and Ranee, and with instructions to lead Hooker to the frontier, though he continued to play games, insisting the border was at nearby Talam (Tallum Samdong). Hooker of course knew the border lay well beyond the village of Thangu, but he played along as Talam lay on the route.

For the time being, Hooker was confined to camp, as the monsoon continued unabated; sadly he lost most of his entomological collections (including many new and beautiful butterflies and other insects) to the pervading damp. Today we know that over 650 species of butterflies, and more than 1,500 species of moth are found in Sikkim.

While Hooker waited, he noticed that daily parties of women and children, brightly decorated with the flowers of *Rosa macrophylla*, passed his tent, on their way to the Zemu valley, to collect *Arisaema* roots for food.

On 11 July 1849, five porters arrived (after a three-week journey) from Darjeeling with rice. They also brought letters from home, and with renewed spirits, Hooker started up the Lachen valley. This valley was certainly far drier than the damp Zemu region and early mornings were bright with good views to the north of Kangchengyao.

Above 3,353 m the valley widened, and the river meandered across a broad bed bordered by a marsh full of yellow and red flowered *Pedicularis* and the dwarf *Myricaria rosea*. They were above the tree-line and alpines appeared in great profusion. It was in this upper part of the Lachen valley that Hooker found the exquisite deep purple-flowered *Cypripedium himalaicum*, the Himalayan lady's slipper orchid, one of the most beautiful of all flowers. With it grew the dwarf Solomon's seal, *Polygonatum hookeri*, then carrying ground-hugging clusters of rose-lilac blossoms.

Rhododendron fulgens, one of Joseph Hooker's favourite species, both for its richly coloured blossoms and its very handsome foliage. Painted by Walter Hood Fitch for *Rhododendrons of the Sikkim Himalaya*.

Polygonatum hookeri, seen here in the Marsyangdi valley, Nepal, between Yak Kharka and Leder, at around 4,300 m (14,108 ft).

On those high alpine grasslands, perennials found shelter among low-growing juniper and rhododendron, including one of his best finds, *Rhododendron campanulatum* subsp. *aeruginosum*. There it formed extensive thickets of dwarfed, dome-shaped bushes smothered in open trusses of bell-shaped blossoms varying in shades of lilac, rose and deep purple, though what impressed Hooker most was the lustre of its young bluish-green foliage, which on emergence displays a glaucous metallic bloom, hence his choice of *aeruginosum*, meaning verdigris-coloured. The underside of leaves is equally striking with a dense rusty-brown indumentum. Hooker is thought to have introduced this showy shrub to cultivation in 1850.

In this valley it grew with one of Nathaniel Wallich's Nepal discoveries *Rhododendron fulgens*, which also clearly impressed Hooker:

> This, the richest ornament of the alpine region in the month of June, forms a very prevalent shrub at the elevations assigned to it, not yielding in abundance to its constant associates, *R. aeruginosum* … and like the former, pushing forth young leaves of a beautiful verdigris-green in July and August. The foliage is perennial, and gives a singular hue to the bleak snowy mountain-faces immediately overhung by perpetual snow, contrasting in August in broad masses or broken clumps with the bright scarlet of the barberry, the golden yellow of the fading birch and mountain ash, the lurid heavy green of the perennial juniper, and the bleak raw brown of the withered herbage. Whether, then, for the glorious effulgence in spring of its deep scarlet blossoms, which

appear to glow like fire in the short hour of morning sunlight, or the singular tint it at other seasons wears, this is among the most striking of plants which lend to these inhospitable regions the varied hues which are denied to the comparatively hospitable but gloomy forests of the temperate zone on the same mountains.

Individual shrubs are of a rounded outline, about four feet high, and twice as much in diameter, and when growing together they compose an impenetrable thicket, as annoying to the traveller as *R. hodgsonii* is at lower elevations.[7]

Describing the species in 1851, Hooker chose the specific epithet *fulgens*, meaning shining, referring to the wonderful lustre of the plant's foliage. One of the seedlings from his 1850 introduction first flowered at Kew in April 1862, where its stunning red blossoms left garden staff awestruck.

New alpine flowers

In the shelter of dwarf rhododendron thickets, Hooker also discovered several new alpine primulas, including the diminutive *Primula pulchra*, bearing violet to rich purple blossoms that appear just as the Himalayan snows are melting. It is widespread across the Himalaya, though scarce and very thinly distributed, and its centre of distribution appears to be in Sikkim's Zemu valley.[8]

Primula hookeri was even smaller, in fact minute, frequenting gravelly avalanche slopes and pushing its white blossoms through the melting snows. Far showier was the cushion-forming *P. concinna*, bearing masses of short-stalked pink, mauve or white, yellow-eyed flowers above its dwarf mealy foliage. It was found by later collectors in E Nepal, W Bhutan and in the Chumbi valley in Tibet, where its preferred habitat is alpine meadows, moraines and screes near glaciers. Perhaps the most exquisite of all these alpine treasures was the tiny *P. sappharina*, bearing several violet-blue *Soldanella*-like pendent flowers on thread-like stems.

With the primulas Hooker was to find other tiny alpines such as the lovely blue-flowered *Eritrichium pustulosum*, *Silene himalayense*, the purple-red flowered *Pedicularis denudata*, *Sanguisorba filiformis* (with creamy-white flowers) and *Vaccinium sikkimense*, a low-growing, dome-shaped shrub, carrying small white, pink-flushed, bell-shaped blossoms in spring followed by delightful light-blue rounded blueberry-like fruits in autumn.

The dominant shade of autumn alpine flowers in the Himalaya is often blue, and this is certainly the case with another of Hooker's finds, the bellflower relative, *Cyananthus incanus*, a gorgeous alpine bearing long-tubed, light blue, star-like blossoms over carpets of ground-hugging silvery foliage.

Orchids occupy almost every habitat in Sikkim, from the steamy tropical valleys to the highest alpine habitats, and Hooker collected and discovered several new species in the high-lying interior valleys. *Calanthe alpina*, for example, is one of very few calanthes to grow in subalpine regions, though it has become rare owing to habitat destruction. Two other alpine orchids from the upper Lachen valley bear their discoverer's name: the tiny green-flowered *Herminium josephi* and *Habenaria diphylla* var. *josephi,* which Hooker found in alpine regions of the Lachen valley and in the mountains above Thangu.

Bergenia purpurascens is one of Joseph Hooker's best-known discoveries. The specific epithet *purpurascens* is very apt. In winter the leaves of the best clones, such as 'Irish Crimson', seen here by the pond at Kilmacurragh, turn a beetroot colour.

Joseph Hooker's best-known alpine discovery must be *Bergenia purpurascens*, which he found in the upper reaches of the Lachen valley on 7 June 1849, and from where he later sent seeds to Kew. From his introduction it was featured (under the name *Saxifraga purpurascens*) in *Curtis's Botanical Magazine* (t.5066) in 1858, and it has remained a firm garden favourite ever since. It is perhaps the most beautiful of all the *Bergenia* species, particularly in winter when the leaves turn a wonderful beetroot colour. The finest selection of this species is *Bergenia purpurascens* 'Irish Crimson', originally raised at the National Botanic Gardens, Glasnevin, and long grown there in the Order Beds; it was popularised by Helen Dillon in recent times through her famous Dublin garden.

In wet places Hooker found the alpine rush, *Juncus sikkimensis*, a species now known to be native to a wide area of the eastern Himalaya and western China, where it grows in rhododendron forest and in alpine swamps at high altitudes. New species of lousewort appeared constantly. On 15 July 1849, as he trekked through the Lachen valley between 3,658 and 4,267 m, he was to discover one of the most beautiful of them all, *Pedicularis elwesii*, named for Sir Henry Elwes. It is particularly showy, with handsome rose-lilac flowers, and is endemic to the alpine meadows of the eastern Himalaya and the mountains of NW Yunnan.

The Lachen Phipun and the Sentam Subah met Hooker at Talam, the latter calling on him daily for friendly conversations. He still insisted, however, that the Lachen River marked the border with Tibet and that Hooker should go no further, even trying to persuade Hooker's men that a person so incorrigibly obstinate must be mad and they would be better advised to flee. To an official used to a better life on the warm hillsides of Sentam, Hooker's ambition to spend weeks in the coldest, highest mountains of Sikkim was unfathomable.

At Talam Hooker discovered several new plants: *Acronema hookeri*, a carrot-relative with handsomely dissected foliage, the lovely rose-lilac flowered *Pedicularis microcalyx* and *Codonopsis foetens*, the stinking bonnet bellflower, a very pretty, low-growing alpine, bearing pendulous blue bell-shaped flowers with a network of darker purple veins running through the petals.

The Sentam Subah finally gave in and proposed to take Hooker to Thangu, a village near the base of Kangchengyao, provided Hooker left behind his ponies and tent, which of course he refused to do. The Subah then disappeared for several days. Hooker worried that he was offended, until one morning his attendant arrived at the campsite looking for medicine for his master who was suffering from severe poisoning, having eaten *Arisaema* root. Relations were restored, and on 22 July a woefully yellow Sentam Subah relented, promising to take Hooker and his men to Thangu and the Kongra La, if he promised to stay only two nights.

Beyond Talam, the valley narrowed and Hooker made a brief diversion up a hanging valley to the west, which, 164 years later, our group would dub 'Hooker Valley'. He soon crossed well above the tree line into a stunted scrub of willow, juniper, birch, barberry and rhododendron, and higher still he met with a rich assemblage of alpine orchids growing in a grassy turf with *Pedicularis, Gentiana, Potentilla, Geranium*, purple and yellow flowered *Meconopsis* and *Artemisia indica*, the Asian mugwort which also grew around Darjeeling.

Thangu, Hooker's Rock and the Kongra La

The little village of Thangu lay in a broad grassy valley at an altitude of some 3,960 m, and was sheltered by precipitous mountains to the north. In the river valley, just below the village, a stupendous rock, over 15 m high, and broken in two, dominated the scene. Hooker pondered this, deciding that ancient glaciers were the only explanation. There were certainly moraines along the mouth of the nearby Thangu Chu, at its junction with the Lachen River. This great rock, with the smaller piece lying close by, suggested that the whole mass had been flung perpendicularly through a crevasse in a glacier. A few wooden houses were built close to it and black tents had been pitched there by nomadic Tibetan farmers. In the woodcut of the

Joseph Hooker's map of Sikkim in *Himalayan Journals* shows that just below Thangu, at Talam, he explored this hanging valley. We christened it 'Hooker Valley'.

rock published in *Himalayan Journals*, Hooker introduced the cliff-like summit of Chomoyummo (Chomiomo), as it appears just north of the rock.

On his arrival, Hooker met the villagers, squatting cross-legged in a circle, listening to a letter from the Rajah, instructing them as to how to treat the young English botanist.

One evening, while based at Thangu, the sick of the surrounding area came to see Hooker with the usual complaints; rheumatism, ophthalmia, goitre and poisoning by *Arisaema*, fungi and various wild vegetables.

While his men pitched his tent, Hooker collected dozens of plants he hadn't previously found, many of which proved to be new species; among them were the tiny dark purple *Pedicularis integrifolia*, the weird and wonderful green-flowered *Parnassia tenella*, the yellow daisy-like flowers of *Nannoglottis hookeri*, purple *Saussurea hieracioides*, and no fewer than three new species of *Astragalus*. In the yak pastures he found a new genus, *Lancea tibetica*, a favourite alpine perennial with visiting plant enthusiasts in the Himalaya, where it studs the turf with its pretty purple blossoms. The genus commemorates Mr J. H. Lance, of the Bengal Civil Service, who contributed plants from Tibet and Kashmir to the Kew herbarium through his friend M. Pakenham Edgeworth.

By streams and on the banks of the Thangu Chu he found the dainty *Iris goniocarpa*, with tiny blue-purple blossoms; it is related to *I. hookeriana* from the western Himalaya and is used in Tibetan folk medicine. Reginald Farrer later re-discovered it in Gansu, and in his book, *The Rainbow Bridge*, he described meeting it in his usual whimsical prose, 'the soft, blue-purple butterflies of *Iris goniocarpa*, an old friend … always welcome'. He also found a white-flowered form and introduced it to cultivation.[9]

By that time the snows had melted and the alpine meadows on barren-looking hillsides over the village were a riot of colour. Here Hooker discovered *Geranium refractum*, with lovely white, pink-tinged, heavily reflexed flowers. This species was described in 1872 by M. Pakenham Edgeworth, based on Hooker's Thangu collection, in the first volume of *The Flora of British India*. Even the main components of the meadows, the grasses and sedges, proved new. Two of these commemorate their discoverer, namely *Trikeraia hookeri* and *Carex daltonii*.

Hooker's Rock at Thangu. This sketch, made on the spot, was later reproduced as a wood engraving in *Himalayan Journals*. The enormous boulder, probably deposited by retreating glaciers, is still the most distinctive feature in this part of the Lachen valley.

Alpine scenery at Thangu. Hooker's Rock, seen here in November 2013. Little has changed in the intervening years and yaks continue to graze the surrounding mountain slopes.

Hooker's new finds were not just botanical curiosities. Many of them were sensationally beautiful, and perhaps one of his finest Thangu collections was the lovely *Onosma hookeri*, a low-growing perennial bearing densely packed cymes of intensely blue flowers. While we admire this for its horticultural potential, natives to the Himalaya rely on it for a multitude of uses; its roots produce a red dye, and when powdered they are given to horses as a cough remedy. As a dye plant it interested Hooker; Kew was always keen to know of plants that could be used for possible economic benefit in other parts of the British Empire.

Two of the prettiest of Hooker's high alpine discoveries above Thangu are the tiny rosette-forming *Saxifraga umbellulata*, which bears cymes of yellow flowers, and the lovely *Pterocephalus hookeri*, which carries rounded heads of scabious-like blossoms in late summer and autumn; Elwes collected it again at Thangu. It is common in the high alpine meadows of the E Himalaya and extends into W China, where one of the best places to see it is in the meadows on Jade Dragon Snow Mountain in NW Yunnan province.

The Asian mandrake is one of the most exciting perennials one encounters when trekking in the Himalaya, and it was another of Hooker's Thangu finds. *Mandragora caulescens* is an altogether more exciting species than its more widely grown European counterpart, *M. autumnalis*. In the wilds of Sikkim and adjacent Tibet it forms a low-growing tap-rooted perennial, carrying bonnet-shaped, purplish-black flowers, with green spreading lobes, on stout 10 cm tall stems; these are followed in autumn by curious orange tomato-shaped fruits.

The great alpine pastures that clothed the sides of the valley made good grazing, particularly for yaks, and having pitched his tent Hooker awoke to a truly Himalayan scene:

On the following morning I was awakened by the shrill cries of Tibetan maidens, calling the yaks to be milked, "Toosh – toosh – toooosh," in a gradually higher key; to which Toosh seemed supremely indifferent, till quickened in her movements by a stone or stick, levelled with unerring aim at her ribs; these animals were changing their long winter's wool for sleek hair, and the former hung about them in ragged masses, like tow. The calves gambolled by their sides, the drollest of animals, like ass-colts in their antics, kicking up their short hind-legs, whisking their bushy tails in the air, rushing up and down the grassy slopes, and climbing like cats to the top of the rocks.[10]

Later that morning Hooker set out for the Kongra La, though he had yet to grasp the equestrian techniques used in this remote Himalayan kingdom:

The Soubah and Phipun came early to take me to Kongra Lama, bringing ponies, genuine Tartars in bone and breed. Remembering the Dewan's impracticable saddle at Bhomsong, I stipulated for a horse-cloth or pad, upon which I had no sooner jumped than the beast threw back his ears, seated himself on his haunches, and, to my consternation, slid backwards down a turfy slope, pawing the earth with his fore-feet as he went, and leaving me on the ground, amid shrieks of laughter from my Lepchas. My steed being caught, I again mounted, and was being led forward, when he took to shaking himself like a dog till the pad slipped under his belly, and I was again unhorsed. Other ponies displayed equal prejudices against my mode of riding, or having my weight anywhere but well on their shoulders, being all-powerful in their fore-quarters; and so I was compelled to adopt the high demi-pique saddle with short stirrups, which forced me to sit with my knees up to my nose, and to grip with the calves of my legs, and heels. All the gear was of yak or horse-hair, and the bit was a curb and ring, or a powerfully twisted snaffle.[11]

A large yak photographed just above Hooker's Rock at Thangu. They are one of the most useful animals in the high Himalayan regions.

Their route took them north along the upper reaches of the Lachen River, past a scrub of rhododendrons, dwarf birch and mountain ash: the most northern woody plants in Sikkim and the last of their kind to be found before the Tibetan Plateau. Alpine plants abounded, particularly *Trollius, Anemone, Arenaria, Draba, Saxifraga, Potentilla* and *Ranunculus*. Many of these proved to be new species, for example, both the dwarf marsh marigold, *Caltha scaposa* and the magnificent yellow-flowered Himalayan cowslip, *Primula sikkimensis*, which gilded the surrounding marshes. When the latter first flowered at Kew in the summer of 1851, having been raised from Hooker's seed collections, William Hooker had it featured in *Curtis's Botanical Magazine* (t. 4597), quoting his son's description: 'It is the pride of all the alpine primulas, inhabits wet boggy places at elevations from 12–17,000 feet, at Lachen and Lachong [sic.], covering acres with a yellow carpet in May and June'.

By the foot of a moraine he found a Tibetan camp of black, yak-hair tents, and in one of these was a young girl making butter and curd from yak milk. The churns were either oblong boxes made from birch bark, or closely worked bamboo wicker, full of rhododendron twigs, in which the cream was shaken. Apart from a few bamboo and copper milk vessels, wooden ladles, tea churns and pots, the tents contained no furniture, but some goat-skins and blankets on which to sleep. In these elevated regions, where timber is scarce, the fire was made of sheep and goats' droppings, lit with juniper wood.

Above there the valley became stony and desolate, with only occasional patches of vegetation, and higher still, at 4,572 m, a small glacier descended almost to the river from the west side of the valley. Eight kilometres further on, he and his men came to the tents of the Phipun, whose wife was preparing to entertain them with Tibetan hospitality. By the tent door, magnificent Tibetan mastiffs created a racket, and yaks and ponies grazed on the alpine turf close by. Inside the tent, Hooker and 11 others were revived with cups of salted butter tea and a number of treats, including roasted maize, which in *Himalayan Journals* Hooker emphasised was 'called "pop-corn" in America', and was 'prepared by roasting the maize in an iron vessel, when it splits and turns partly inside out, exposing a snowy-white spongy mass of farina. It looks very handsome, and would make a beautiful dish for dessert'.[12]

The sound of avalanches and rockfalls reverberated through the valley as they sat cross-legged in the tent. The Phipun then stood up and announced it was time to move on, since the rock falls meant it would soon rain. The moist vapours travelling up the Lachen valley met their final obstacles, the great cliff face of Chomoyummo (6,829 m), and Kangchengyao (6,889 m), condensing on their rock faces, which being loosened, precipitated avalanches of rocks, boulders and snow. Hooker and his team proceeded up the valley through dense mist, with the roar of these rocks falling on either side sounding like deafening peals of thunder. The mist cleared, allowing transient views of black rock and great sheets of ice towering 1,524 m above him, producing an almost overpowering feeling of awe for this wild, majestic landscape. Though the valley was engulfed in a gloomy grey mist, in front lay a high blue arc of absolutely cloudless sky, between the overhanging cliffs that formed the portals of the Kongra La pass.

The wife of the Lachen Phipun treated Hooker to butter tea, a favourite Tibetan beverage. This wood engraving, published in *Himalayan Journals*, depicts a Tibetan teapot with a block of brick tea.

Hooker reached the pass in the early afternoon. The boundary between Sikkim and Tibet was a low, flat spur running east from Kangchengyao to Chomoyummo, and was marked by a simple stone cairn decorated with colourful prayer flags. A powerful wind blew up the valley, making it a bitterly cold visit, and they had been thoroughly soaked by the earlier rains.

Cushion plants, being well adapted to the extreme climate, found a suitable habitat on the pass, and Hooker managed to collect over 40 species; a rich flora for such a high altitude. Several members of the Caryophyllaceae formed hemispherical balls on the bare mountain slopes, while others grew in matted tufts, level with the soil. He regretted that there were no woolly species of *Saussurea*, common at the same altitudes on the wetter mountains further south, but there were plenty of rewards such as the blue poppy, *Meconopsis horridula,* and a curious yellow *Saxifraga* with long runners and the strongly scented spikenard, *Nardostachys jatamansi.*

Hooker was to discover four new cushion-forming species of *Arenaria* or sandwort on the Kongra La, all described by his friend, M. Pakenham Edgeworth. All of these, *Arenaria ciliolata, A. monticola, A. polytrichoides* and *A. pulvinata*, were white-flowered species, and much sought-after by keen growers of alpine plants.

Arenaria polytrichoides seen here on the Goecha La in north Sikkim. It is generally found near the highest mountain passes in the eastern Himalaya and western China.

Tibetan nomads on the Phalung Plains, one of the few plateau areas in this mountainous landscape. Tibetans crossed the border into Sikkim and grazed their yaks there during the summer months.

On the same screes, he found *Eutrema hookeri*, a tiny, alpine member of the Brassicaceae, with pretty white flowers. Its bedfellows included the minuscule yellow-flowered *Saxifraga lychnitis*, the intensely blue *Corydalis polygalina, Ligularia hookeri* and *Potentilla articulata*. This last was later described by the French botanist Adrien René Franchet (1834–1900) from collections made decades after Hooker, by Père Jean Marie Delavay (1834–1895) in western China.

Franchet would also later describe another of Hooker's Kongra La alpines, *Androsace strigillosa*, a hummock-forming species at high altitude, bearing loose umbels of small white flowers on 30 cm tall stems. Louseworts abound in the Himalaya, especially on alpine slopes, and one of the most gorgeous of all is the tiny rose-purple flowered *Pedicularis bella*; Hooker was obviously charmed by this lovely annual, which he found in alpine meadows close to the pass, choosing the epithet *bella*, when naming it in 1885. Of his many Kongra La discoveries, the tiny *Astragalus kongrensis* bears the name of the pass.

These meadows were tinted a rose colour by the flowers of *Primula tibetica*, a tiny species described by the English botanist David A. P. Watt (1830–1917); he based his description on the frontier collections of Hooker and of Lt Col. Sir Henry Strachey ((1816–1912) and his travelling companion, James Edward Winterbottom (1803–1854). It is one of the highest-growing of all the Himalayan primulas.

After spending two hours on the pass, Hooker began to feel the effects of altitude. He was stiff, cold and dizzy and had a throbbing headache. Having walked a good 21 km botanising, he was glad to ride back down. They reached the Phipun's tent by 6 p.m., and after some Tibetan butter tea, continued in the darkness towards Thangu. On arrival there he wrote a letter to the great Prussian explorer Alexander von Humboldt, opening the letter with 'I have carried out my point, and stood on the table-land of Thibet, beyond the Sikkim frontier, at an elevation of 15,500 ft., – at the back of all the snowy mountains!'[13]

Women on the Phalung Plains. The outer figures are Lepcha girls, in conversation with Tibetan women.

Despite the darkness, Hooker's pony knew the route instinctively, and so as he rode he reflected that an ambition he had held since childhood had finally been realised. He was the first European explorer to gaze into the forbidden land of Tibet from the kingdom of Sikkim, having surmounted every obstacle that had been placed in his way. One has to admire his determination, stubborn energy and drive. Few other explorers endured the hardships and privations he constantly encountered, and he must have felt a warm satisfaction as he rode down those bleak slopes, having earlier that day stood on the roof of the world:

> The night was fortunately fine and calm, with a few stars and a bright young moon, which, with a glare from the snows, lighted up the valley, and revealed magnificent glimpses of the majestic mountains. As the moon sank, and we descended the narrowing valley, darkness came on, and with a boy to lead my sure footed pony, I was at liberty uninterruptedly to reflect on the events of a day, on which I had attained the object of so many years' ambition. Now that all obstacles were surmounted, and I was returning laden with materials for extending the knowledge of a science which formed the pursuit of my life, will it be wondered that I felt proud, not less for my own sake, than for that of the many friends, both in India and at home, who were interested in my success?[14]

East to the Phalung Plains

On 26 July, the Phipun, who now treated Hooker with great respect, offered to take him to an encampment of Tibetan nomads, at the base of Kangchengyao. They mounted their ponies and followed the banks of the Thangu Chu eastwards. Above Thangu it was a rapid, turbulent river, confined by a narrow gorge, between the grassy slopes of the surrounding hills, on which large herds of yaks were grazing, tended by women and children, whose black tents were scattered across the slopes.

The boisterous young yak calves immediately left their mothers to run alongside Hooker and the Phipun's ponies. Chased by the grunting mass, the situation became unmanageable and Hooker was thrown by his pony.

The rocks were covered with *Ephedra gerardiana* ssp. *sikkimensis* (a new subspecies), indicating a dry, rather barren climate; it grew with *Onosma hookeri*.

The moist ground here was liberally scattered with the great Himalayan cowslip, *Primula sikkimensis*, its yellow blossoms making a splendid contrast with sheets of the spectacular blue-flowered *Anemone trullifolia*, one of William Griffith's Bhutan discoveries. In the high alpine meadows, he was to discover gems such as the tiny blue-flowered annual, *Gentiana micans*, and the diminutive *Aconitum naviculare*, with large, violet, hooded blossoms. In the Himalaya this alpine monkshood is an important medicinal herb.

Beyond the meadows the valley opened out to the north, offering a superb view of the astounding perpendicular face of Kangchengyao, studded with immense icicles, from which it gained its name *gyao* or *jhow*, meaning bearded, and *kang chen*, or big snow. From it, a long jagged spur stretched south, the Changmekang (Chango-khang) or eagle's crag, from whose flanks glaciers descended to form the source of the Thangu Chu.

Kangchengyao, seen here from Yume Samdong in north-east Sikkim, just a few miles from the Tibetan border. Hooker visited glaciers and hot springs on its southern flank. Its flat ice-covered summit makes it one of the most distinctive mountains in Sikkim.

Another glacier descended from the southern flank of Kangchengyao, hidden to the east by an enormous, barren terminal moraine that lay obliquely across the valley; the turquoise waters from this glacier fed into Sebu Lake. From its top a succession of smaller parallel ridges indicated the retreat of the glacier over past millennia, and on one of several excursions, Hooker found a hot spring bursting from granite rocks just a mile below its icy tongue. Several plants in the immediate vicinity of the hot spring exploited it as a little oasis in an otherwise cold and sterile landscape; these included the cress-relative *Rorippa elata*, which he had earlier discovered growing near cottages and in fields near Thangu.

Above there he ascended a series of ancient moraines to the plains of Phalung (Palung), an elevated grassy plateau, 3 by over 6 km broad, extending south from the flank of Kangchengyao, with a mean height of 4,877 m. It was curious to stumble across a high plain in the depths of the Sikkim Himalaya, and it was there that the Tibetans had made their camp.

The enormous Tibetan mastiffs guarding the campsite began barking savagely, as the ponies carrying Hooker's party galloped into the enclosure of stone dykes that protected the cluster of black yak-hair tents; thankfully the dogs were individually chained to large stones.

Primula sikkimensis, discovered by Hooker near Thangu. He sent seeds to his father at Kew where it flowered alongside *Inula hookeri* and *Primula capitata* in May 1851. It is abundant in alpine bogs in Sikkim, Bhutan and parts of western China.

These nomads were natives of the cold, arid, windy plateau area of Kampa Dzong in nearby eastern Tibet, from which the source of the rivers that flow into Nepal, Sikkim and Bhutan originated, and that of the Yarlung Tsangpo to the west. The same families migrated annually to Phalung to graze their animals, paying a small tribute to the Sikkim Rajah in return, arriving in June after the snow had melted, and returning to Tibet in September, where they stall-fed their herds on grass, cut on the banks of Tibetan rivers.

They lived in a harsh environment, and smoke and exposure had blackened their faces; the women applied a pigment of grease as a protection from the incessant winds. The tents were pitched in depressions about 75 cm deep, encircled by a wall of similar height, and at the centre of each tent was a clay arched fireplace, on which cauldrons were placed. Immense heaps of dried goat and sheep droppings were the only source of fuel; when Hooker boiled water in one of the tents, its temperature of 85°C indicated an altitude of 4,836 m.

These nomadic farmers kept a fine flock of Changra goats, which were then being sheared for their Pashmina wool, and during their short stay at Phalung they grew turnips, the most alpine crop in Sikkim. Above the campsite Hooker climbed the bare, rocky hills overlooking the Lachen valley, where at 5,181 m he found several minute arctic plants growing alongside another of his Sikkim rhododendron discoveries, the tiny *Rhododendron nivale*, one of the world's highest-growing alpine shrubs. It is also one of the smallest of

Hooker's forget-me-not, *Chionocharis hookeri*. Joseph Hooker described the blossoms as 'looking like turquoises set in silver'.

The forget-me-not relative, *Chionocharis hookeri*, one of the most beautiful alpine plants. Hooker made his collections on Chomoyummo, on the Phalung Plains and on the Kongra La. It is seen here on the Beima Shan in north-west Yunnan, China.

all the Sikkim *Rhododendron* species, the epithet *nivale* meaning 'the snow rhododendron'. In Sikkim it occupies the highest altitudinal zone between the tree-line and perpetual snow, forming straggly, cushion-like plants up to 30 cm tall, smothered in funnel-shaped purple blossoms between June and August when the snows have melted. Hooker was thoroughly impressed by this tough little plant:

> The hard woody branches of this curious little species, as thick as a goose-quill, straggle along the ground for a foot or two, presenting brown tufts of vegetation where not half a dozen other plants can exist. The branches are densely interwoven, very harsh and woody, wholly depressed; whence the shrub, spreading horizontally, and barely raised two inches above the soil, becomes eminently typical of the arid stern climate it inhabits. The latest to bloom and the earliest to mature its seeds, by far the smallest in foliage, and proportionally largest in flower, most lepidote in vesture, humble in stature, rigid in texture, deformed in habit, yet the most odoriferous, it may be recognised, even in the herbarium, as the production of the loftiest elevation on the surface of the globe, – of the most excessive climate, – of the joint influences of a scorching sun by day, and the keenest frost at night, – of the greatest drought followed in a few hours by a saturated atmosphere, – of the balmiest calm alternating with the whirlwind of the Alps. For eight months of the year it is buried under many feet of snow; for the remaining four it is frequently snowed and sunned in the same hour. During genial weather, when the sun heats the soil to 150°, its perfumed foliage scents the air; whilst to snow-storm and frost it is insensible, blooming through all, expanding its little purple flowers to the day, and only closing them to wither after fertilization has taken place. As the life of a moth may be indefinitely prolonged, whilst its duties are unfulfilled, so the flower of this little mountaineer will remain open through days of fog and sleet, till a mild day facilitates the detachment of the pollen and fecundation of the ovarium. This process is almost wholly the effect of winds; for though bumble-bees and the "blues" and "fritillaries" (Polyommatus and Argynnis) amongst butterflies, do exist at the same prodigious elevation, they are too few in number to influence the operations of vegetable life.
>
> The odour of the plant much resembles that of "Eau de Cologne." … This singular little plant attains a loftier elevation, I believe, than any other shrub in the world.[15]

Rhododendron nivale was introduced to cultivation by Roland Edgar Cooper in 1915 from Bhutan, first flowering at the Royal Botanic Garden Edinburgh in March 1920, though it has never been common in the gardens of Britain and Ireland.[16]

With it grew *Thylacospermum caespitosum*, one of the Himalaya's most prominent high alpine cushion plants, forming moss-like mounds across arid rubble-strewn slopes, reminding Hooker of another curious cushion plant, the balsam bog, *Bolax glebaria*, which he had collected on the Falkland Islands, during his Antarctic voyage.

Wild animals were scarce on the lofty borderlands between the Kongra La and Thangu, though there was plenty of vegetation to sustain mammals. Apart from wild sheep, these mountains were home only to Himalayan marmots, which migrated in vast numbers from Tibet as far as Thangu. Hooker put this lack of mammal life down to the cold, moist climate, comparing the great paucity of land

animals with the damp west coast of Tasmania, New Zealand and Tierra del Fuego, areas he had explored. Just a few miles north, on the dry, high Tibetan Plateau, the alpine landscape abounded with wild horses, antelopes, foxes, marmots and hares. The laws governing the distribution of large mammals seemed to Hooker, as a naturalist, to be intimately connected with climate.

He returned to Thangu, spending a week there with the villagers, who were extremely kind and hospitable towards him. On 28 July there was a major fall of snow on the mountain slopes above 4,267 m, while rain fell at Thangu. Hoping to reach the line of perpetual snow, he headed for the mountainous mass of Chomoyummo the following morning, and there made one of his most spectacular discoveries, *Chionocharis hookeri*, describing it as large silky cushions of forget-me-nots, looking like turquoises set in silver. It grew with another equally exciting alpine: the diminutive *Delphinium glaciale*, which, despite its beautiful orbs of sky-blue flowers, exhaled a rank smell of musk. Dense fog and sleet prevented Hooker reaching higher than 5,107 m, and there was no perpetual snow to be found; that was of little matter, as he had just discovered two of the most beautiful alpine plants to be found in the entire Himalayan range.

Hooker departed from Thangu on 30 July 1849, reaching the little village of Lachen just two days later. Along the route he made a rich collection of plants he had missed on his journey up the valley, including *Berberis concinna*, a striking low-growing barberry whose leaves are vividly white beneath. He sent seeds of this to Kew from here in 1849, though despite being highly garden-worthy, it has never received the popularity it deserves.

While still at high altitude Hooker discovered and collected several new alpines, including the diminutive *Gentiana phyllocalyx*, smothered in glorious blue, trumpet-shaped flowers, and the curious dwarf shrub *Diplarche multiflora*, a new genus in the heath family (Ericaceae), with clusters of inflated, bell-shaped, rose-pink blossoms. Further down the valley he found *Allium macranthum*, a highly ornamental perennial bearing masses of rose-lilac flowers on 38 cm tall stems.

The Himalayan marmot, *Marmota himalayana*, an inhabitant of alpine grasslands throughout the Himalaya.

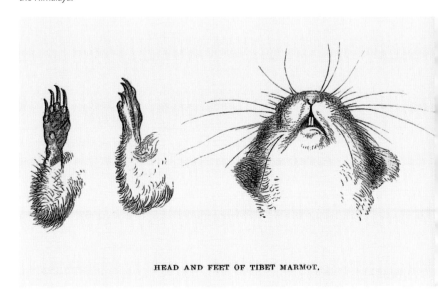

HEAD AND FEET OF TIBET MARMOT.

Rhododendron campanulatum ssp. *aeruginosum*, seen here in the Chopta valley above Thangu. It is one of the highest-growing rhododendrons in the Sikkim Himalaya. Native to large tracts of the eastern Himalaya from Nepal to Arunachal Pradesh in NE India, its young growths are a striking metallic blue-green.

Rhododendron campanulatum ssp. *aeruginosum* was first described in 1851 by Joseph Hooker in *Rhododendrons of the Sikkim Himalaya*. The image above is Walter Hood Fitch's re-working of Hooker's field sketch.

The area where he had camped at Lachen in May and June was now a marsh, so he stayed for two days instead in a house in the village. Tropical cicadas ascended into the fir and hemlock forests above the village, making their shrill song in the heat of day, while glow-worms lit the woods by night. By 5 August, he had reached the mellow climate of Chungthang; bridges and paths had been swept away by the torrential rain and landslides, making their return journey extremely difficult. Along the river plains, by then covered in dense, rank vegetation, they waded knee deep in mud, swarming with leeches, making long detours to avoid the dangerously swollen Lachen River.

Great care had to be taken when leaving the path, since the venomous black cobra was so common there, often basking in the sun in open glades. Harmless green snakes, up to 1.2 m long, glided silently through the bushes. Hooker remained at Chungthang until 15 August, sorting and arranging his collections, before proceeding

up the nearby easterly Lachung River valley, where he planned to cross the Dongkya La (pass) and the source of the Teesta River at the Cholamu Lake.

Despite all this, he still made time to write up descriptions of the new rhododendrons he encountered during his travels, which he posted to his father at Kew. He had earlier received the first part of *Rhododendrons of the Sikkim Himalaya* from Sir William, which no doubt acted as a stimulus for further work. Joseph was ecstatic about the quality of the publication, delighting in the wonderful plate of *Rhododendron grande*, writing to his mother from Chungthang, that 'all the Indian world is in love with my *Rhododendron* book, and extracts from my Tonglo journal, which I sent to the Asiatic Society Journal, have been praised in all the public papers'.[17] His rhododendron book would appear in successive parts between 1849 and 1851, and remains, even to this day, one of the most beautiful florilegia ever produced.

CHAPTER 12

THE LACHEN VALLEY

The greatest highlights of any visit to Sikkim are the Lachen and Lachung valleys, which run parallel to one another in NE Sikkim and ultimately reach passes leading into Tibet. The more westerly of the two is Lachen, which we were to explore twice in autumn when the monsoon mists had cleared, and again in spring during the flowering season.

Our approach to the valley during our first trip was northeast from Darjeeling, and again a year later, from the State capital Gangtok via the busy market town of Mangan. The Darjeeling route brought us through the outskirts of the hill station, through vast tea gardens, with scattered trees of *Cryptomeria japonica*, *Albizia procera* and *Schima wallichii* (in bloom at the time) fringed by thickets of the wonderfully scented East Indian lemongrass.

Beyond the tea gardens our route brought us to the junction of the Great Rangit and Teesta Rivers, where Joseph Hooker and his companion, Charles Barnes, had explored in May 1848. We visited this famous viewpoint on several occasions, latterly in May 2015, sheltering from the scorching sun beneath a forest of sal, *Shorea robusta*, then still bearing a few late blooms.

A scene from the Buddhist Wheel of Life, painted as a mural by the entrance to the Lachen monastery. It shows scenes of hell, the realm where beings endure unimaginable suffering and torment for the evil deeds they have committed.

Tree ferns, *Cyathea spinulosa*, in the upper tropical Teesta valley. They are seen here in May 2015, expanding their fronds at the beginning of the pre-monsoon season.

From there we began an exciting journey through extensive teak plantations before plunging into the tropical valley of the Teesta River, past cardamom plantations cultivated beneath native forest trees. *Callicarpa arborea* was common in the lower valleys, forming trees to 15 m tall, with enormous cymes of showy purple blossoms. It grew alongside the red silk-cotton tree, *Bombax ceiba*, which I had last seen in the tropical rainforest area of Xishuangbanna in southern Yunnan province, in SW China. It was lovely to see it again here, this time bearing masses of showy scarlet blossoms. *Bauhinia variegata* was abundant on the banks of the Teesta, and in its crown were masses of large showy, fragrant blossoms that varied in shades of magenta, to indigo and, very occasionally, pure white.

We had previously visited the valley in late November 2012 when we stopped to study a wild population of *Magnolia hodgsonii*. Not a common tree in Sikkim, it grew at an altitude of 1,210 m and formed a striking evergreen tree bearing enormous obovate-oblong leaves and great globular cone-like seed capsules. We met it again in May 2015, when the valley lay wreathed in mist, and trees sported wonderful copper-coloured new growths and fat flower buds, just days away from expanding their wonderfully spicy-scented blossoms. At that time of year, several epiphytic orchids were giving a spectacular floral display, particularly the superb *Dendrobium nobile*, Sikkim's national flower, and *Coelogyne nitida*, then laden with sprays of showy white blossoms.

Dendrobium nobile, the national flower of Sikkim, growing wild as an epiphyte in the Teesta valley.

Magnolia hodgsonii is still relatively common in the Teesta valley. Frustratingly, we were a few days too early to see it in bloom. It is an extremely handsome foliage plant however.

Sikkim has one of the richest orchid floras in the world. Seen here is Coelogyne nitida, perched in the branches of Magnolia hodgsonii. It was abundant in the tropical Teesta valley,

At 805 m the forests became more jungle-like in appearance, with native screw pine, *Pandanus furcatus*, giant bamboo, *Dendrocalamus giganteus* and graceful tree ferns such as the slender-trunked *Cyathea spinulosa*. Along our route (during autumn visits), blazing red Mexican poinsettias, *Euphorbia pulcherrima,* lined the roadsides forming small trees to 4 m, reminding us that despite the welcome heat of this tropical area, Christmas in Ireland was just weeks away.

Our journey took us to the Temi tea gardens, the only tea-growing district in Sikkim. The tea gardens cover 177 ha and were established in 1969 under the supervision of Keshab Pradhan, for the king of Sikkim and his government. All the plants growing there were propagated from Darjeeling stock, and are thus descended from Robert Fortune's original introduction for the British East India Company. We arrived to see *Prunus cerasoides*, an autumn-flowered cherry, at its spectacular best, carrying a dazzling display of blossoms above a sea of thousands of tightly clipped tea bushes.

The Temi tea gardens with trees of *Prunus cerasoides* in bloom in November 2013. Sadly, this species is not suited to British and Irish gardens.

Beyond Temi our route continued up the tropical Teesta River valley, through forests of *Toona ciliata, Terminalia myriocarpa* and *Pandanus furcatus*. To reach Chungthang, and ultimately, the little village of Lachen, meant passing Singhik (Singtam of Hooker), one of the most famous viewpoints in the Himalaya, where Joseph Hooker sketched Kangchenjunga and the Talung River valley.

This is perhaps the most iconic (and most published) of Hooker's Himalayan scenes, and, descending a flight of steps beneath the view point, through an avenue of Buddhist prayer flags, I was absolutely thrilled to stumble across the flat place where Hooker had pitched his tent. We returned to the same spot two years later, when a few energetic members of the party climbed down to the ridge, to find a tiny cluster of wooden houses with thatched roofs, with a further two miles of steps reaching to the river below.

Singhik, overlooking the Talung River. It was on this flat ridge, in 1849, that Joseph Hooker pitched his tent and sketched one of north Sikkim's best-known views.

A long steep flight of steps lined by prayer flags led us down to Joseph Hooker's campsite. In November 2012 clouds obscured the views, but we were luckier on return visits.

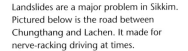

Landslides are a major problem in Sikkim. Pictured below is the road between Chungthang and Lachen. It made for nerve-racking driving at times.

Chungthang — cobra lilies and wild bananas

The road approaching Chungthang from the south, particularly via the town of Mangan, is a major landslide area, and like Hooker we encountered enormous slips where entire mountain faces had slid into the Teesta River. This is especially problematic after the monsoon, when forests, hillsides and roads can disappear overnight. On the plus side, this can provide opportunities to botanise in warm-temperate areas that we would normally pass over in our desire to reach the temperate flora higher up.

On one such occasion, in May 2015, we were forced to botanise just a few kilometres south of the village on the banks of the Teesta River. With the sweltering heat and humidity, it felt like being in the Great Palm House at Glasnevin, and we found some exciting and very showy plants.

The freezing waters of the Teesta River made a strange contrast with enormous clumps of the Darjeeling banana, *Musa sikkimensis*, lining its banks. Like Joseph Hooker, we were impressed that while we stood in this subtropical area, glaciers and icy peaks lay less than 30 km away. On the river bank we also found a single clump of black, or Bengal, cardamon, *Amomum subulatum*, with spectacular semi-translucent cream and orange starfish-like flowers. Like true cardamon, *Elettaria cardamomum* from southern India, the seeds of this cardamon are used in Indian cuisine, particularly as a spice in meat stews.

The Darjeeling banana, *Musa sikkimensis*, wild on the roadside between Singhik and Chungthang. It is a relatively hardy species, surviving out of doors at Kilmacurragh until the arctic winter of 2010.

The cardamon grew in a thicket of *Impatiens arguta*, a perennial balsam discovered by Hooker in September 1848. It is probably the loveliest species of balsam in Sikkim, with large, pale purple blossoms carried on long, slender peduncles.

Overhanging the Teesta we found an enormous leaning strangler fig, and suspended from its trunk was a superb plant of the epiphytic *Aeschynanthus parviflorus*, the Sikkim blushwort. This is occasionally grown as a conservatory plant in cooler parts of Europe and North America for its clusters of spectacular crimson tubular flowers, and it was wonderful to see it here.

The Bengal cardamom, *Amomum subulatum*. We found it in bloom on the banks of the Teesta River.

Impatiens arguta, a perennial species with flower colour ranging from lavender-blue to almost blue. One of the loveliest of the Sikkim balsams.

The flowers of this trailing epiphyte are extremely showy and all *Aeschynanthus* species make good glasshouse plants.

We stopped briefly at Chungthang, where the Lachung and Lachen Rivers join to form the Teesta. Immediately below the confluence, an enormous dam was under construction, one of many built on the river, and much debated by environmentalists in Sikkim. Following the 1962 Sino-Indian war there is now a large army base nearby, placed there to guard the Indo-Tibetan border.

Chungthang marks the transition from the tropical and subtropical floras of the Teesta valley to the temperate flora of the Lachen and Lachung valleys. Hooker discovered *Rhododendron maddenii* on the hillside here, and we were delighted to find it growing alongside *R. lindleyi* on roadside cliffs, and the charming *Aster sikkimensis*, which Hooker discovered in E Nepal in 1848, introducing it to cultivation via Kew, where it first flowered in October 1850.

We saw several plants of both *Rhododendron maddenii* and *R. lindleyi* flowering on riverside cliffs just south of Chungthang in May 2015. Cobra lilies were abundant in the damp roadside thickets and forests, particularly *Arisaema speciosum* and the whipcord cobra lily, *Arisaema tortuosum*, an altogether more upright plant with a very distinctive green, whip-like spadix emerging from its spathes.

The Teesta valley was incredibly hot and humid as we botanised this stretch. Cascading from the trees above we found the shrubby epiphyte *Aeschynanthus parviflorus*. Here its blossoms are suspended above the icy, racing waters of the Teesta River.

Aster sikkimensis, another of Hooker's Himalayan discoveries. It is particularly common around Chungthang.

Arisaema tortuosum, another of Sikkim's several cobra lily species. It is also native to Myanmar and parts of China.

Looking up the Lachen River towards the modern-day village of Chungthang. The flora at this point takes on a more temperate aspect.

Good woodlanders abounded there, including the graceful *Disporum cantoniensis*, carrying masses of lemon-yellow, bell-shaped blossoms on slender, upright branching stems over a metre tall.

Here also was *Begonia flaviflora*, whose large, bold leaves, beautifully marbled with chocolate-brown and various shades of green, made a splendid contrast with other fine foliage plants such as *Setaria palmifolia* and *Elatostema hookerianum*.

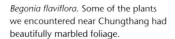

Begonia flaviflora. Some of the plants we encountered near Chungthang had beautifully marbled foliage.

Immediately beyond Chungthang, *Alnus nepalensis*, draped in yellow catkins, lined the banks of the Lachen River. At 1,605 m *Mahonia acanthifolia* was common, with dense racemes of yellow blooms above handsome pinnate leaves. It grew among *Anemone vitifolia*, *Hoya polyneura*, *Schefflera venulosa* (with densely packed panicles of cream and purple-red blossoms), and the beautifully scented *Luculia gratissima*.

The Lower Lachen valley

Beyond Chungthang we soon entered the precipitous lower Lachen valley, whose cliffs and waterfalls, with rounded mountain summits above clothed in pristine, primeval forest, were just as Hooker had seen it.

It was late autumn and the forest was just taking on autumnal tones, though in some places the tree canopy was splashed pink with the flowers of the giant pseudovine, *Wightia speciosissima* and the delightful and conspicuous autumn-flowering cherry, *Prunus cerasoides*. Several other trees were also in full flower in this warm, low-lying part of the Lachen valley, including one of my favourites, *Leucosceptrum canum* (Lamiaceae), which bears long upright spikes of yellow-white flowers with protruding stamens, looking rather like bottle brushes as the inflorescence ages.

Several species of *Lindera* inhabit this valley, including the November-flowered *L. neesiana*. It was at its spectacular best during our visit, when branches were wreathed with densely packed small star-shaped, yellow flowers.

Following the Lachen River we soon reached the hamlet of Menzithang at 1,840 m. Here we passed trees of *Acer sikkimense* wreathed in epiphytic orchids such as *Coelogyne cristata*. The roads were skirted by wild walnuts, *Juglans regia,* and *Rhus hookeri*, with pinnate leaves of a wonderful claret-red, while the feathery plumes of *Astilbe rivularis* abounded in damp banks and gullies by the roadside.

Rising in altitude, hardy trees appeared, including large specimens of *Acer campbellii* (with striking yellow autumn colour) and two birches, *Betula cylindrostachya* and *B. alnoides,* the latter with wonderfully scented bark. *Tetracentron sinense* is also locally common at one point in this valley, a disjunct Himalayan population of this rare Chinese relict species, its orange-yellow autumn colour making it easy to identify on the mountainsides

On the forest floor lay a dense thicket of the lovely *Woodwardia unigemmata*, the jewelled chain fern, with lush fronds well over 2 m long. This lovely fern is perfectly happy in Irish gardens, and one of my travelling companions, Helen Dillon, grows it to perfection in her garden in Ranelagh, in Dublin, where it paints a very pretty picture every spring when its dramatic copper-coloured fronds unfold. Near the fern we spied the feathery heads of *Miscanthus nepalensis* and the superb *Rhododendron griffithianum*.

The mountainside ridges below Lachen bristled with tall trees of *Tsuga dumosa*, while by the roadside grew an astonishingly rich array of good garden plants such as *Corylus ferox* (draped in yellow-purple catkins), *Populus ciliata* (with yellow autumn foliage), *Rhododendron arboreum* (red flowered forms, we were told), *R. maddenii, Hydrangea heteromalla, Zanthoxylum oxyphyllum* and *Ampelopsis semicordata*. On a nearby hillside *Himalayacalamus falconeri* and various *Lindera* species covered the slopes, obviously pioneer species that had 'moved in' after a recent fire.

The roads approaching the village were bounded by thickets of exciting garden plants, including a young tree of *Quercus oxyodon*. I was thrilled to see one of my favourite garden shrubs, *Prinsepia utilis*, a thorny member of the rose family, which bears masses of showy white flowers in winter and early spring. It is unaccountably rare in gardens, though there are fine old specimens at Glasnevin and in the garden of another of my travelling companions, Robert Wilson-Wright from Coolcarrigan in Co. Kildare. *Stachyurus himalaicus* grew nearby, forming a tall, upright shrub, which, in its best forms, carries drooping densely packed racemes of lovely rose-pink blossoms.

The autumn-flowered pseudovine, *Wightia speciosissima*, seen here in the lower Lachung valley by the village of Lemna. We also found it several times in the Lachen valley.

Though rarely seen in cultivation, the linderas are often spectacular in flower and many species have very showy foliage. Seen here is *Lindera neesiana*, one of the more commonly encountered Sikkim species.

Tetracentron sinense, some authorities place the Himalayan population in var. *himalense*. The genus has been found in Iceland's fossil record.

Griffith's birthwort, *Aristolochia griffithii*, seen here scrambling through *Acer campbellii* in the Lachen valley. It was one of the most exciting vines we found during our travels in north Sikkim.

Several *Rhododendron* species also grew here, including the low-growing hummock-forming *R. glaucophyllum*. This pretty little shrub was originally discovered by William Griffith in Bhutan, and re-collected by Hooker in the Lachen valley in the spring of 1849; it was introduced by him from seeds collected in Sikkim in the autumn of that year, flowering for the first time in cultivation under glass at Kew in March 1853. Flower colour varies from pink, to old rose, to in some cases, pure white. On the cliffs above we found *R. virgatum*, another of Hooker's introductions from this valley, although, as with *R. glaucophyllum*, this species had previously been found in Bhutan by Griffith. *R. virgatum* grows in relatively small groups in the wild and it has an extremely restricted altitudinal range between 2,400 and 2,700 m. I must admit that I find it one of the least interesting of the Sikkim rhododendrons, preferring the larger-flowered species.

In May 2015 we explored this stretch of valley at a time when many of these rhododendrons were in bloom. The star performer was the very lovely *R. edgeworthii* that grew epiphytically on many of the surrounding trees. There it formed small shrubs no more than a metre tall, and one pure-white flowered plant, growing on a cliff-side tree of *Schefflera rhododendrifolia*, scented a considerable area with its powerfully sweet aroma. Named for the Co. Longford

botanist, M. Pakenham Edgeworth (a friend of Hooker), we, as an Irish party, were particularly pleased to meet such a wonderful plant, named for one of our most notable botanists. It was everywhere to be seen in the forests below Lachen that day, even 36 m up, in the canopy of venerable old moss-covered *Tsuga dumosa*, crowning the mountain ridges.

During that same spring visit we were thrilled to find dozens of vines of *Aristolochia griffithii* carrying masses of pendulous purple-splashed, mustard-yellow flowers shaped like a Dutch-man's pipe, hence another popular common name. Hooker also collected it here, and named it for William Griffith who had previously found it in Bhutan; it commonly clambered through trees of *Acer campbellii*. In more open areas the Himalayan strawberry, *Fragaria nubicola*, was carrying both flower and fruits. This pretty little alpine strawberry is widely distributed through the Himalaya, its specific epithet *nubicola* meaning 'dweller among the clouds'.

Rhododendron triflorum, first found by Griffith in Bhutan, re-discovered by Hooker near Chungthang in 1849, and introduced by him to Western gardens in 1850, was also in full flower in this part of the valley in May 2015. While not as showy as many of the other Sikkim rhododendrons, it is particularly pretty, with flat, saucer-shaped lemon-yellow flowers.

The Himalayan strawberry, *Fragaria nubicola*, a pretty little carpeting species. It is abundant in the Lachen and Lachung valleys of north Sikkim.

Rhododendron triflorum, a very distinctive Sikkim species with blossoms in shades of lime and cream. Though not quite as showy as other species, it possesses a quiet charm.

Paris thibetica, which we found on the approach to Lachen village. It was by no means common and is quite distinct from the commoner *Paris polyphylla*, which also grows in this valley.

It grew with *Viburnum erubescens*, a common and very variable plant in Sikkim, and the forms we encountered in this part of the Lachen valley were probably the finest in this corner of NE India, with long white tubular blossoms, variously marked rose-purple at their extremities. *Paris polyphylla* is a common plant in the temperate woodlands of western China and the Himalaya. In this part of the valley we found the even more spectacular *P. thibetica*, bearing whorls of long, narrow leaves, bluish green on their upper-surface and of a striking purple hue beneath.

Over the course of four expeditions we stayed in the little town of Lachen on several occasions, though I think few of us will ever forget the first visit. In retrospect, seeing our drivers covering the engines of our four-wheel drives with several layers of carpet and rugs should have been a sign of what was to come, but for many of us that first night was to be the coldest and most uncomfortable any of us spent anywhere, even beating a night spent out in the open in SE Tibet.

We reached Lachen late in the evening of 24 November 2012 and stayed in an ice-cold hotel near the monastery. Our rooms were surrounded by thickets of *Daphne bholua*, *Pieris formosa* and the

The brilliant white stems of Rubus biflorus became a familiar sight around Lachen village. If only our native brambles looked this good!

The stunning *Berberis virescens*, seen here in the Lachen valley, where it is common. The species was described from flowering and fruiting material sent to Kew by Thomas Acton from the Kilmacurragh Estate, Co. Wicklow.

Ireland, but to see the white flowered form was a real treat. There was also a young tree in bloom by our hotel, and while the white-flowered form is relatively rare in Western gardens, it is not at all uncommon in the wilds of Sikkim.

In his 1849 sketch of the village, Hooker, normally a stickler for accuracy, exaggerated the scale of the rocky, pyramidal-shaped peaks that loomed over the hamlet, though he did capture the spectacular position on the very edge of the looming Himalaya. It is still home to the Lachenpas, a Bhutia group, who are a hybrid mix of Tibetans, mostly from the Chumbi valley, and native Lepcha peoples.

We climbed above the village to 2,850 m, passing 10 m tall trees of the Himalayan buckthorn, *Hippophae salicifolia*, absolutely laden with bright orange fruits. The plant that stole the show however was a barberry, with stunning purple-red stems. On my return to Kilmacurragh I identified it as *Berberis virescens*, and while in the rare book room at Glasnevin was surprised to learn from *Curtis's Botanical Magazine* that Hooker described it, not from the Lachen valley where he discovered it in May 1849, but from flowering and fruiting specimens cultivated by Thomas Acton at Kilmacurragh!

In late autumn *Hippophae salicifolia* is one of the showiest trees in the Lachen valley. The vitamin-rich fruits are eaten by local inhabitants.

ghostly-white stems of *Rubus biflorus*, one of the most spectacular plants in the valley at this late season.

One of the most abundant plants around the village was the rhododendron look-alike, *Daphniphyllum himalayense,* and it was in this valley that Hooker discovered it on 27 May 1849. Another common inhabitant on grassy slopes above the village was *Roscoea auriculata*, which Hooker discovered in nearby Chungthang in May 1849, and also collected at Lachen on 1 June 1849.

Lachen has grown slightly since Hooker's visit, though many of the traditional cottages have been replaced with large modern houses, and a large ornately decorated gompa has replaced the old late 19th-century monastery that was badly damaged by recent earthquakes. On our second visit we met young French and German women involved in the older gompa's restoration, and were told that specific timbers were employed in the monastery's reconstruction and restoration, *Abies densa* providing enormous columns to support ornate ceilings, and only *Magnolia doltsopa* considered worthy to provide altar timber.

Magnolia campbellii grows wild around Lachen, and in May 2015 we were lucky to stumble across a tree carrying enormous white blooms in the valley just above the village. To see this species blooming in its native habitat was itself worth the journey from

Though it was introduced to cultivation by Hooker in 1850 to Kew, where it first flowered (as an undescribed species) in July 1855, it received little attention. It surfaced again in June 1887, when Thomas Acton sent material to Kew for naming. H. J. Elwes, it seems, collected this species when he re-traced Hooker's route in 1870 and presumably sent seeds to David Moore at Glasnevin, which were forwarded to Thomas Acton, who sent specimens to Hooker at Kew. It had also been raised at the Royal Horticultural Society's garden at Chiswick from Elwes's Sikkim collections.

It seems that Hooker had been unlucky; while trekking up the valley he had made no note of the plant's flower colour, and as he was travelling during the rainy season, he lost many of his specimens, hence the description from cultivated material.[1] *Berberis virescens* was later found in Bhutan by Ludlow and Sherriff in June 1933, and five years later they would see it again in SE Tibet.

This region of the valley was home to a rich array of interesting herbaceous plants, many of them familiar, including the thistle-like *Morina longifolia* and another great garden favourite, *Primula capitata*, which was abundant by the roadside and still bore a few flowers. Near it, *Elaeagnus parvifolia* carried wonderfully sweet-scented blooms, and cascading through its wiry branches were the clambering stems of *Celastrus hookeri,* carrying capsules of bright-red seeds. Among other climbers, *Clematis montana* and *Jasminum humile* rambled their way through a shrubby tree of *Corylus ferox*, which carried masses of nuts enclosed in a large spiny, husk.

It is always exciting to encounter *Cardiocrinum giganteum* in the wild, and the Lachen plants were no exception. We saw this wonderful plant several times during repeated visits to the valley, much to the delight of one of our party, Helen Dillon, whose earliest memory, as a child growing up in Perthshire, was of standing in her parent's garden, awestruck by the sight of the enormous giant lilies; in her own words she was 'hooked', and has grown plants with the greatest of passion ever since.

Primula capitata, one of William Griffith's Bhutan discoveries. Hooker found it growing on gravel banks near Lachen village and sent seeds to Kew, where it flowered in October 1850.

Returning to the same area in late autumn of 2013, one of the most conspicuous plants in bloom on the forest's edge and in open mountain moorland was the glorious *Daphne bholua* var. *glacialis*, the high-altitude, deciduous variety of *D. bholua* originally named from Sikkim material. Near it we found dense carpets of the lovely *Primula scapigera*, not flowering of course , but with leaves covered in a dense, mealy silvery-white farina. Hooker discovered this little rose-purple or pink flowered species on the summit of Tonglu. It appears to have a very limited range of distribution in western frontier regions of

The modern Lachen monastery seen here in autumn 2013. Trees of *Larix griffithii* in full autumn garb may be seen on the ridge of the hill above.

The interior of the newly built monastery at Lachen. Colourful thangkas are suspended from the ceilings. The timber employed is *Abies densa*, though only *Magnolia doltsopa* is considered sacred enough for altar work.

Daphne bholua var. *glacialis*, the deciduous high-altitude variant of this widespread Himalayan shrub. It was just coming into flower as we drove up the Lachen valley in mid-November 2013.

Sikkim, so our plants were something of an anomaly. It is extremely rare in Western cultivation, though was grown successfully for a time during the 1950s at the Royal Botanic Garden Edinburgh.

Near the giant lily we saw *Philadelphus tomentosus*, *Sarcococca hookeriana* and *Sambucus adnata*, a Sino-Himalayan elder, with wide, flat-topped umbels of orange-red fruits. This species is pretty in the wilds of W China and the Himalaya, though it is best seen there. In cultivation, probably due to humus-rich soil, it becomes an aggressive thug, and never flowers or fruits as well as it does in its native homeland. Another plant growing near it, with a similar distribution, was the handsome *Onychium japonicum*, the carrot fern, which I've previously seen wild in China and Taiwan.

Our aim during that first visit had been to drive to the head of the Lachen valley and continue to the small village of Thangu, but our plans had to be abandoned following a single night of extremely heavy snowfall, rendering the roads impassable.

Despite this, we managed to walk a short way, beneath enormous trees of *Picea spinulosa* and *Tsuga dumosa,* venerable giants up to 55 m tall, that must have been several centuries old, and on our way down were happy to spot a number of birds, among them the hill pigeon, *Columba rupestris*, yellow-billed blue magpies, *Urocissa flavirostris* and, occasionally, majestic golden eagles, *Aquila chrysaetos*.

We were bitterly disappointed not to have made it on to the Tibetan borderlands at Thangu, though the snowy landscape and ancient trees had more than compensated our loss. Lachen had been wonderful, but it was time to push on.

In November 2012 heavy snow thwarted our plans of reaching the upper Lachen valley and Thangu. Though frustrating, we did see some of the big trees of the valley, including these mammoth specimens of *Picea spinulosa*.

Sikkim lorry drivers love to decorate their trucks. Colourful vehicles like this are a familiar sight on mountain roads.

The upper Lachen valley and Thangu

We returned to Lachen in the autumn of 2013, and again in May 2015, determined to reach the Zemu River, Thangu and the spectacular Chopta valley. Weather conditions, thankfully, were much improved during 2013, when we successfully made it up the valley. By the Zemu River we spied a well-worn track, the same taken by Joseph Hooker all those years before us, though very frustratingly, were told (whether true or not) that that particular river valley was off limits to Western travellers. What a pity. Undoubtedly a vast array of plants and an exciting landscape lay along the upper course of the Zemu River. But it was not to be for us and a glimpse would have to suffice, and so we pressed on further still along the upper course of the Lachen River.

The May visit was particularly rewarding and spectacularly colourful, particularly for *Rhododendron* species. One of the showiest of these was *R. cinnabarinum*, another of William Griffith's discoveries from Bhutan, named and introduced from Sikkim by Joseph Hooker in 1850 from seeds collected in the Lagyap valley, in east Sikkim, in November 1849. The forms we encountered had tubular bell-shaped flowers, typically in orange-yellow shades, though we were also happy to encounter the striking *R. cinnabarinum* Roylei Group, with thick, waxy rose-red blossoms.

R. cinnabarinum is highly variable in flower colour throughout Sikkim, and Hooker introduced several shades, including *R. cinnabarinum* Blandfordiiflorum Group, bearing larger than normal bi- or even tricoloured flowers. He also found and introduced *R. xanthocodon* (syn. *R. cinnabarinum* var. *pallidum*), a species described from one of Frank Kingdon Ward's 1924 Tibetan collections. Hooker sent seeds to Kew and, plants flowered there under glass in May 1854.

Rhododendron cinnabarinum, one of the showiest of the Sikkim species, but also an extremely poisonous plant.

Rhododendron cinnabarinum Roylei Group, seen here at Mount Usher gardens in Co. Wicklow. We found occasional plants in the upper Lachen valley.

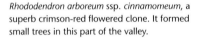

Rhododendron arboreum ssp. *cinnamomeum*, a superb crimson-red flowered clone. It formed small trees in this part of the valley.

We were equally impressed by *R. arboreum* ssp. *cinnamomeum*, a superb clone with vivid red-scarlet blooms, held in tight trusses and sheltered by trees of *Abies densa* and the abundant Himalayan bird cherry, *Prunus cornuta*.

Beyond there *R. campanulatum* became more abundant, particularly on the banks of the Lachen River, where it grew with the catkin-covered *Salix daltoniana*. *R. campanulatum* is one of the loveliest of all Sikkim rhododendrons and is highly variable in the wild. The plants we saw ranged from rose, to lavender-blue, to, in some cases, almost white. Soon after its introduction to cultivation its horticultural potential was realised, particularly by David Moore, who, in 1860, crossed one of Hooker's seedlings at Glasnevin with a white-flowered form of *R. arboreum* ssp. *cinnamomeum* to create the superb 'Thomas Acton'; the original tree still grows at Kilmacurragh.

On the humus-rich forest floor exciting woodlanders abounded, including yet another cobra lily, *Arisaema propinquum*, again a Hooker discovery, and one of the most exciting species to grow in cultivation, with blackish-purple spathes. Even more thrilling were the many colonies of the lovely *Bryocarpum himalaicum*, which,

Rhododendron campanulatum on the banks of the Lachen River. An abundant species in north Sikkim.

The rather sinister-looking *Arisaema propinquum*, one of the species in this remarkable genus that can truly claim to be a 'cobra lily'.

when Hooker found it in these same woods in 1849, proved to be a new genus in the Primulaceae. It must be one of the most gorgeous of all Asian woodlanders, with its delicate pendent yellow, *Soldanella*-like blossoms. Sadly, it is virtually impossible to grow in cultivation and very rarely seen in alpine collections. Its bedfellows included the charming *Clintonia udensis*, just about to open racemes of pale violet-blue flowers carried above tight rosettes of handsome elliptical leaves.

Bryocarpum himalaicum, a rare woodlander from the alpine valleys of Sikkim. The best place to see it is in its native homeland. It resists all efforts in cultivation.

Clintonia udensis, another good woodlander. It grew in leaf litter with *Bryocarpum himalaicum*.

A bridge across the Thangu Chu (the upper Lachen River) leading into the tiny hamlet of Thangu. Hooker's Rock is in the valley beneath. An Indian army camp lies nearby. The Chinese border is just a few miles away from this position.

Beyond there, the valley opened out into a spectacular Himalayan landscape inhabited by herds of yaks, with their Lachenpa shepherds. Moving on, we met *Rhododendron barbatum*, a form with 'unbearded' petioles. During autumn visits to this part of the valley we often saw *Rosa macrophylla* draped in bright orange-red pendulous fruits and 6 m tall trees of *Hippophae salicifolia*, heavily laden with fruits.

Above Talam, we stopped to photograph a snowy fir-lined valley to the west, which Hooker explored on his approach to Thangu in 1849. Here we reached the tree-line, and a little further on arrived at the village of Thangu at 4,267 m. Beneath lay Hooker's Rock, surrounded by a few small ploughed fields, the only sign of cultivation in the region. On the edge of these fields we spied the dead flowering stems of thousands of the lovely yellow-flowered *Primula obliqua*.

In his sketch (p. 184) Hooker included the scrub-like vegetation on the flat surface of the rock, and this we climbed to catalogue, recording the following: *Rhododendron campanulatum, R. setosum, Juniperus pseudosabina, Cotoneaster glacialis,* and dwarf trees of *Betula utilis.* Thangu can't have changed much since Hooker visited. The local inhabitants were very curious about Hooker's sketch, exclaiming with delight when they recognised the giant rock in the valley. Many

Villagers from Thangu. In their hands is the sketch of Hooker's Rock from *Himalayan Journals*. They were amazed when they recognised the scene.

The author with Hooker's Rock looming overhead. Beside me is *Juniperus pseudosabina* while above, on the rock itself, is a thicket of *Rhododendron campanulatum.*

of the people we met that day were probably descendants of the Lachenpa characters portrayed in the sketch; they have never been inclined towards migration. We returned in the spring of 2015, when the meadows surrounding the rock were filled with thousands of flowering plants of *Primula denticulata*.

Also scattered through the grass were several thousand Tibetan dandelions, *Taraxicum tibetanum*, all extremely dwarfed as a result of altitude. This species was first found by Hooker there on the Tibetan borderlands, and was described from material also gathered about the same time in Tibet by Hooker's friend Thomas Thomson, and from collections made in 1893 on the mountains above Tengchong, in Yunnan by the French missionary Père Jean André Soulié (1858–1905).

Taraxicum tibetanum, the alpine dandelion, another of Hooker's Himalayan discoveries.

Lachenpa girls at Thangu. We always received the warmest of welcomes wherever we travelled in Sikkim, particularly in the remote northern valleys.

May 2015 and heavy drifts of snow still linger in the Chopta valley. Seen here are my travelling companions, left to right: Gráinne Larkin, Averil Milligan, Kristin Jameson, Daphne Levinge Shackleton, Lesley Fennell, Bruce Johnson, Derek Halpin, Orlaith Murphy, Bruno Nicolai and our ever-dependable guide, Thupden Tsering.

A view down the Chopta valley, the haunt of many rare alpines like the giant Sikkim rhubarb, *Rheum nobile*.

A scene from the upper Lachen valley in November 2012. Waterfalls stream like silver threads down the steep valley walls into the racing turquoise waters of the Lachen River far below.

Above Thangu lies the spectacular Chopta valley (in *Himalayan Journals*, Hooker simply refers to this region also as the Lachen Valley), another place of soaring peaks, streaked with glaciers, frozen waterfalls and ice fields. It was interesting to observe how much drier the atmosphere becomes as one approaches the Tibetan Plateau. We recorded a relative humidity of 48% at Chungthang (Darjeeling would have been far higher), and by the time we reached the Chopta valley this had dropped to just 22%, particularly dry by Himalayan standards.

The upper part of the main valley was crisscrossed by ancient moraines on which *Rhododendron campanulatum* ssp. *aeruginosum* had formed enormous colonies, and scattered in the shelter of the rhododendrons were alpines such as *Fritillaria cirrhosa*, *Rhodiola fastigiata* (the latter introduced to cultivation by Edward Madden through Glasnevin in the 1840s), the huge *Gentiana stylophora*, with enormous bell-like yellow flowers, and two Himalayan poppies, the lovely blue-flowered *Meconopsis simplicifolia* and the yellow-flowered *M. paniculata*.

Rhododendron setosum turned the mountainsides an ochre hue, with its aromatic autumnal foliage. Its companions included the equally aromatic *R. anthopogon*, *Diapensia himalaica* (sporting intense claret-purple autumn colour), *Cassiope fastigiata*, *Juniperus pseudosabina* (all burned as incense in local monasteries) and *Berberis angulosa*. The latter was introduced to cultivation by Colonel Charles Ball-Acton (1830–1897), an Irish soldier in the British East India Company's army, who sent seeds from his base in Kashmir to his brother, Thomas Acton (who first flowered it at Kilmacurragh in 1888). This widespread barberry was common in the Chopta valley where it had assumed a fiery-orange autumnal hue, and I departed the area pleased to see a 'Kilmacurragh plant' thriving in its native Himalayan home.

CHAPTER 13

THE LACHUNG VALLEY

Having sent his Lachen collections to Darjeeling, Hooker set off again, up the Lachung valley. He must have exasperated the Sikkim officials; the Sentam Subah had thought his troubles were over, but hadn't realised just how headstrong and stubborn the young English botanist was:

> On August 15th, having received supplies from Dorjiling, I started up the north bank of the Lachoong, following the Singtam Soubah, who accompanied me officially, and with very bad grace; poor fellow, he expected me to have returned with him to Singtam, and thence gone back to Dorjiling, and many a sore struggle we had on this point. At Choongtam he had been laid up with ulcerated legs from the bites of leeches and sand-flies, which required my treatment.[1]

The Lachung River was crossed by means of a solidly constructed cane-bridge 36 m long, spanning a river 26 m across. It was at this point that Hooker lost his constant companion, Kinchin, his Tibetan Mastiff hybrid. The dog had the frustrating habit of running for several yards across the round bamboo culms supporting cane-bridges, on which it was impossible for him to retain his footing. At this point the poor fellow was terrified and lay down on the bamboo with his legs dangling above the waters. Hooker had managed to save him on several previous occasions, though this was to be Kinchin's last bridge crossing.

The Lachung River in the Yumthang valley, one of the most spectacular and accessible landscapes in north-east Sikkim. This area of the valley lies in the coniferous zone, with trees of *Larix griffithii*, *Juniperus recurva*, *Abies densa*, *Picea spinulosa* and *Tsuga dumosa*.

A Tibetan Mastiff hybrid, pictured here at Chungthang. It gives a good idea of what Hooker's ferocious hound, Kinchin, may have looked like. This one was a little more placid.

Joseph Hooker had become extremely fond of his dog Kinchin and missed him greatly after he accidentally fell from a bridge at Chungthang and drowned in the Lachung River. Kichin is here portrayed by Hooker in a playful sketch in a letter written to Archibald Campbell, dated 6 June 1849.

Hooker was botanising beneath the bridge when he suddenly heard the dog's shrill, short barks ringing out, but it was too late. Long before he made it to the bridge, Kinchin lost his footing and was swept away by the raging rapids. It was a bad start to the second part of his expedition, and Hooker continued in a melancholy mood, missing him by his side by day, and at his feet by night. Kinchin had grown into a beautiful dog, with glossy black hair, jet-black eyes and a glorious tail curling over his back.

The lower Lachung valley was far more open and grassy than that of the Lachen, and the vegetation differed considerably. To the east lay Tangkra (Tunkra) mountain, rising in a superb open sweep of dark fir and hemlock woods, and cliffs with great snowy peaks that ultimately reached 5,700 m. A pass crossed the mountain's steep slopes, and from there a route continued through the valley of the Thangka River, into the Chumbi valley in Tibet, where the Sikkim Rajah kept a summer residence.

The lower part of the valley before the village was densely forested and full of exciting new exotics. The best of these was *Brassaiopsis mitis*, a spectacular member of the *Aralia* family, forming a small tree bearing enormous leaves (used as fodder plants locally) rather like those of the rice paper plant, *Tetrapanax papyrifer*. It is extremely rare in Western gardens, but if ever promoted will cause quite a stir

among lovers of bold foliage plants and should prove hardy in the sheltered coastal gardens of Britain and Ireland and similar regions in the United States. It almost defies recognition in the temperate forests of Sikkim, looking more suited to the jungle-like vegetation of the tropical valleys.

Bamboos are a common component of the Himalayan flora, and in the lower parts of the Lachung valley they formed impressive specimens, particularly *Dendrocalamus hookeri*, whose enormous rounded culms soared to over 20 m tall. In neighbouring Nepal, where it also occurs, it is sometimes grown for its edible shoots, while its stems are used in basketry and for scaffolding.

The balsams, a group in which Hooker was particularly interested, were well represented in the lower, warmer, Lachung valley, especially in forest understorey in damp, shaded places. He must have drawn a breath on first finding *Impatiens cymbifera*, a spectacular annual, bearing myriad large faded blue-purple blossoms with stunning dark-purple spurs. His herbarium specimens record that he found it on the banks of the Lachung River; all the balsams are water lovers.

Another of Hooker's Lachung balsams was the annual, golden-yellow flowered *I. falcifer*. Though he collected it in 1849, he didn't name and describe it as a new species until 1903, when it was cultivated at Kew from seeds collected in Sikkim by Major Prain, the

Impatiens falcifer, discovered by Hooker on the banks of the Lachung River on 2 October 1849. Commonly known as the sickle balsam, the specific epithet *falcifer* refers to the upper sickle-shaped spurs. From *Curtis's Botanical Magazine* t. 7923 (1849).

Superintendent of the Royal Botanic Garden, Calcutta. *Impatiens* are a difficult group to dry and press, and it was by having living material on hand, which flowered in frames at Kew in May 1903, that Hooker could determine it as a valid new species.

Another annual species, the yellow (sometimes pink) flowered *I. laxiflora*, was named more promptly by M. Pakenham Edgeworth in 1851, based on Hooker's collections from the Lachung valley, and those Edgeworth himself had made in nearby north-west Bengal.

Rising higher up the valley a more cold-temperate flora appeared, and Hooker collected *Acer stachyophyllum*, a species Griffith had found in Bhutan a decade previously. In the Lachung valley it formed small trees with young shoots and petioles beautifully flushed scarlet. Below Lachung village he entered a 'pine wood', a vast forest of fir and hemlock, indicating that he was approaching alpine regions. By the river bank, he found the tiny *Rubus fockeanus*, a charming little woodlander that carpeted the mossy forest floor, and whose small, orange, raspberry-like fruits nestled in corrugated, strawberry-

like foliage. It's a delightful little plant, not at all like its invasive cousin, the common bramble of the English and Irish countryside, and though rare in gardens, it makes a lovely plant in a peat bed or woodland garden.

Stems of the climbing bellflower, *Codonopsis affinis*, twined through the surrounding vegetation, its gorgeous bell-like flowers deep purple-brown near their mouths and fading into green along the upper parts.

Modern-day travellers in the eastern Himalaya will be familiar with *Mulgedium bracteatum*, a perennial lettuce-relative, bearing short spikes of showy purple-blue blossoms in autumn; Hooker discovered it beneath a canopy of *Tsuga dumosa*. These same forests also provided a habitat for a previously unknown tiny subalpine orchid, *Neottia karoana*, one of the smallest flowered of all the Himalayan orchids. While the latter is probably of botanical interest only, another orchid, *Neottianthe cuccullata* var. *calcicola*, is a superb, showy little subalpine variety, with dense spikes of fragrant, rose-purple blossoms; it was first described, not from Hooker's Lachung collections, but from those made by George Forrest in north-west Yunnan, China in September 1910.

The village of Lachung (Lachung means small pass) was large by Sikkim standards: a cluster of about a hundred well-built wooden houses, raised on posts, and positioned close together without any particular order. A little cultivation was carried out in the fields, mainly barley, wheat, peas, radishes and turnips. Rice was also grown, though at this altitude (2,438 m), the crop was rather uncertain. In little gardens by the houses, the inhabitants grew hollyhocks, introduced through Tibet from China. They had also planted *Pinus wallichiana*, a recent introduction from nearby Bhutan, alongside peaches, walnuts and weeping willows. Soon after setting up his tent, Hooker was visited by the Lachung Phipun, the village headman, who greeted him, bringing the usual present, and requesting he visit his sick father.

Just beneath the village of Lachung, on the mossy forest floor, by the banks of the Lachung River, Hooker found the diminutive *Rubus fockeanus*. It is photographed here in a similar setting near the Jele Dzong in western Bhutan.

Hooker's sketch looking south over the village of Lachung, then a cluster of traditional Bhutia houses sitting on a gravelly plain on the left bank of the Lachung River.

The Tangkra La and the Chumbi valley

The Sentam Subah was also laid up with illness, as a consequence of his many leech bites, so while waiting for him to recover, Hooker set off to explore the Tangkra La, the pass leading into the Chumbi valley in Tibet. It was said to have the deepest snows of any of the Sikkim passes, and Hooker was to find this true. His route would also take him to the Kanko La, a minor pass on Mount Kanko and the glaciers below the Tangkra La.

He followed the course of the Shargaphyu River, whose banks were luxuriantly clothed in a range of herbaceous plants and showy shrubs. *Clematis, Thalictrum, Anemone*, a scandent species of *Aconitum, Berberis, Deutzia, Philadelphus, Habenaria, Fritillaria, Aster, Euphorbia, Hypericum, Halenia, Codonopsis, Enkiantus, Sorbus, Prunus, Pieris* and *Daphyniphyllum* were among the many showy genera he collected along the way. Large monkeys inhabited the edges of fir and hemlock forest, as did the shy red panda.

The forests there contained some remarkable giants. One of the trees of *Tsuga dumosa* Hooker measured was a whopping 8.5 m in girth and over 36 m high; and as if that wasn't impressive enough, the fat barrel-like trunk of a Himalayan fir, *Abies densa*, measured an astounding 11 m in girth, with an unbranched trunk for 12 m. Sadly, no trees of a comparable size survive in Sikkim today, though in neighbouring west Bhutan there remain some mighty hemlocks.

Emerging from the forest, he and his team crossed some vast moraines into a shallow mountain valley choked with rhododendron scrub. Magnificent gentians grew there, with *Aconitum luridum*, a new species whose root was as virulent as the closely allied *A. ferox*, whose roots supplied poisons for the spear tips of forest-dwelling groups in the Himalaya. He was surprised to stumble across begonias at this altitude, plants normally confined to hothouses in the gardens of Western Europe. He had, in fact, found a new species, *Begonia sikkimensis*, a lovely plant with deeply serrated foliage that's wonderfully splashed with silver and brown, and further enhanced with distinctive red veins. It is also relatively hardy and grows out of doors in a number of Irish woodland gardens.

This was a cold, exposed landscape, though Hooker was a well-seasoned traveller, and fairly immune to low temperatures. On this particular trip he had no tent and slept beneath an enormous boulder, as the rain poured down. The following morning, he trekked with his men through the upper reaches of the valley, where the river had cut a narrow, deep gorge, with the snowy peak of Tangkra in the near distance.

He was back, once again, in an alpine landscape, and saxifrages grew there in cushion-like tufts, smothered in tiny golden-yellow blossoms, with bladder-headed *Saussurea*, whose putrid-smelling flowers were enclosed by inflated bracts, protecting them from the cold, searing mountain winds. Sheets of *Primula* also gave a spectacular display alongside golden potentillas and *Pleurospermopsis sikkimensis*, a carrot relative, carrying masses of ripe seeds on broad, handsome umbels.

The village of Lachung in May 2015. It has changed little since 1849 and many of the features from Hooker's sketch are easily recognised. The river bed of the Lachung River lies below the low cliffs in the centre.

One of his most charming discoveries from this region was *Saxifraga palpebrata*, a tiny species bearing solitary golden-yellow flowers above tight rosettes of foliage. Even tinier was the gorgeous cushion-forming *S. stella-aurea*, whose golden blossoms were held on slim stems above dense hummocks of foliage.

Sikkim is considered the megadiversity centre for louseworts, and throughout his travels there Hooker found a multitude of them. On the Kanko La he discovered the tiny white-flowered *Pedicularis albiflora*, one of the very few white-flowering species.

The cremanthodiums are another group often sought by alpine enthusiasts visiting the Himalaya. Essentially a group of alpine daisies, they have a refined, graceful habit, with lovely pendent bell-like flowers in a range of beautiful colours. On the Kanko La he discovered no fewer than three species, naming the finest of these *Cremanthodium thomsonii*, with gorgeous rose-pink flowers that bow their heads to avoid the monsoon rains, for his friend and fellow explorer Thomas Thomson. *C. palmatum*, another of his discoveries, is just as beautiful with graceful purplish-red nodding blossoms, and the final species in this trio was *C. pinnatifidum*, which bore yellow flowers above deeply lobed foliage.

At this altitude the woody flora was reduced to thickets of very dwarf shrubs, the most charming being *Diplarche pauciflora*, a tiny heather-like bush reaching no more than 7 cm tall. At flowering time this little alpine shrub is smothered in tiny white, pink-tinged flowers. Through it trailed the delicate prostrate stems of *Polygonum nummularifolium*, and another of his finds, the sedum-like *Rhodiola fastigiata*, a low-growing handsome perennial that is gaining popularity with keen gardeners in Britain and Ireland.

Hooker's field sketch of the village at Lachung was later reproduced as a wood engraving in *Himalayan Journals*.

In the damp alpine turf, he made further discoveries. This habitat was perfect for the fleshy golden saxifrage, *Chrysosplenium carnosum*, though Hooker must have been distracted by no fewer than three alpine primulas, all of which were later to prove new to science. All were tiny, though delightful, especially *Primula klattii*, a creeping species with nodding cup-shaped lavender-blue blossoms borne on wiry stems. In the Mount Everest region there is a white-flowered form, though the plant resists all attempts at cultivation, and so primula aficionados must travel to the highest passes in the Himalaya to see this jewel. Even tinier was *P. muscoides*, a delightful little species with minute solitary lilac-pink blossoms above moss-like foliage.

The most beautiful of the three, however, was *P. soldanelloides*, a plant that makes the most avid alpine enthusiast swoon. In the Himalaya this species has a range from central Nepal, east to Sikkim, and across the border into the Chumbi valley in Tibet, before meeting its most eastern limit of distribution in western Bhutan. Nowhere across that area is it common, growing on rock faces and cliff ledges at 4,300–5,000 m. It is spectacular, a tiny creeping plant, in some places forming considerable patches and smothered with nodding, white funnel-shaped blooms, reminiscent of another alpine, the European *Soldanella alpina*.

The ascent to the pass (as is often the case) was steep and difficult, up a stony valley boxed in by cliff-like walls. Above 4,572 m, Hooker and his men finally left extensive snowbeds behind them and began to climb along a glacier for several miles. A gently sloping saddle, absolutely bare of snow, succeeded the glacier, and formed the pass, which by boiling water, Hooker reckoned to be 4,919 m. Nothing could be more different from the two slopes approaching the pass, that from Sikkim being precipitous and snowy, while the Tibetan side was a steep broad grassy valley with not a particle of snow, and yaks grazing around a lake just 305 m below. This, of course, was owing to the southerly winds that ascend the Tibetan valley being drained of their moisture by intervening mountains before they reached the pass, while in Sikkim warm currents carried a mass of moisture up the Teesta and Lachung valleys.

Pagri (Phari), one of the highest towns in the world at 4,300 m, lay at the head of the Chumbi valley in Tibet, a two days' trek to the east from Hooker's position. Next to Darjeeling, Pagri was the greatest Tibetan, Bhutanese, Sikkimese and Indian entrepôt along the whole Himalayan range, east of Nepal. Several other mountain passes led over the eastern frontier of Sikkim, into the Chumbi valley, a high region, bounded on the Sikkim side by Pauhunri and the Chola

Snowy peaks on the approach to the Tangkra La. One of Hooker's many unpublished field sketches preserved in the archives at the Royal Botanic Gardens, Kew.

Rhodiola fastigiata (formerly *Sedum fastigiatum*), one of Joseph Hooker's Lachung valley discoveries. It is still common there. This plant grows at Hunting Brook Gardens in Co. Wicklow, Ireland.

range, and on its eastern side, by the mountains of Bhutan. By mule, Lhasa, the ultimate goal of many a Western traveller, lay just fifteen days away. How Hooker's heart must have raced at the thought.

A few plants grew among the stones on the pass, and these were quite different from those he had collected on the Kongra La and on the Phalung Plains. The showiest plants were a pink-flowered *Arenaria*, two species of *Corydalis*, a cottony *Saussurea* and diminutive primroses. Of the *Corydalis* species, one proved new: the sensational *C. latiflora*, a true alpine, its finely dissected bronze-flushed grey foliage pushing through the surrounding screes, capped by smoky-blue fragrant blossoms. At this point even buttercups were reduced to lilliputian species, and on the pass itself Hooker discovered the minuscule *Ranunculus sikkimensis*

Returning by way of the glacier, Hooker stayed for a further two days, sleeping by night beneath an enormous rock overlooking the valley. The night was misty and miserably cold; his fire was of wet rhododendron wood, especially the dwarf aromatic species, *R. anthopogon* and *R. setosum*, which, being full of resinous glands, blazed with fury. The same dwarf rhododendron thickets provided shelter for another of his many discoveries, *Pleurospermum hookeri*.

It was a desolate place, but commanded incomparable views across the Lachen and Lachung valleys, and the massive Kangchenjunga massif, which proved to be valuable for geographical and map-making purposes. He returned the following morning to Lachung, where he remained for the next few days botanising.

During his stay in the village, he was woken at night several times by screaming voices, groans, the beating of drums, firing of guns and flaming torches gleaming through the trees, as figures flitted from house to house. These were night-time exorcisms, he was told, intended to drive out demons, who stole from the village during their frequent visits.

On the higher peaks above the settlement, Hooker discovered a wide range of subalpine flowers on exposed cliff faces and ledges, and in the high-lying pastures. In the latter he found gems such as the delightful *Cyananthus inflatus*, with its lovely mid-blue blossoms, perched above a curiously 'inflated' calyx. Higher still, he saw *Primula glabra*, a tiny, though showy little species bearing drumstick-like umbels of pinkish-purple to bluish-violet blossoms with a central white eye.

Lachung is also the type locality for Hooker's rockfoil, *Saxifraga hookeri*, a pretty yellow-flowered species, that is found in the subalpine forests of Sikkim, Nepal, E Tibet and Bhutan, where it generally prefers damp habitats and grows on the banks of rivers and streams.

He also collected specimens of *Aconitum lethale*, one of Griffith's 1836 discoveries from the Mishmi Hills of Arunachal Pradesh in north-east India. Its name leaves little to the imagination, or anyone in doubt of its virulence, and according to Griffith, this was the source of the celebrated 'bhi' or 'bis' poison of the Mishmi people. On the moraines above the village he found a second new species, *A. elwesii*, a slender climber, which was also later collected in Sikkim by its namesake, Sir Henry Elwes.

The Yumthang valley

On 29 August, Joseph Hooker set off to the Yumthang (Yeumtong) valley, a northerly extension of the Lachung valley. The road ran along the river terrace, across extensive beds of unstratified gravel and sand, and from that point through fir, juniper and hemlock forest. Above there the valley contracted into a narrow gorge, and landslides and old glacial moraines covered its floor. The broad flats were richly clothed with rhododendrons.

Yumthang was a small summer yak-station by the Lachung River at an altitude of 3,633 m. It took just two days to reach, and reminded Hooker of the Swiss valleys. The west flank had precipitous cliff-like walls, and narrow gullies still choked with the previous winter's snow at 3,810 m. The eastern side gradually sloped up towards Pauhunri and its snowy peaks. *Abies densa* ascended as far as 3,962 m, above which it was replaced by *Juniperus recurva*, which formed sizeable trees over 18 m high.

Hot springs occurred near the banks of the river, and were popularly used as therapeutic baths, the patient remaining in them for three days, only taking occasional breaks to eat in a shed close by. The temperature at their source was a pleasant 45°C , whereas that of the Lachung River, just 9 m away was a chilly 8°C.

Hooker had once again reached the alpine Himalaya, and on the glacial moraines and steep slopes made a rich collection of plants, including two discovered by the French collector Victor Jacquemont. Like William Griffith, Jacquemont was a brilliant botanist of astounding ability, but his life was cut short in 1832 when he died of cholera in Mumbai (then Bombay) aged just 31. He collected widely in the western Himalaya, particularly for the Jardin des Plantes in Paris, and achieved more in his short life than most accomplished botanists would wish for in an entire lifetime. On 6 September 1849 Hooker collected material of *Lonicera purpurascens,* which Jacquemont had found in Kashmir many years previously. The alpine grass *Poa pagophila* was described at Kew, a century after Hooker collected it, from collections he made at Yumthang at 4,267 m and from those made by Jacquemont in western Tibet.

While the latter are really only of botanical and historical interest, Hooker was to find dozens of showy new plants, including dense thickets of *Spiraea arcuata*, a low-growing shrub with arching sprays of pink blossoms in mid-summer. It is common on the highest mountains of the E Himalaya, Myanmar and W China.

In scrub at the edge of alpine forest he found *Codonopsis subsimplex*, a delightful perennial carrying pendulous, bell-shaped blossoms on wiry, thread-like stems. In Yumthang the predominant

The Yumthang valley above Lachung. The scenery reminded Hooker of Switzerland. The Lachung River, a furious torrent, is at this point spanned by a narrow pedestrian bridge. The surrounding flora is amazingly diverse.

Juniperus recurva is commonly found in the higher valleys of Sikkim. Here, Kristin Jameson puts a tree in scale in the upper Lachen valley. In Ireland there are good cultivated specimens at Birr Castle and Kilmacurragh.

a perfect habitat for the cushion-forming, yellow-flowered *Saxifraga aristulata*, which would later be collected by other plant hunters in the mountains of Tibet, Nepal and Bhutan.

The Sentam Subah had plodded along in a foul temper since leaving Chungthang, and once again Hooker found it virtually impossible to purchase supplies. Matters were made worse by the fact that Hooker had reprimanded him for sanctioning the flogging of one of his more civil and industrious men, a huntsman from Chungthang. Neither had he managed to find much for Hooker, except a large monkey, that almost bit the head off his best dog.

Yume Samdong and the Dongkya La

The Lachung Phipun visited on 7 September; he had crossed the border to see what Tibetan people felt of Hooker's intended journey to the Dongkya La (which then marked the border with Sikkim), and finding them quite indifferent, offered to be a guide. Luckily, a month's provisions had just arrived from Darjeeling, and the following day Hooker set out for Yume Samdong (Momay Samdong), the loftiest yak-grazing station in all of Sikkim, just a day's trek from the Dongkya pass.

Higher up the valley, below the glaciers of Changmekang, the ancient moraines were impressive, reaching prodigious heights as they piled on one another, creating a chaotic landscape clothed in places by scattered, stunted firs. Above 4,267 m even shrubs ceased. There the valley opened out into a bleak, stony plateau, with high mountains on either side. Yume Samdong was just 10 km SE of Kangchengyao, and its rounded dome-shape summit, capped with a heavy layer of snow dominated the scene. At this point the Lachung was a stream just 11 m wide, cutting a shallow channel through the gravel deposits to the bedrock beneath. Hooker recorded the altitude of Yume Samdong as 4,682 m.

colour of the corolla is blue, though in E Sikkim, at the Kyongnosla Alpine Sanctuary, there is a stunning yellow-flowered form.

Gentians, so evocative of flowery alpine valleys, appeared in great numbers, including two new species: *Gentiana infelix*, a tiny species never reaching much more than 5 cm tall, with dark-blue trumpet-like blossoms in autumn, and the showier *G. elwesii*, a tufted perennial spreading by means of stolons and bearing clusters of urn-shaped, bluish-purple flowers in September. It is still common in the Yumthang valley, and in the upper Lachen valley, where it was also collected by Elwes in the 1870s.

The minuscule *Primula tenuiloba* was one of Hooker's most stunning discoveries from the Himalaya. In the highest valleys of Sikkim, it occurs between 4,500 and 5,800 m, on rocky hillsides and mossy boulders, forming extensive cushions of foliage, above which rise glorious violet-blue blossoms with curious y-shaped petals, giving the flowers a frilly appearance.

In the dry, boulder-strewn screes he also found the stonecrop relatives *Rhodiola humilis* and *R. crenulata*, both short, multi-stemmed perennials bearing terminal clusters of red flowers. *R. crenulata* is used in the Himalaya to treat acute mountain sickness, though its effectiveness has recently been disputed. The same screes provided

Yume Samdong lies just a few miles from the Tibetan border. Hooker must have been greatly excited at the prospect of being so close to the frontier. The Lachung River at this point breaks its way through a field of ice and snow.

The Lachung River at Yume Samdong. Here it is little more than a fast-flowing alpine stream. An interesting flora is found on the surrounding mountain ridges; many are Hooker plants.

That evening he pitched his tent close to a small shed, by the gently sloping base of a mountain that divided the Lachung River from a western tributary. His men occupied a hut nearby, just vacated by a party of young Tibetan girls, who moved on with their yaks to fresh grazing ground.

It was a wild, exposed, dramatic landscape with razor-edged mountain ridges, stupendous precipices, distant snowy peaks, moraines, glaciers and rounded boulders and rocks. There wasn't a bush over 15 cm high, the few woody plants consisting of minute creeping willows, dwarf rhododendrons, prostrate junipers and *Ephedra*.

Hooker was the first Western botanist to reach this elevated region, so it is no surprise that the bulk of his collection proved new to science when later studied at Kew and other scientific institutions in Europe. One of his most charming finds was the annual *Microcaryum pygmaeum*, a borage relative, forming tiny ground-hugging rosettes covered in sapphire-like blossoms set on long dark-purple peduncles. The gravelly slopes were covered in sheets of the gentian-relative *Jaeschkea microsperma*, another tiny annual that paints the higher alpine valleys of the eastern Himalaya purple-blue in autumn.

Many of the grasses proved new too, including *Calamagrostis nivicola*, a pretty little alpine. When naming it in 1896, Hooker chose the epithet *nivicola*, meaning growing near snow. When our group visited Yume Samdong we found the region buried beneath a heavy carpet of snow.

Aconitum hookeri, an exquisite little plant, reached no more than 10 cm high, bearing individual blue (sometimes purple-blue) bonnet-like blossoms above finely dissected foliage. However lovely Hooker's monkshood may be, it could in no way surpass the beauty of his finest alpine discovery at Yume Samdong, *Silene nigrescens*, a stunning little perennial carrying masses of solitary lantern-like blooms in autumn. The individual flowers are like little inflated pendent, lime-green balloons, heavily striped with dark purple longitudinal veins. It grows across much of the E Himalaya and the mountains of W China, where I first saw it over 20 years ago growing on limestone clefts in NW Yunnan. It is a beautiful perennial, whose flowers have evolved to withstand the monsoon and are a lovely sight when seen blowing on the wind and bedewed by mountain mists.

To the east, tributaries fell into the Lachung River, having originated near the enormous Tista Glacier that descended from Pauhunri. A girdle of snow-clad mountain peaks swept around to enclose the headwaters of the Lachung, which found its source in the innumerable streams arising from the surrounding glaciers, lakes and perpetual snows. Hooker was mightily pleased with his progress. Despite the obstacles he was on the very borderlands of Tibet and this journey would prove important in completing his map of Sikkim, a map that would be of immense importance over the following 50 years, and repay the Admiralty every penny they had invested in Hooker's travels:

> The spot appearing highly favourable for observations, I determined to remain here during the equinoctial month, and put my people on "two-thirds allowance," i.e., four pounds of rice daily for three men, allowing them to send down the valley to cater for what more they could get. The Singtam Soubah was intensely disgusted with my determination: he accompanied me next day to the pass, and having exhausted his persuasions, threats, and warnings about snow, wind, robbers, starvation and Cheen sepoys, departed on the 12th for Yeumtong, leaving me truly happy for the first time since quitting Dorjiling. I now had a prospect of uninterruptedly following up my pursuits at an elevation little below that of the summit of Mont Blanc, surrounded by the loftiest mountain, and perhaps the vastest glaciers on the globe; my instruments were in perfect order, and I saw around me a curious and varied flora.[2]

A double rainbow, perhaps an omen of good luck, spanned the easterly horizon as Hooker and his men set off towards the Dongkya La on the morning of 9 September 1849. Following the Dongkya River (an extension of the Lachung), they headed in a north-easterly direction along the lower slopes of Pauhunri (7,125 m). Shallow valleys, choked higher up by glaciers, descended the rearward mountains, in which there were several lakes. In the distance a broad opening to the west led to Gurudongmar (6,715 m), the easterly peak of Kangchengyao.

At 5,334 m he passed several shallow lakes, and beyond there lay his final obstacle: a steep rocky crest, 152 m high, rising between two precipitous snowy peaks, a final exhausting ascent before reaching the pass at a numbing 5,533 m. Some stone cairns, decorated with prayer flags, had been erected there, and Hooker found a mani stone, inscribed with the mantra *'Om mani padme hum'*, which he took away as a memento of his climb. No snow lay on the pass itself, though glaciers descended its northern slopes to 5,182 m, and it was exactly at that dizzying altitude that he found *Saussurea hookeri*.

Adaptating to the extreme conditions, many plants formed great cushion-like mounds; these included the aptly named, yellow-flowered, *Saxifraga hemisphaerica*. *Cortiella hookeri* was another of Hooker's discoveries here, a stunning short-stemmed perennial, with flat, compact rosettes of finely dissected foliage, closely appressed to the gravelly surface, surrounding a dense, flattened umbel of white, purple-tinted flowers.

On this high mountain pass, plants were absolutely dwarfed by their environment. *Pedicularis daltonii*, named for Hooker in *The Journal of the Asiatic Society of Bengal* in 1889, is one of the tiniest of the Himalayan louseworts. On river banks, in wet gravel, he found the tiny *Juncus minimus*, growing alongside *Polygonum hookeri* and the gorgeous *Gentiana emodii*, the latter then smothered in lovely blue-purple, trumpet-like flowers. Hooker's specimens of this gentian, beautifully pressed, and preserved in the herbarium at the Royal Botanic Gardens, Kew, still display their rose-purple, blue-striped blossoms to this day.

The summit of the forked Dongkya, a pass leading into the forbidden plateau of Tibet: an irresistible temptation to Joseph Hooker who threw caution to the wind and made an unauthorised entry into this mysterious land.

A modern-day view of the forked Dongkya. Hooker employed a measure of artistic licence, though his field sketches of Sikkim are generally very accurate.

His most exciting discovery on the Dongkya La, however, was the pale blue-flowered *Corydalis ecristata*, a species he initially confused with *C. cashmeriana*, though the Sikkim plant is far shorter, and more alpine in all respects than its cousin from Kashmir. The latter is also sometimes confused with yet another of Hooker's discoveries, *C. casimiriana*, from the Lachung valley. The tiny *Saussurea donkiah* was named after its type locality 151 years after Hooker collected it on the Dongkya La; no doubt, new species still lurk in Hooker's Sikkim collections.

Tibet and the Cholamu Lake

During this visit to the pass, the views were all but obscured, except on the Tibetan side where Hooker obtained a magnificent panoramic sweep of Ü-Tsang Province. Just 457 m (1,500 ft) below, lay the Cholamu Lake, a substantial sheet of blue water, hemmed in by jagged, snow-clad mountain spurs. Cholamu lay in a broad, scantily grassed, stony valley, and towards its head, snowbeds and sheets of ice dipped sharply towards the lake, whereas on its west side, a lofty brick-red coloured spur rose from its shores. The closest Tibetan village of any significance to the lake was Kampa Dzong (Kambajong). It lay on the banks of the valley of the Arun River, on the road from Sikkim to Shigatse.

Hooker marked all these places on his map, and over half a century later, in June 1903, Britain's ultimate imperial adventurer, Lt Col Francis Younghusband (1863–1942) travelled to the village's remarkable fortress from the Nathu La in Sikkim, while on a 'diplomatic mission' to Tibet. Needless to say, he followed Hooker's maps, and it is hard to underestimate the value of his cartographic work during his visits to the frontier regions. His map was published by the Trigonometrical Survey Office of Calcutta, and for the next century it was to become the standard map of Sikkim.

Years later, in November 1873, Hooker wrote to his friend Charles Darwin, 'I am in a state of temporary inflation – a book just published on the military operations in Sikkim says of my travels: "never was the officer commanding a force favoured with a fuller, more able, or more lucid report of a country and its inhabitants than I was by the study of Dr. Hooker".'[3]

Hooker must have been astounded, when in 1903, at the grand old age of 86, he received a telegram from British officers of the Sikkim–Tibet Boundary Commission, congratulating him on his map:

Khambajong, Thibet: 'Major Prain, Colonel Younghusband and officers of the Thibet Mission desire to send you their felicitations by telegraph from Khambajong and express their high admiration of that zeal displayed by you fifty-five years ago, which has enabled them to follow in your steps and has inspired them to emulate your devotion to science and to your country.'[4]

Major, later Sir David Prain (1857–1944), was then Director of the Royal Botanic Garden, Calcutta, and in 1905 was appointed to the role of Director of Kew. Younghusband visited Hooker during the Christmas period of 1904. Hooker, then elderly, was still greatly respected as a pioneering explorer and Younghusband was highly amused to see the telegram framed and hung on a wall in Hooker's house.

The Lachen River rose at this point, flowing from the lake's northerly extremity before turning westwards and running through a barren, desert-like valley. Beyond there it passed between the red, stony mountain of Bamcho (Bhomtso) and the northern slopes of Kangchengyao, before running south towards Thangu and Lachen, joining the Teesta, and then the Ganges.

To the north of Bamcho lay the broad valley of the Arun River and in the extreme north-west of Hooker's view some immense snowy peaks, reduced to mere specks on the horizon. Hooker sketched the scene, not realising he had produced yet another pioneering depiction of Mount Everest.

A view of Tibet and the Cholamu Lake from Hooker's vantage point at 5,639 m. Several stone cairns, decorated with prayer flags, mark the Dongkya pass. The Lachen River arises in the Cholamu Lake and may be seen exiting from the broad sheet of water. The tallest peak in the distance is thought to be Mount Everest. This is W. H. Fitch's watercolour of Hooker's original sketch.

The archives at the Royal Botanic Gardens, Kew preserve several of Joseph Hooker's Sikkim notebooks. They are tiny and very fragile. Some contain wonderful sketches like this view of the far-distant mountains of Tibet and the ink-blue waters of the Cholamu Lake as seen from the Dongkya La.

The river valley was bounded to the north by black, precipitous, snow-capped mountains, and behind lay other ranges including the Qang La (Kiang-La), which formed the watershed of this meridian. Hooker estimated the mean elevation of this district to be 6,069 m, and in salt lakes beyond there nomadic traders dug salt from the edges of the lakes.

This contradicted the common belief that beyond the Dongkya pass the landscape of Tibet sloped back in descending steps to the Tsangpo River, and was more or less a plain. Had he trusted his eyes only Hooker might have believed this assumption. However, his instruments, particularly his theodolite, recorded a vastly high country of plateaus and massive mountains.

Hooker was greatly impressed by this landscape and its colours, which contrasted so brilliantly with the black, rugged heavily snowed peaks of the Sikkim Himalaya. All the mountains between the Dongkya La in Sikkim and the Arun in Tibet were relatively gently sloped, and of a yellow-red hue, rising and falling in long undulations, like dunes, and bare of permanent snow or glaciers.

The upper slopes of Pauhunri rose above the pass, and although his view of the summit was obscured he continued his ascent along the slopes before finally reaching 5,883 m, where he found enormous sheets of blue-ribboned ice capping the ridges.

It was at this elevation, as he hung over a tremendous precipice, that he witnessed the 'Brocken spectre', an atmospheric phenomenon in which the magnified shadow of the observer, looking down from a ridge or peak, with the sun shining from behind, is cast against the upper surface of clouds. The magnification of the shadow is an optical illusion and the head in the shadow is often surrounded by a glowing halo of different coloured rings of light.

A violent southerly wind constantly blew over the pass, often carrying snow. Hooker and his men camped on the north face, bringing fuel with them for campfires. He gathered a pink-flowered woolly *Saussurea* and *Delphinium glaciale*, both common at altitudes

between 5,334 and 5,486 m, and lichens, one of his life-long interests, were quite common near the summit, including *Thamnolia vermicularis*, a brilliantly silver species found around the world at high altitudes. It is particularly showy and its specific epithet *vermicularis*, or 'worm-like', is very apt. He also found the tiny orange-coloured *Lecidea miniata (Lecanora miniata)*, a species he had previously collected in Antarctica:

> To-day I went up the flanks of Donkiah to 19,300 feet, amongst the knot of snowy peaks west of Chumulari, and such gulfs, craters, plains, and mountains of snow are surely nowhere else to be found without the polar circles. Of course I have seen nothing to compare for mass and continuity with Victoria Land, but the mountains, especially Kinchin-jhow, are beyond all description beautiful; from whichever side you view this latter mountain, it is a castle of pure blue glacial ice, 4000 feet high and 6 or 8 miles long. I do wish I were not the only person who has ever seen it or dwelt among its wonders. Now I have been N., S., E., and W. of it, up it, down it, to 16,000, 17,000 and 18,000 feet; and every view enchants me more than another.
>
> I was greatly pleased with finding my most Antarctic plant, *Lecanora miniata*, at the top of the pass, and to-day I saw stony hills at 19,000 feet stained wholly orange-red with it, exactly as the rocks of Cockburn Island were in 64° South, is not this most curious and interesting? To find the identical plant forming the only vegetation at the two extreme limits of vegetable life is always interesting; but to find it absolutely in both instances painting a landscape, so as to render its colour conspicuous in each case five miles off, is wonderful.[5]

From the Dongkya pass, Hooker and his men returned to their campsite at Yume Samdong, reaching his tent as darkness fell. He complained that when the dogs were let loose at night they 'howled dismally'. Little has changed in the intervening years: modern-day visitors to both Sikkim and Bhutan will have had a similar experience. The yaks were equally bad sleepers and Hooker was woken on several occasions when they pushed their heads into his tent and announced their presence with snorts and blasts of their hot, moist breaths. Yume Samdong lay on a relatively busy route to the Tibetan passes and several groups passed by Hooker's campsite, headed for Tibet:

Throughout September various parties passed my tent at Momay, generally Lamas or traders: the former, wrapped in blankets, wearing scarlet and gilt mitres, usually rode grunting yaks, which were sometimes led by a slave-boy or a mahogany-faced nun, with a broad yellow sheep-skin cap with flaps over her ears, short petticoats, and striped boots. The domestic utensils, pots, pans, and bamboos of butter, tea-churn, bellows, stools, books and sacred implements, usually hung rattling on all sides of his holiness, and a sumpter yak carried the tents and mats for sleeping. On several occasions large parties of traders, with thirty or forty yaks laden with planks, passed, and occasionally a shepherd with Tibet sheep, goats and ponies.[6]

On 18 September 1849 Hooker and his men climbed the lofty mountain range that divides the Lachung valley from the neighbouring Lachen valley, to the Sebu La (Sebolah), a mountain pass leading into the southern ranges. A yak track led across the glaciers beneath Kangchengyao, along the shores of Sebu Lake and then westwards along a steep spur, with sheets of glacial ice and snow. Few plants grew there above 4,877 m. He ultimately reached 5,360 m and his party began to suffer from the effects of altitude.

The view from the summit of the ridge was dominated by the precipitous face of Kangchengyao, the soaring cliffs of Pauhunri to the east, the grassy dunes on the Phalung Plains and of course, by the mass of Kangchenjunga in the distance. By then the Phalung Plains were yellow with withered grass and the yaks were gone, but the black tents of the Tibetans remained.

Yaks in the Yumthang valley below Yume Samdong. They made nocturnal visits to Hooker's tent when he camped at Yume Samdong in 1849.

Ever the inveterate explorer, Hooker crossed the Sebu La and from a nearby ridge gained sweeping views of Pauhunri, the Phalung Plains and the immense snow-capped mass of Kangchengyao. The latter is seen here from Yumthang in November 2012. At the base of its cliff face he trekked across glaciers. To the right is Kangchenyao's jagged east top, of which he made a superb sketch.

Having completed his alpine travels and ascended the Tista Glacier, Hooker descended the Lachung valley for Chungthang where he met Archibald Campbell. Their plan was to make a circuitous journey through the Lachen valley, cross into Tibet from the Kongra La, and return to Sikkim across the Dongkya La. The scene above would have been familiar to him: the upper Lachen valley with traditional Bhutia houses.

Two days later, Hooker ascended the Tista Glacier (he called it the Donkia Glaciers) to the east of his base at Yume Samdong. This is the most extensive glacier in NE Sikkim and lies along the immense upper slopes of Pauhunri. Between 4,877 and 5,182 m it contained four marshy lakes, alternating with several transverse moraines that had dammed the river. Below Pauhunri, the valley widened and a terminal moraine rose like a mountain to a staggering 244 m high.

From the summit, the scene was striking: a vast river of ice filling an immense basin several kilometres broad and long, contained by serried mountain ranges, all over 6,096 m, separating large tributary glaciers. Hooker walked for over 1.6 km across the glacier, until he reached the mountain's red cliffs at an altitude of 5,668 m. By 3 p.m. temperatures plummeted to 1°C and it began to snow heavily. At these heights the weather was never fine for more than an hour at a time, and relentless sleet, followed by heavy snow, drove down on him during both his visits.

On 26 September he climbed to 5,7921 m onto a scarped precipice that fell sheer for 914 m. He had been attracted to the cliff area by its bright orange-red colour, which he found had been caused by peroxide of iron. He made several sketches of this wild landscape.

Two days later the Sentam Subah travelled up the valley from Yumthang, requesting leave to return to his wife, who was apparently

At 5,486 m (18,000 ft) Joseph Hooker stopped to sketch the view of the east top of Kangchengyao, and of Tibet, which appears in the distance. Tibetan argali, *Ovis ammon* subsp. *hodgsoni*, appear in the foreground. This peak appears on the right-hand side of the image opposite.

ill, but also to inform Hooker that Archibald Campbell had left Darjeeling with the Chebu Lama, on the authorisation of the Sikkim Rajah. Hooker left to meet him at Chungthang, stopping at Yumthang on the way.

By then the deciduous trees of the Yumthang valley were displaying beautiful autumn colour, and at Lachung the following day *Larix griffithii* was shedding its rust-coloured needles, though many perennials were still in bloom, including the blue-flowered climbing gentian, *Crawfurdia speciosa*.

Hooker reached the cane-bridge crossing the Lachung River into Chungthang on the second day of October 1849, pausing to remember the sad loss of his dog; he reached his camping ground by the river at 10 p.m., after an exhausting twelve-hour trek. The porters arrived two hours later with his tent and bed, though not before leeches and mosquitoes had made a savage attack.

Chungthang, was, and still is, landslide country, and next morning Hooker found the valley buried beneath a torrent of mud and boulders, and for some way the Teesta River was dammed into a deep lake. Just beneath the landslide he met an exhausted Campbell, looking the worse for wear, having travelled through the dangerously unhealthy, hot, humid Teesta valley. There he had almost been killed when the pony he was riding had been dashed to pieces when it fell over a precipice into the river.

According to Hooker, Campbell was visiting Sikkim to establish better relations with the Rajah and his court. Hooker believed that his treatment, and the behaviour of the Sikkim court towards the representatives of the British East India Company, was a violation of good faith on the part of the government of Sikkim (i.e. the Dewan), towards the Company, for which the Rajah, however helpless, was yet responsible.

Campbell, meanwhile, claimed to have been given authorisation from the Deputy-Governor of Bengal (Lord Dalhousie was absent at the time) to visit the area to enquire personally into the matter, and to familiarise himself with the kingdom, which by Treaty, the British East India Company was bound to protect, but from which they had been so forcibly excluded.

An unauthorised visit to Tibet

Diplomacy, however, did not seem to enter into Campbell's plan. On 6 October he and Hooker left directly for the Kongra La, planning a circuitous route around Cholamu Lake and re-entering Sikkim over the Dongkya La. Hooker claimed the frontier area was uninhabited, so the Chebu Lama saw no difficulties with their travelling through the area, providing the Lachen Phipun and the Tibetans did not object. The Sentam Subah, who had happy plans of returning to nurse his wife in the warm foothills, must have been appalled to learn that he had to travel once more into the cold, high border regions. Hooker claimed that the Rajah had ordered the poor Subah to travel with them, but the latter was furious and didn't try to hide his anger from Campbell, to whom he was sulky and rude.

As they retraced their route back up the Lachen valley they found the village of Lachen totally deserted, as its inhabitants were still grazing their herds higher up, though crops of buckwheat and turnips were maturing near their houses.

Hooker's depiction of Tibetan man with his prayer wheel. Prayer wheels are still widely in use in Tibetan areas of the Himalaya.

Their journey meant retracing the route Hooker had taken earlier that summer, though by then autumn colour had transformed the valley, and at Zema he feasted on the large, succulent scarlet hips of *Rosa macrophylla*. *Berberis*, which was abundant on the stony hillside, had turned scarlet, giving a warm glow to the slopes above the tree-line.

Hooker generally purchased goats and kids from local farmers to supplement food rations, and it was in this part of the valley that he lost several from their grazing on the foliage of *Rhododendron cinnabarinum*, one of the most poisonous of all the species growing in Sikkim. The Lepchas knew it as *kema kechung* (*kema* meaning *Rhododendron*). When used for fuel, it caused the face and eyes to become swollen and inflamed, and Hooker was to find his poor goats and kids foaming at their mouths and grinding their teeth. We know today that *R. cinnabarinum* contains the compound andromedotoxin, a serious cardiac poison, and honey produced by bees that have visited it is also very poisonous.

On the morning of 16 October 1849, the sun rose over Hooker's campsite near the Kongra La at 7.30 a.m. He and Campbell set off with the Chebu Lama to hold a parley with the Tibetans, instructing the porters and the rest of the team to follow at their leisure. As the trio approached the pass Tibetan soldiers appeared on the scene, Hooker panicked, diplomacy was forgotten, and he bolted, quite illegally, and rather arrogantly, over the pass and into Tibet:

> We had not proceeded far when we were joined by two Tibetan Sepoys, who, on reaching the pass, bellowed lustily for their companions; when Campbell and the Lama drew up at the chait of Kongra Lama, and announced his wish to confer with their commandant.
>
> My anxiety was now wound up to a pitch; I saw men with matchlocks emerging from among the rocks under Chomiomo, and despairing of permission being obtained, I goaded my pony with heels and stick, and dashed on up the Lachen valley, resolved to make the best of a splendid day, and not to turn back till I had followed the river to the Cholamoo lakes. The Sepoys followed me a few paces, but running being difficult at 16,000 feet, they soon gave up the chase.[7]

A short ride to the north-east brought him back to the Lachen River, there a stream, running through a broad, open stony valley. Though dismally barren, some low-growing plants did exist in that high-altitude alpine desert on the Tibetan side of the frontier. One of Hooker's letters to Kew, written in Lachung village on 25 October 1849, after the border crossing describes their behaviour:

> What do you think – we spent four days in Thibet! In spite of Chinese guards, Dingpuns, Phipuns, Soubahs, and Sepas. It was a serious undertaking, and required a combination of most favourable incidents, together with my previous acquaintance with the country, and a most indomitable share of resolution and boldness. Campbell has behaved splendidly, and diverted me by throwing all the sage precepts he sent me to the winds … he had not been two days with me before he was storming and bullying right and left. The unfortunate Singtam Soubah, with whom at C[ampbell]'s intercession I had kept such good friends, he gave no peace to, blackened his face, and sent him to the Durbar in disgrace.[8]

The vegetation here consisted of stunted, scrubby *Ephedera*, cushion-forming *Arenaria*, *Artemisia* (including *Artemisia minor*, one of Victor Jacquemont's Himalayan discoveries), *Astragalus*, *Androsace* and the most alpine of rhododendrons, *R. nivale*. Few botanists had ever collected plants on the Tibetan Plateau, and despite his excitement, Hooker gathered and pressed many plants.

Some of his Tibetan collections proved new, including *Lonicera rupicola*, a dense low-growing shrub in that part of Tibet, where its preferred habitat is boulder slopes and screes, hence *rupicola*, meaning 'a dweller among rocks'. It is a showy little shrub, bearing clusters of scented, rose-pink trumpet-shaped flowers in summer.

In his account of this area in *Himalayan Journals*, Hooker mentions finding a *Myosotis*, or forget-me-not, in the surrounding area. In fact, he had found a new genus, *Microula tibetica*, a tiny perennial growing to less than a centimetre tall, with lovely forget-me-not like flowers. It was described in 1876 by George Bentham in his classic work with Hooker, *Genera Plantarum*. Though tiny, this plant is tough, growing on the shores of alpine lakes and on gravel banks in the mountains of Nepal, Tibet and north-east India at altitudes of up to 5,300 m.

Another of his Tibetan discoveries was the glorious *Dracocephalum tanguticum*, a low growing perennial to about 45 cm, which in summer bears masses of blue-purple tubular blossoms, beloved of bees. It was described in 1881 from the collections of the famous Russian soldier and naturalist, Nicolai Mikhailovich Przewalski (1839–1888), from material he found in Gansu, China (hence *tangutica*, from the Tangut people of Gansu). Przewalski's Asian collections were made between 1870 and 1885 however, long after those of Hooker's, who claims precedence as the discoverer of this beautiful perennial.

I have seen it on sun-baked slopes in SE Tibet, where it is truly one of the most beautiful high-altitude perennials. Helen Dillon grew it in her garden in Dublin, from a plant I gave her, though it eventually succumbed to visiting cats, who found it even more delicious than catmint. Speaking of which, the aromatic Asian catmint, *Nepeta coerulescens*, has a similar history. Hooker found this pale blue-flowered perennial on similar dry slopes in Tibet, though it was described by the St Petersburg-based botanist, Carl Ivanovich Maximowicz (1827–1891), from another of Przewalski's Gansu collections.

Behind Kangchengyao, at an altitude of 5,480 m, Hooker came across some marshy flats, and in the waters found some shells belonging to a species of freshwater snail, which proved to be a new species, named, in 1850, by the mollusc expert Lovell Augustus Reeve (1814–1865) as *Orientogalba hookeri* (syn. *Lymnaea hookeri*). It is found today, generally in slow-running freshwater streams, across east Asia from Russia to NE China.

Rounding a great easterly spur of Kangchengyao Hooker sighted the blue waters of Cholamu Lake, with Pauhunri rearing its great precipices of rock and ice to the east. His pony wasn't at all well, nor was he himself; altitude was again taking a heavy toll, and both

Dracocephalum tanguticum, one of Hooker's Tibetan discoveries. Photographed by the author in SE Tibet in the autumn of 1997.

were glad to rest by the rippling waters of the lake. Even then Hooker hadn't grasped the gravity of his incursion into Tibetan territory. He reflected on the events of the day:

> ... and there, with the pleasant sounds of the waters rippling at my feet, I yielded for a few moments to those emotions of gratified ambition which, being unalloyed by selfish considerations for the future, become springs of happiness during the remainder of one's life.
>
> The landscape about Cholamoo lakes was simple in its elements, stern and solemn; and though my solitary situation rendered it doubly impressive to me, I doubt whether the world contains any scene with more sublime associations than this calm sheet of water, 17,000 feet above the sea, with the shadows of mountains 22,000 to 24,000 feet high on its bosom.[9]

Despite the aridity of the landscape, two species of antelope abounded. The largest of these, the Tibetan antelope, *Pantholops hodgsonii*, was commonly known in the Himalaya as '*chiru*'. This handsome creature, named for Hooker's friend Brian Hodgson, is now sadly endangered owing to illegal poaching. Endemic to the Tibetan Plateau, in the 20th century their numbers plummeted from an estimated one million to fewer than 75,000, being hunted for their valuable wool.

Pantholops hodgsonii, the Tibetan antelope or chiru. The images here, reproduced in *Himalayan Journals*, were sketched by Lieutenant Maxwell of the Bengal Artillery from a pair brought to Darjeeling.

The second species, the Tibetan gazelle or goa, *Procapra picticaudata* (named and described by Brian Hodgson), fed on sedges surrounding the shores of the lake, alongside the Tibetan antelope. For such a seemingly arid landscape, mammal life abounded, with tailless rats, slate-coloured hares and marmots, a wonderful contrast to the Sikkim side of the Dongkya pass just 8 km away, where animal life was very rare.

By 1 p.m. there was still no sign of Archibald Campbell. Hooker wondered whether or not he had followed, and began to speculate on the possibility of sleeping in the open, by the warmth of his pony; not a pleasant proposition at this altitude. Two hours later he arrived at Yum Cho (Yeumtso), exhausted and with a pounding headache. Waiting there, to his enormous relief, were the Chebu Lama and the Lachen Phipun, who were dreadfully worried about Hooker's whereabouts and fearful that he could freeze to death on these high lands.

They told Hooker that after a long conversation with the commander or Dingpun of the Tibetan Sepoys, the party had been allowed to pass. The Sepoys had guided his porters, who were close behind, but no one had seen Campbell, and so the Chebu Lama went in search of him. By 5 p.m. the sun had sunk behind Chomoyummo's icy peak, the wind dropped and temperatures sank rapidly. There was no way that Campbell could survive a night in the open, but, as luck would have it, the Chebu Lama found him by some lakes below Kangchengyao, which he had mistaken for the Cholamu Lake, where he presumed he'd find Hooker. Campbell had also thought about passing the night under a rock with his servant; had this happened, they would most likely have died.

Campbell related to Hooker that after he had 'bolted' over the pass, the Dingpun and 20 of his men absolutely forbade the remaining party to cross the frontier, but they eventually persuaded the commander to allow them through. The porters soon caught up. Hooker's men, by then well accustomed to travel at high altitudes were all fine, but Campbell's Nepalese Gorkhas were suffering desperately from altitude sickness. The Tibetan border guards and their commander were the last to arrive at the campsite at Yum Cho (5,123 m), and they made a formidable sight:

> The Tibetan Dingpun and his guard arrived last of all, he was a droll little object, short, fat, deeply marked with small-pox, swarthy, and greasy; he was robed in a green woollen mantle, and was perched on the back of a yak, which also carried his bedding, and cooking utensils, the latter rattling about its flanks, horns, neck, and every point of support: two other yaks bore the tents of the party. His followers were tall savage looking fellows, with broad swarthy faces, and their hair in short pig-tails. They wore the long-sleeved cloak, short trousers, and boots, all of thick woollen, and felt caps on their heads. Each was armed with a long matchlock slung over his back, with a moveable rest having two prongs like a fork, and a hinge, so as to fold up along the barrel, when the prongs project behind the shoulders like antelope horns, giving the uncouth warrior a droll appearance. A dozen cartridges, each in an iron case, were slung round the waist, and they also wore the long knife, flint, steel, and iron tobacco-pipe, and purse, suspended to a leathern girdle.[10]

The night at Yum Cho stayed fine with a dark sky and blazing stars; flashes of lightning lit up the sky. What they were seeing was the reflection of a massive storm raging on the plains of India, far away on the Terai, 240 km in the distance, to the south of their remote campsite. The temperature fell that night to –15°C.

The Dingpun departed the following morning, having written a letter on daphne paper to his superiors at Kampa Dzong, explaining his reasons for allowing Hooker's party to travel so far across the border; he also allowed the group a day to rest.

Above the campsite lay the lake Campbell had visited the previous day, and both men climbed the slopes above Yum Cho to reach it. Gurudongmar Lake (Yeumtso Lake) is fed by glaciers to the north of Kangchenjunga, and is famous throughout Sikkim as one of the world's highest-altitude lakes, the second highest lake in India after Cholamu (the border between Sikkim and Tibet has moved and Cholamu is nowadays in Sikkim). Perched at 5,430 m, Gurudongmar Lake is sacred to both Buddhist and Sikhs.

Fording the river, Hooker and Campbell crossed to the northern side of the valley from their campsite to climb to the summit of Bamcho (5,666 m) and visit the pass at Bamcho La (5,352 m), which led to the imposing fortress at Kampa Dzong in Tibet.

The south-facing side of the mountain consisted of alternating steep rocky slopes and broad shelf-like flat areas, and Hooker found many more plants growing there than he had previously expected. All were perfectly adapted to inhospitable growing conditions: woolly *Leontopodium*, cushion-forming *Androsace* and tufts of the pretty lavender-purple flowered *Astragalus chiliophylla*. The latter is quite common across the highest, more arid parts of the Himalaya, particularly Tibet. As Hooker recorded, this species forms great circular tufts on the ground, the centre eventually decaying, and annual shoots growing outwards from a deep-reaching tap-root, thus constantly enlarging the circle.

Up to 5,486 m, he found a scattered mix of alpine inhabitants including *Gnaphalium*, *Stipa*, *Sedum*, *Salvia*, *Draba*, *Pedicularis*, *Sibbaldia*, *Gentiana* and *Erigeron*. He also re-encountered an earlier discovery, the lovely ink-blue *Meconopsis horridula*, its bristly seed capsules now full of ripe, viable seeds. He also discovered a new species of wormwood, a strongly scented species with wonderfully silky foliage – *Artemisia campbellii*, which he chose to name for his adventurous travelling companion.

The summit of Bamcho commanded spectacular views of the mountains of Sikkim, north Bhutan and Tibet. To the north lay the broad sandy valley of the Arun River, while to the south, deep in the valley below, lay the azure waters of Cholamu Lake. Tibet, that forbidden land, lay directly in front of him, and Hooker spent the bulk of his time on the summit, gazing into the far horizon, studying the plateau, its peaks and deep river valleys. Lhasa, Tibet's ancient capital, with the great Potala Palace, the seat of the Dalai Lama, its convents, monasteries and network of narrow streets, aroused his curiosity. No doubt he yearned to see this holy of holies, though that would mean losing his head, so he had to content himself instead with a view of this mysterious land.

The view from Bamcho looking towards the Cholamu Lake. The Lachen River may be seen exiting the lake. Wild asses and Tibetan antelopes appear in the foreground. Hooker's illustration later appeared as a lithograph in *Himalayan Journals*.

Hooker's Bamcho field sketch was re-worked and appeared as a coloured lithograph by W. L. Walton in *Himalayan Journals*.

The range of mammals and birds of this dry Tibetan region impressed Hooker, once again in stark contrast to adjacent Sikkim:

> No village or house is seen throughout the extensive area over which the eye roams from Bhomtso, and the general character of the desolate landscape was similar to that which I have described as seen from Donkia Pass … The wild ass grazing with its foal on the sloping downs, the hare bounding over the stony soil, the antelope scouring the sandy flats, the fox stealing along to his burrow, are all desert and Tartarian types of the animal creation. The shrill whistle of the marmot alone breaks the silence of the scene, recalling the snows of Lapland to the mind; the kite and raven wheel through the air, 1000 feet over head, with as strong and steady a pinion as if that atmosphere possessed the same power of resistance that it does at the level of the sea. Still higher in the heavens, long black V-shaped trains of wild geese cleave the air, shooting the glacier-crowned top of Kinchinjhow, and winging their flight in one day, perhaps, from the Yaru, to the Ganges, over 500 miles of space, and through 22,000 feet of elevation.[11]

The wild ass or kiang he mentions was *Equus kiang*, the largest of the wild asses, found native almost exclusively to the alpine grasslands of the Tibetan Plateau, with scattered populations in Sikkim, Nepal and Kashmir. It is untameable; efforts to domesticate the *kiang* have always met with failure. By the Yaru, he referred to the great Yarlung-Tsangpo River, which he correctly claimed entered Assam as the Brahmaputra River, a fact only confirmed following exploration during the 1880s.

Hooker and Campbell climbed Bamcho twice. The Tibetan Sepoys, there to guard the border, were left bewildered and frustrated by the second visit. Campbell broke off to race ahead on his pony,

leaving the border guard chasing behind him, totally out of breath. When Hooker reached the summit, he found Campbell resting against a wall he had just built to shelter from a piercingly cold wind. Surrounding him was the border guard, totally puzzled by his admiration of the views beyond.

They descended that same afternoon, though it was dark before they reached Cholamu Lake; it was freezing hard, there was absolutely no sign of their campsite fires and in despair they began shouting until the echoes of their voices were heard by the Chebu Lama. Their tent was pitched on the shores of the lake, on a broad gravelly plain at 5,151 m. A cold wind howled down from Pauhunri, and that night the minimum temperature fell to –10°C.

A return to Sikkim via the Dongkya La

The following morning they set off for the Dongkya La with Campbell's porters, who were suffering quite badly from the altitude. The previous day it had taken them a full day to travel from Yum Cho to Cholamu Lake, a journey of just 10 km. Though beginning to tire of Hooker and Campbell's delays, the Tibetan Sepoys showed great patience, carrying Campbell's sick porter to the top of the pass on the backs of one of their yaks. Nothing however could persuade them to cross the frontier into Sikkim, which Hooker claimed they considered to be 'Company's territory'.

Before crossing the pass, Hooker first turned off to the east, re-ascending Pauhunri and reaching a peak overhanging a yawning precipice at almost 5,791 m. At its base were small lakes and glaciers, the source of the Lachung River. Among the snow and rocks he found the golden moonglow lichen, *Dimelaena oreina*, an Arctic European placodiod lichen. Beyond there he took a northern route to the Dongkya pass, across vast piles of rock and ancient glacial debris, reaching the pass by 3 p.m., thoroughly exhausted, but finding two of his best Lepcha porters sheltering behind a rock with his theodolite and barometers.

Having taken a series of heights and bearings, Hooker set off for Yume Samdong, not before giving the two Lepcha lads a small bottle of brandy and some biscuits. It was as well he did. The pony carrying his geological specimens soon afterwards became ill from exhaustion and altitude, and they spent the night by the warm flank of the beast, under a rock at 5,486 m, without fire, food or shelter.

At Yume Samdong Hooker found that Campbell had pitched his tent where he had camped the previous September. Campbell had had a difficult journey down the mountains with his very ill porters, who had thrown their packs off several times, and, lying on the ground, had to be shaken from a lethargy that would soon have led to death. One has to question Hooker and Campbell's lack of responsibility in bringing ill-equipped and wholly untrained porters into such dangerous terrain.

Trekking further down the river valley, Hooker reached the yak camp at Yumthang by the following evening. There he found herbaceous plants that were withered, brown and seeding at his previous base at Yume Samdong, green and unripe. Yumthang was a mere 610 m lower down the mountain slopes, and there, owing to lower altitude, the growing season was longer. He was keenly observant and recorded the following fruiting times for *Rhododendron* species at various altitudes in Sikkim:

16,000 to 17,000 feet, *Rhododendron nivale* flowers in July; fruits in September = 2 months.

13,000 to 14,000 feet, *Rhododendron anthopogon* flowers in June; fruits in Oct. = 4 months.

11,0000 to 12,000 feet, *Rhododendron campanulatum* flowers in May; fruits in Nov. = 6 months.

8,000 to 9,000 feet, *Rhododendron grande* flowers in April; fruits in Dec. = 8 months.

This apparent inversion of the order of nature and earlier seed set at higher altitudes, Hooker put down to the brighter skies and more powerful and frequent solar radiation.

Still further down the valley at Lachung, the weather was beautifully warm and mild. Cotoneasters were covered in scarlet berries and the brushwood was silver with the feathery seed heads of *Clematis montana*.

It was at Lachung that Hooker found he had lost a valuable thermometer used to measure high temperatures. It seems there was a hole in the bag in which his Lepcha porter, Cheytoong, carried his most regularly used instruments. It had last been used in the hot springs of the Kangchengyao Glacier, and Cheytoong was so worried about its loss that he set off, with a blanket on his back, and a few handfuls of rice, to retrieve it.

Hooker worried for his safety. There was now, at this late season, not a single person between Lachung and Kangchengyao, even the nomadic Tibetans having moved on. Three days later Cheytoong reappeared, beaming with happiness, having found the lost instrument. The first evening, he searched in vain, and rather than wasting time travelling back down the mountain, spent a freezing October night in a hot water spring, at 4,877 m.

His second day was equally fruitless, though crossing the Lachen River at Yume Samdong, he spotted the brass case glistening between two planks of the bridge. It had almost fallen into the torrent below, and Hooker was mightily impressed at his porter's bravery:

The Lepchas have generally been considered timorous of evil spirits, and especially averse to travelling at night, even in company. However little this gallant lad may have been given to superstition, he was nevertheless a Lepcha, born in a warm region, and had never faced the cold till he became my servant; and it required a stout heart and an honest one, to spend the night in so awful a solitude as that which reigns around the foot of the Kinchinjhow glacier.[12]

Cheytoong caught up on Hooker's party in the lower reaches of the Lachung valley, near the little village of Khedum on 26 October 1849. The villagers were busy harvesting crops of millet, maize and rice. Peaches had also been gathered, and though only half ripe, were far better than those at Darjeeling. Even as far north as the lower Lachung valley, monsoon rain and mists made ripening any of the European fruits difficult.

Hooker reckoned *Exbucklandia populnea* to be one of the most beautiful trees in Sikkim, and near Khedum he found a veteran specimen with a girth of 6.4 m at 1.5 m above ground level, with an unbranched trunk for 12 m. Epiphytic ferns and orchids wreathed its thick branches, including an exquisite purple-flowered species identified by Hooker as *Coelogyne wallichii*, while *Clematis* and *Stauntonia* climbed its trunk. Hooker reminisced about the great names brought before the traveller's notice: William Buckland (1784–1856), the English palaeontologist, theologian and Dean of Westminster; Sir George Leonard Staunton (1737–1801), the noted Irish diplomat and naturalist; and Nathaniel Wallich, former Superintendent of the Royal Botanic Garden, Calcutta.

The orchid Hooker saw, however, was *Pleione praecox*, an autumn-flowered species common in the warmer valleys and foothills in Sikkim. The error had initially been made by his father, Sir William, when he published *Coelogyne* 'wallichii' instead of *C. wallichiana*, an older name for *P. praecox*.

William Hooker had seen it growing in the great glasshouses at Chatsworth, where it blossomed alongside the gigantic *Victoria amazonica*. The *Pleione* had been collected by John Gibson (1815–1875) who was sent to India in 1835 by the sixth Duke of Devonshire to collect plants for his enormous glasshouse range. Gibson collected in the Khasi Hills of NE India (where he had been advised to go by Nathaniel Wallich).

After seven months of travel Gibson returned to Calcutta in February 1837, sending on 30 cases of plants to Chatsworth. He set sail a few weeks later, with 12 Wardian cases on the deck of the ship, while his cabin was crammed with orchids, many of them still attached to branches of trees and suspended from the ceiling. His Indian orchids made the Chatsworth collection the largest and most important in Europe, and it was by this means that the lovely *Pleione praecox* came into cultivation.

A scene from Yumthang, with an entire mountain slope covered by *Rhododendron campylocarpum*. Joseph Hooker recorded the duration between the times that various species flowered and set seed. For the high-altitude plants this was very brief indeed.

The following day Hooker and his party reached Chungthang. By then the rhododendrons growing on the hills behind the village had ripened their seeds, and it was from his collections there that *Rhododendron dalhousieae, R. maddenii, R. edgeworthii, R. grande, R. griffithianum* and *R. virgatum* were introduced to cultivation, and from where *R. arboreum*, probably the red-scarlet flowered variant, was re-introduced.

The Sikkim court's Lepcha messenger, Meepo, was waiting for Hooker at Chungthang, with instructions to take him to the Cho La and Yak La passes in east Sikkim. The road lay past the Rajah's palace at Tumlong (Tumloong), and Campbell hoped to have a meeting with the Rajah as they passed.

Below Chungthang the valley of the Teesta River was still flooded, with trees of *Alnus nepalensis* standing with their trunks submerged beneath 3.6 m of water, though Hooker presumed the dam would probably be washed away at the beginning of the rainy season the following year.

Beyond there he left behind the temperate flora of N Sikkim, and as he ventured further south the surrounding vegetation became increasingly tropical. On 30 October they reached the village of Sentam, to find that the road had fallen into the river far below as a result of landslides.

Travelling from Chungthang had been long and exhausting, but at last the monsoon was over and the dry, sunny season arrived. The remaining undamaged roads were dry and hard and, mercifully, there were no leeches at this late season, though on the hills above the river Hooker and Campbell encountered the vicious elephant nettle, *Dendrocnide sinuata*. Known to the Lepchas as *mealumma*, it forms a small tree up to 5 m tall. Contact with this species can cause fever, chills, painful itching and hives that may reappear every ten days to six months. Hooker had great difficulty in getting his Lepcha collectors to help cut a plant down. He gathered several specimens, avoiding contact with his skin, though the invisible, scentless effluvium was so strong that his eyes and nose poured for the rest of the afternoon. The severest warning from a Lepcha parent was to be punished with *mealumma*.

CHAPTER 14

OUR JOURNEYS THROUGH THE LACHUNG VALLEY

The village of Chungthang is a small, though busy marketplace, from which we generally stocked up with provisions before setting off to explore either the westerly Lachen valley, or its easterly counterpart, the Lachung River valley.

Lachung, like Lachen, is a major centre of biodiversity in northeast Sikkim, and is without doubt the more spectacular of the two; we explored the area on several occasions, both in late autumn, once the monsoon season had ended, and again in late spring when the valley floor was painted every shade imaginable by the many spectacular spring flowers that find a refuge there.

From Chungthang we followed the more easterly course of the Lachung Chu, heading for the little village of Lachung, a stronghold of the Bhutia and Lachungpa peoples. Departing Chungthang we crossed a modern bridge that spanned the lowest reaches of the Lachung River. It was on this very spot, in August 1849, that Hooker's dog Kinchin fell from a cane-bridge and was swept away and drowned in the icy waters beneath. We drove this route several times, and on each occasion, as we crossed the bridge, Hooker's hybrid Tibetan mastiff came to mind.

Cultivation in the lower Lachung valley. Traditional Bhutia houses, like those that appear in Hooker's 1849 sketch of Lachen, still survive in the Lachen and Lachung valleys. People in this valley belong to the Lachungpa ethnic group.

Brassaiopsis mitis is a variable species in Sikkim and around Darjeeling District. At its best, like the specimen shown here, which we found above Chungthang, it is a superb foliage plant. Plants around Darjeeling tend to be less heavily lobed.

The lower Lachung valley

The lower reaches of the Lachung valley are less precipitous than those of the neighbouring Lachen valley and thus better suited to agriculture, particularly arable farming, which is carried out on a small scale in forest clearings. Here, farmers still live in traditional Bhutia houses, much like those sketched by Joseph Hooker in 1849.

When we botanised the lower valley in early May 2015, men and women were busy in small fields harvesting rye, and the potato crop was in full flower. This was in total contrast to the village of Thangu, where, owing to the alpine climate, the potato crop is not planted until early May.

The little fields of rye and potatoes formed a pretty picture around the colourfully painted Bhutia houses, with steep, densely forested peaks rising immediately behind. It was a scene of unspoiled rural beauty, subsistence farming, and man living in complete harmony with the surrounding countryside. If we all lived this way what a happier place our planet would be.

We should have taken it as a good omen for the days to come, when, driving up the lower valley, we passed two trees of *Magnolia doltsopa*, flowering spectacularly. This lovely evergreen species is a firm favourite in British and Irish woodland gardens, though only in coastal settings. It is slightly tender, suffering in occasional exceptionally cold winters, but where it thrives it forms large, shapely trees, as at Mount Congreve in Co. Waterford. It's not fussy about pH either, growing (albeit slowly) on the shallow alkaline soils of the National Botanic Gardens, Glasnevin, in Dublin, and flowering so heavily there that its handsome foliage is almost hidden. The trees we saw were wreathed in pure white, heavily scented blossoms, and it was marvellous to see it in its Himalayan setting.

Autumn is equally interesting in this valley, particularly in November when the cherry blossoms come into bloom. Near the village of Lemna, at 2,175 m, where some of the farmers from Lachung live during the winter months, the warm, sun-baked, densely forested slopes of the valley were painted pink during our successive visits by the blossoms of the autumn-flowered Himalayan cherry, *Prunus cerasoides*, a glorious sight. *Himalayacalamus falconeri*, Falconer's bamboo, was abundant and grew with other exotics such as *Brassaiopsis mitis*, *Woodwardia unigemmata*, *Pittosporum floribundum* and *Lindera neesiana*, a small tree smothered in sprays of golden-yellow flowers.

Above there, in May 2015, we stumbled across one of the loveliest shrubs in this valley, *Deutzia bhutanensis*, bearing masses of large pale lilac-pink blossoms on gracefully arching stems. Strangely, Hooker

Prunus cerasoides, one of the loveliest autumn-flowering trees of Sikkim and Darjeeling. It is at its best in November when the post-monsoon rains clear and skies are gentian blue.

Dorjee and Sonam, our north Sikkim drivers, seen here feasting on the berries of *Rubus ellipticus*. These are extremely good and full of flavour, but this species has proved invasive in several parts of the world.

missed this, and it wasn't described until 1973 from a collection made by Frank Ludlow (1885–1972) and Major George Sherriff (1898–1967), on the Yonpu La, in east Bhutan in June 1947. It was later collected by Adam Stainton (1921–1991) in Nepal in 1980, so it was exciting to find it, fully in bloom, in the wilds of the Lachung valley.

A little further we stopped to gorge on the fruits of the golden Himalayan raspberry, *Rubus ellipticus*. It formed enormous, viciously armed, arching shrubs and was common on steep, roadside embankments. It is native to a large area of Asia, from India across China, and as far south as the Philippines, though it has proved to be a noxious pest outside its native terrain and is listed as one of the world's worst 100 invasive species.

Several other plants were in fruit, including Himalayan ivy, *Hedera nepalensis,* smothered in dense umbels of rounded berries. It made a great contrast with the very striking purple-blue pendulous bloomy fruits of *Prinsepia utilis*. I have never seen this rare rose-relative set fruits in cultivation, though in May 2015 roadside bushes near Lachung bore masses of bloomy blue-black pendulous fruits, on elegant arching branches, a lovely sight. By the edge of the road and on open hillsides it formed mounded, densely branched 3 m tall shrubs. The genus commemorates James Prinsep (1799–1840), Secretary of the Asiatic Society of Bengal, and the specific epithet *utilis*, meaning useful, refers to the fact that an oil is expressed from

Deutzia bhutanensis, a showy species with beautifully coloured flowers. We found it in a roadside thicket in the lower Lachung valley.

Prinsepia utilis, a particularly showy Himalayan shrub, though rarely seen in cultivation. In Irish gardens it flowers in February and March.

Elaeagnus parvifolia ripens its fruits in May. While extremely handsome, they also apparently have the potential to become a culinary fruit. From my experience silverberries may be an acquired taste.

Berberis sikkimensis was just opening its blossoms as we travelled up the Lachung valley. It grew in roadside thickets with other superb garden plants including Daphne bholua.

Disporum cantoniensis. The specific epithet relays the fact that it is also native to China. We found it in bloom around Chungthang and in the lower reaches of the Lachung valley.

the seed. This is used for cooking and medicinal purposes, while a deep-purple dye obtained from the fruits is used to paint the windows and walls of homes in the Himalaya.

Elaeagnus parvifolia completed this trio of fruiting shrubs. On previous autumn visits to Sikkim and Tibet I had seen this species in bloom, and been lucky enough to enjoy the powerfully sweet fragrance of its small silvery-white blossoms. Commonly known as the Himalayan silverberry because of the silvery speckles across its orange-red fruits, these are relished as a dessert in the Himalaya and the species has some potential as commercial fruit in temperate regions.

During our May visit we spotted several good garden plants in bloom. On the damp cliffs by waterfalls Bergenia pacumbis (syn. Bergenia ciliata f. ligulata) luxuriated, giving a good indication of the sort of conditions it requires at home. Lyonia ovalifolia, forming small trees, was then in full bloom and bore slender racemes of lovely white, bell-shaped blossoms. I can't understand why this lovely species is so rare, even in good woodland gardens in Britain and Ireland. It is certainly worthy of wider cultivation.

Berberis sikkimensis was also common in the roadside thickets. Hooker discovered it on 27 August 1849, and his specimens in the herbarium at Kew record that he collected it in the Lachung valley at an altitude of 9,000 ft, meaning he found it above the village. It was just coming into bloom at the time of our visit, when plants were laden with short racemes of lovely yellow blossoms.

Its bedfellows included Disporum cantoniensis, Pteris wallichiana (unfolding its delicate fronds), and the Sikkim spurge, Euphorbia sikkimensis, a stout multi-stemmed perennial, covered in cymes of lime-green blossoms surrounded by showy, long-lasting bracts. The spring foliage of this spurge is perhaps its finest feature. Many of the plants we saw had striking red stems and foliage with ruby-rose midribs and a leaf margin of a similar hue. In autumn, in the Lachen and Lachung valleys, its foliage assumes wonderful amber

RIGHT Pteris wallichiana, a very distinctive Himalayan fern. We found it twice during our travels in north Sikkim. It commemorates Dr Nathaniel Wallich.

The Sikkim spurge, *Euphorbia sikkimensis*, a first-rate perennial both in terms of its rose-purple tinted foliage, which assumes burgundy autumn colour, and its striking cymes of lime-green blossoms.

tones. The Sikkim spurge was discovered by Hooker at Chungthang in May 1849, and he made further collections in the Lachung valley in August of the same year.

By the roadside approaching Lachung village, grew a medley of superb plants including *Magnolia globosa*, *Magnolia campbellii*, *Philadelphus tomentosus*, *Prunus mira* (a wild peach, possibly naturalised in this region), huge tree-like plants of *Hydrangea heteromalla*, *Rhododendron arboreum*, *Alnus nepalensis* and *Prinsepia utilis*, which, in our November 2012 visit, were laden with starry white blossoms.

The village, perched by the banks of the Lachung Chu at 2,400 m, hasn't grown significantly since Hooker sketched it in August 1849, and we were able to recognise many features: dark trees of *Abies densa* still bristle the steep-sided mountain ridges that enclose the village, as they did in Hooker's time and snow lay on the highest peaks.

In the spring of 2015 we arrived at the village in the early evening and so, with time and daylight to spare, decided to explore a valley just to the north. The surrounding forests had just come back to life after the long winter, and we witnessed an incredible sight, hundreds upon hundreds of plants of the giant Himalayan lily, *Cardiocrinum giganteum*, then expanding wide rosettes of glossy heart-shaped leaves. Obviously there are no slugs in this valley!

Thousands of *Arisaema nepenthoides* pushed their inflorescences through the giant lilies, growing alongside occasional plants of the Himalayan gingseng, *Panax pseudogingseng*, *Arisaema propinquum* (with black, hooded cobra-like 'flowers') and the very lovely *Trillium tschonoskii*, with small violet-coloured flowers. Another bold foliage

Panax pseudogingseng, an important medicinal herb. It grew on the edge of a forest in a copse of *Corylus ferox* with dozens of giant Himalayan lilies.

plant, *Ligularia fischeri*, was also emerging. It is a common plant in the swampy meadows and stream banks of the Himalaya and much of the temperate Far East.

Beneath all these woodlanders was a carpet of *Fragaria nubicola*, fully in flower, though we made a hasty retreat when we realised that one of our party was covered in masses of leeches, the dread of every Himalayan explorer.

Bushy trees of *Corylus ferox*, with delicate copper-flushed foliage, covered in a soft silver down beneath, formed a canopy overhead, alongside trees of *Rhododendron arboreum* ssp. *cinnamomeum* (all red-flowered). The plants we saw there, as on Tonglu near Darjeeling, bore short stout young stems and leaves of a striking crimson-red hue, similar to the new growth of *Pieris formosa* (which also grew nearby). We couldn't determine if this was a physical abnormality caused by insects or disease, but it was certainly eye-catching.

An abnormality on the foliage of *Rhododendron arboreum* ssp. *cinnamomeum* caused new growths to be strikingly red-flushed.

The staminate flowers of *Daphniphyllum himalayense*. All species are dioecious, that is, male and female flowers are carried on different plants.

The silky rose, *Rosa sericea*, was one of Nathaniel Wallich's Nepal discoveries. Hooker sent seeds to Kew about the same time as when Sir Richard Strachey sent seeds to England.

The rhododendron look-alike, *Daphniphyllum himalayense*, was in flower and abundant on the forest margin. Trees are dioecious, with male flowers predominating and bearing showy purple stamens contained in a rose-pink tinged bract. The weeping *Picea spinulosa* was also common at this altitude, with its branchlets supporting freshly formed rose-purple cones.

Rosa sericea formed enormous upright shrubs, some up to 4 m tall, with branches smothered in snow-white single blooms. This lovely Himalayan rose makes a wonderful plant for the woodland garden and is equally effective in autumn when it bears masses of crimson and yellow, edible pear-shaped fruits. There are very good plants in the Earl of Rosse's garden at Birr Castle, Co. Offaly, in the Irish midlands.

The evergreen laburnum, *Piptanthus nepalensis*, was absolutely abundant along the gravelly, boulder-filled banks of the river, though the form growing there was far finer than those generally found in cultivation, with flowers at least twice normal size and of a more pleasing lemon-yellow. What's more, many had foliage clothed with silvery hairs, which contrasted beautifully with the very showy blossoms.

Our last find that evening, as the light was fading, was on the river bank, where we spotted a single plant of the rather dainty *Rhododendron camelliiflorum*, then in full bloom, though early and out of season. Discovered by Hooker in the Yalung valley, the most easterly valley in Nepal, in the autumn of 1848, he found it again in the Lachen valley in 1849, and introduced it to cultivation from seeds collected in the Lagyap valley in E Sikkim, in November 1849. Though it is more normally seen as an epiphyte, the plant we found grew terrestrially, on a sharply drained river bank formed of gravel and boulders: it bore creamy-white flowers spotted with lime-green.

Lachung, our hotel in November 2012. Several new hotels have been built in the meantime and much grander accommodation is available, though we found this place charmingly rustic.

The upper Lachung valley

Travelling up the valley, beyond the village, conifers dominate the scene and the landscape becomes increasingly alpine, the steep slopes densely clothed with *Abies densa* and moss-laden trees of *Tsuga dumosa*. Beneath these vast forests lives a rich understorey of woodland plants, while in open glades and on the valley floors rhododendron species abound, in greater diversity than in any other part of Sikkim.

The gorgeous *Rosa macrophylla* is common here, and during our autumn visits it was very conspicuous, carrying a fine crop of large pendulous flask-like fruits. *Abies spectabilis*, one of the most beautiful of all the firs, was also common there, reminding me of the gigantic old specimen in the Deer Park at Kilmacurragh. *Berberis hookeri*, growing in thousands, formed a low-growing spiny ground cover beneath the firs. Near it, under a canopy of *Tsuga dumosa*, grew *Dryopteris wallichiana*, a few scattered plants giving a wonderful shuttlecock effect.

Rhododendrons, to our great delight, abounded. We stopped first at the Yakchey reserve, which, with the nearby Shingba Rhododendron Sanctuary, is home to 24 different species. On the edge of a forest almost entirely composed of *Abies densa*, *Tsuga dumosa* (draped with 1 m long strands of the pendulous lichen *Dolichousnea longissima*) and *Picea spinulosa*, grew several species, including the widespread *Rhododendron lepidotum* (forming dwarf shrubs), *R. virgatum* (a Griffith discovery/Hooker introduction, whose flowers may vary from white, to pink, to purple) and the rather rarer, plum-purple flowered *R. baileyi*.

Rhododendron lepidotum is a variable plant in Sikkim. Hooker described both *R. salignum* and *R. elaeagnoides* from plants he found on the hills above Chungthang, though today these are regarded as variants of *R. lepidotum*. In Sikkim certainly, its flowers may vary in shades of greenish yellow, white, pink, red, rose, purple or crimson. Hooker introduced it to cultivation in 1850 from Sikkim.

During our visit in May 2015 we were astounded by the colour provided by the many species growing in this valley, none more so than *R. niveum*, the State tree of Sikkim and the rarest rhododendron species in the Himalaya. We had timed our visit perfectly. It was optimum flowering time and we were left in awe of small dome-shaped trees bearing densely packed trusses of blossoms varying in shades of deep magenta to purplish-lilac. It's a curious colour that some of us love, and others loathe, and, during the 1920s and 1930s, the shade became so deeply unfashionable that plants in large collections in Britain and Ireland were grubbed up and destroyed,

Piptanthus nepalensis, the evergreen laburnum. The form growing in the upper Lachung valley is far superior to those in cultivation.

Rhododendron virgatum is native to the mountains of Sikkim, Bhutan, Tibet and north-west Yunnan. Hooker introduced it to cultivation in 1850.

the Cho La pass in 1849, and sent seeds back to Kew from Lagyap in the same year. From this collection a plant bloomed for the very first time in cultivation at Kew in May 1853. Visiting plantsmen must have been bowled over by the strange hue of its blooms.

While Hooker has been inadvertently credited with the discovery of a large number of Sikkim rhododendrons, David Long, an expert on the flora of Bhutan and Sikkim, based at the Royal Botanic Garden Edinburgh, has pointed out that of these, Griffith was the discoverer of *R. cinnabarinum, R. dalhousieae, R. edgeworthii, R. grande, R. griffithianum, R. glaucophyllum, R. triflorum* and *R. virgatum,* though *R. falconeri* cannot be claimed as a Griffith discovery since it had been introduced to cultivation by Col. Sykes in 1830. For his part, Joseph Hooker always gave credit, where possible, to Griffith's earlier collections.

Hooker didn't immediately recognise *R. niveum* as a new species. He never saw it in bloom in Sikkim (it had finished flowering by the time he reached Lachen and Lachung), and he described it later

Tsuga dumosa, on a large rock cleft in the upper Lachung valley. In Britain and Ireland it is liable to be damaged by late spring frosts, though there are good trees at Fota and Kilmacurragh.

with the result that it is no longer common in gardens. What a pity. In my opinion, it is one of the most beautiful of all Himalayan rhododendron species, and the best specimens in cultivation, like those at the Royal Botanic Garden Edinburgh, are breathtakingly beautiful when fully in bloom.

We were certainly impressed by the many plants we saw at Yakchey, framed by a suitable backdrop of snowy peaks, and flowering alongside red-flowered forms of *R. arboreum* ssp. *cinnamomeum.* First found by William Griffith in Bhutan in 1838,[1] Hooker re-discovered *R. niveum* growing in both the Lachen and Lachung valleys and by

Rhododendron baileyi, a rare inhabitant of the upper Lachung valley. It is named for F. M. Bailey, Britain's ultimate double agent, a spy who also discovered the fabled blue poppy, *Meconopsis baileyi.*

Rhododendron niveum, the national tree of Sikkim. It is pictured here at the Shingba Rhododendron Sanctuary in the Yumthang valley above Lachung. This species is one of the rarest *Rhododendron* species in Sikkim. A new population was discovered in Kangchenjunga National Park in 2010.

instead from its foliage, distinguishing it from the allied *R. arboreum* ssp. *cinnamomeum* by means of its white woolly indumentum. Instead it was described in the *Journal of the Horticultural Society* (later the RHS) in 1852 from one of his Sikkim seedlings. Hooker chose the very apt specific epithet *niveum*, meaning 'snow-like', on account of the conspicuous snow-white tomentum on the undersurface of leaves and on the petioles.

In the upper Lachung valley *R. niveum* grows in incredibly small numbers, and the populations in the Lachen valley and near the Cho La pass from where Hooker made his collections no longer exist, perhaps as a result of landslips. Such is the delicate balance of nature in the Sikkim Himalaya that a rare endemic, restricted to single valleys, can be destroyed, in an instant, by landslides. In more recent times another Sanctuary designated for the preservation of *R. niveum* (near Yakchey) was washed away in an avalanche in 1970.[2] It is endemic to Sikkim and adjacent Bhutan, and a new population was discovered in Kangchenjunga National Park as recently as 2010.

Rhododendron niveum, an illustration that perfectly depicts the rather strange and unique colour of the inflorescence. The gardens at Stonefield Castle, in Argyll, Scotland, received many of Hooker's seedlings and created several hybrids using this species.

Rhododendron niveum in bloom at Yakchey in the upper Lachung valley in north-east Sikkim. It is seen here in May 2015, peak flowering season for most rhododendron species in this part of Sikkim.

Rhododendron niveum, a superb plant in the collections at the Royal Botanic Garden Edinburgh. Seen here flowering in April 2013.

Close by, in open moorland beyond the forest margin *R. glaucophyllum* abounded, forming low, densely branched mound-like shrubs to not more than a metre tall, and smothered with masses of lovely rose-pink, bell-shaped blossoms. At Yakchey it created a sea of colour on the boggy flats, though of the smaller-flowered species growing there, my personal favourite was the lovely *R. ciliatum*, another of the many species first named and described by Hooker in his classic work *Rhododendrons of the Sikkim Himalaya*.

He discovered it in the Lachen valley in June 1849, sending seeds to Kew in November of the same year, from which six plants bloomed at Kew in March 1852. It made a lovely sight on the same boggy moorland, its white, flushed-pink campanulate flowers delighting our group. The species, when crossed with *R. edgeworthii*, created the wonderfully fragrant hybrid *R.* 'Lady Alice Fitzwilliam',

Rhododendron glaucophyllum, an abundant species on the boggy flats at Yakchey, is one of the smaller-growing Sikkim species, though absolutely charming and with richly coloured blossoms.

Rhododendron ciliatum, whose flowers are proportionally very large for the size of the plant. One of the loveliest of the Himalayan species, it is the parent of many good hybrids.

named for the daughter of the sixth Earl Fitzwilliam, whose Irish seat was the extensive Coolattin Estate in Co. Wicklow. The same Earl was a Yeomanry Aide-de-Camp to Queen Victoria's Viceroy in India between 1884 to 1894.

Beyond Yakchey, in the upper Lachung valley at 3,148 m, we drove further north into the famous Shingba Rhododendron Sanctuary. Steep mountain cliffs, streaming with waterfalls, rose overhead, while snow capped the higher peaks. I had been to the Yumthang valley in previous years, and the reason for a return visit was to take another look at the 43 km² sanctuary, where 24 of Sikkim's 40 rhododendron species are found, including a further population of *R. niveum*.

In a damp boggy area at the forest's edge, in carpets of *Chrysosplenium nepalense* and near fruiting spires of *Cardiocrinum giganteum*, we spied *Primula capitata* interspersed with *P. spathulifolia*, a few plants of the latter carrying lilac-blue blossoms in November 2013; *Daphne bholua* is also very common here, with an altitudinal range of between 3,048 and 4,575 m. Higher up, on the same autumn trip, we found deciduous plants of *Daphne bholua* var. *glacialis* with flowers varying from white to candyfloss pink on stems bare of foliage.

Another common shrub here was *Berberis virescens*, a plant with stunning red stems that seemed to glow in the low November sunlight. As previously stated, this species was described from material, sent to Hooker at Kew, by Thomas Acton of Kilmacurragh, Co. Wicklow.

RIGHT *Rhododendron arboreum* ssp. *cinnamomeum*, a crimson-flowered plant seen here in bloom at Yakchey with the spectacular backdrop of the upper Lachung valley.

Claire Mullarney from the National Botanic Gardens, Kilmacurragh, seen here crossing the turbulent waters of the Lachung River. For any keen plantsperson Sikkim is a superb place to botanise.

Viburnum nervosum seen here forming trees at Shingba. This species is much underrated and its autumn colour, in shades of deepest claret, is spectacular. A widespread species in the eastern Himalaya.

In the surrounding forests we found fine trees of *Acer caesium, A. sikkimense, Enkianthus deflexus, Betula utilis* (with wonderful peeling cherry-brown bark) and *Pieris formosa* in several forms. *Clematis montana* draped its way through the surrounding trees, in autumn covered in silky, silvery seed heads, and by the edge of thickets we saw the wonderfully architectural *Cirsium eriophoroides* (another of Hooker's 1849 finds) and the woolly rosettes of *Meconopsis paniculata*.

We had reached the tree-line at this point and the vast forests of *Abies densa* began to peter out, to be replaced by ancient trees of *Viburnum nervosum*, which during our May visit bore broad, rounded corymbs of white flowers above copper-flushed foliage. *Prunus cornuta*, the Himalayan bird cherry (previously encountered in the Lachen valley), was equally abundant and it too sported freshly emerged coppery-red foliage.

Prunus cornuta, the Himalayan bird cherry, pictured here in the Shingba Rhododendron Sanctuary. Its dark copper-coloured foliage provided a striking contrast to the snowy peaks that enclosed the valley.

Another deciduous tree, *Magnolia globosa*, 12 m tall, attains the highest altitude of any of the Sikkim magnolias, though it rarely reaches such heights in cultivation. It is also native to Yunnan in W China from where it was introduced by George Forrest in 1919. It is a lovely species, closely allied to the Chinese *M. wilsonii*, to which it bears a resemblance, particularly in its lovely summer-produced, pendulous, fragrant white flowers. It was just coming into leaf at the time of our visit, and its expanding foliage and fat, plump flower buds were heavily clothed in a dense, woolly fawn-coloured tomentum. Hooker found it in 'the interior ranges', meaning the Lachen and Lachung valleys, where it grows in *Abies* forest.

Magnolia globosa, one of Joseph Hooker's lesser-known discoveries. It is seen here in May 2015 expanding a plump, flower bud. We were a week too early to see it in bloom.

Rhododendron pendulum, one of the epiphytic species. This plant grew on rocks, though elsewhere at Shingba and the Yumthang valley we found it on *Betula utilis*. In west Sikkim it grows high up on the trunks of *Abies densa*.

Not far from the entrance to the Sanctuary we were very pleased to encounter the high-altitude epiphytic *Rhododendron pendulum*, springing from the trunks of *Betula utilis* and *Tsuga dumosa,* and also from nearby rock faces. Hooker first found it in June 1849 in the adjacent Lachen valley, where its pendulous branches hung from the limbs of *Abies densa* and *Tsuga dumosa*, or very rarely on rocks, and he went on to state that it was very often covered in the spectacular lichen *Dolichousnea longissima*. Its epiphytic nature means it is difficult to cultivate and it has always remained rare in British and Irish gardens. It is, as Hooker pointed out, a difficult plant to spot in the gloomy, almost impenetrable forests of Sikkim, though the plants we found were very spectacular, forming dense, dwarf, very compact mounds, smothered in small, snow-white blossoms. It was another of his 1850 *Rhododendron* introductions via Kew.

In open glades and on the edge of rhododendron thickets cobra lilies abounded, particularly the spectacular *Arisaema utile*, with sinister brown-purple spathes. The cobra lilies are of course not true lilies, but aroids with a spectacular inflorescence. Of 180 species worldwide, 60 are found in the E Himalaya, with 78 in China, and these regions are thought to be the centre of diversification and origin for the genus.[3]

Cobra lilies grow with amazing vigour in the Yumthang valley, especially within the area of the Shingba Rhododendron Sanctuary. Seen here is *Arisaema utile* with flowering plants of *Primula denticulata*.

Rhododendron hodgsonii is commonly found in the higher valleys in north Sikkim, including the Yumthang valley, where this plant was photographed in May 2015. There it is invariably found beneath the canopy of forests composed of *Abies densa*, and it can form enormous populations. Hooker named it for his hospitable friend at Darjeeling, Brian Hodgson.

Beyond there big-leaved *Rhododendron hodgsonii* formed impenetrable thickets. I can remember how pleased we were to see this in the Shingba Rhododendron Sanctuary in November 2012, and so it was exciting to meet it in its native homeland. It was even more exciting to catch it in full flower in May 2015, bearing tightly packed clusters of flowers. This species was first found by Hooker in 1848 in the remote valleys of E Nepal, and later in Sikkim, from where he introduced it to cultivation (see chapter 5, page 90). He rightly considered it to be the characteristic understorey shrubby tree in the higher valleys of Sikkim:

> Nowhere can the traveller wander, in the limits assigned to the present species, without his attention arrested by its magnificent foliage, larger than that of *R. falconeri*, and remarkable for its brilliant deep green hue. In summer the leaves are broad, and spreading all around the plant; in winter rolled up, and pendulous from the tips of the branches. It is alike found at the bottom of valleys, on the rocky spurs or slopes and ridges of hills, in open places, or in gloomy pine-groves, often forming impenetrable scrub, through which the explorer in vain seeks to force his way. Nor is this a thicket merely of twigs and foliage, that will fall under a knife or cutlass, but of thickset limbs and stout trunks, only to be severed with difficulty, on account of the toughness and unyielding nature of the wood.
>
> … Of the wood, cups, spoons, and ladles are made by the Bhoteas [*sic*], and universally the little "Yak" saddle, by means of which the pack-loads are slung on the back of that animal. Easily worked, and not apt to split, it is admirably adapted for use in the parched and arid climate of Thibet. Nor is the foliage without its allotted use. The leaves are employed as platters, and serve for lining baskets for conveying the mashed pulp of *Arisaema* root … and the accustomed presents of butter or curd is always made enclosed in this glossy foliage.
>
> Such are the characteristics of this *Rhododendron*, which I desire to dedicate to my excellent friend and generous host, B. H. Hodgson, Esq., of Darjeeling, formerly the Hon. East India Company's Resident at the Court of Nepal; a gentleman whose researches in the physical geography, the natural history, especially the zoology, the ethnology, the literature of the people, &c. &c, of [the] Eastern Himalaya, are beyond all praise.[4]

Rhododendron thomsonii. Unlike *Rhododendron hodgsonii,* this species prefers to grow in the open, in full sun, out of the canopy of *Abies densa*. It hybridises readily in the wild and has been widely used for similar purposes in cultivation.

Rhododendron campylocarpum pictured here flowering in the Shingba Rhododendron Sanctuary in the Yumthang valley. In the Lachen and Lachung valleys of north Sikkim it forms natural hybrids with *R. thomsonii*, which grows alongside it.

Hooker's 'gloomy pine groves' were in fact the Sikkim fir, *Abies densa*, and wherever the modern-day traveller finds *R. hodgsonii*, they can be sure to find *Abies densa* close by. The *Arisaema* he refers to is *A. utile*, to which he gave the specific epithet *utile* meaning useful, when naming it many years later.

We were photographing cobra lilies when we heard a sudden cry from one of our travelling companions. Pushing our way through dense rhododendron thicket, we soon realised what the commotion was about! She had stumbled her way into the Yumthang valley, above the rhododendron sanctuary, and the view we saw that day amazed us all: on the valley floor beneath, the landscape had been transformed into a sea of red by thousands of plants of *Rhododendron thomsonii*.

I had been here before, though never at flowering time, and to see thickets of this rhododendron in bloom, draped in millions of blood-red waxy bells, was one of life's high points. This species is justly popular in Irish gardens (where it forms small trees) on account of the loose trusses of waxy, bell-shaped flowers that smother plants in April and May and also for its wonderful smooth plum and cinnamon-coloured bark.

In this valley it grew alongside the yellow-flowered *Rhododendron campylocarpum*. Where both species overlapped the hybrid offspring *Rhododendron × candelabrum* occurred. Hooker described this plant as a species in 1851, from his collections made in fir and hemlock forest near the village of Lachen, though it is far more abundant at Yumthang, where with fellow rhododendron enthusiast Averil Milligan, Head Gardener at Rowallane Gardens in Co. Down, we spent a considerable period admiring its lovely pale-rose and apricot-rose blossoms. It also occurs in Bhutan, from where it was introduced to gardens by Ludlow and Sherriff in 1937. Rhododendrons are very promiscuous, some species more than others, and higher up the Yumthang valley we found natural hybrids of *R. thomsonii × R. wallichii* producing pleasing, large flowered rose-magenta forms.

Beneath *R. thomsonii* and its hybrids grew a ground flora of blue-flowered gentianellas, *Osmunda japonica* (then expanding its crozier-like fronds) and extensive areas of *Arisaema griffithii*. If any of the arisaemas have a claim to being 'cobra lilies', this species certainly has precedence. The enormous plants in the Shingba Rhododendron Sanctuary and in the Yumthang valley were slightly unnerving when seen though a camera lens!

The perfect wild garden. Averil Milligan and Bruno Nicolai pictured here framed by *Rhododendron campylocarpum* and *Rhododendron thomsonii*, with *Viburnum nervosum* overhead.

Rhododendron thomsonii appears to be a highly promiscuous species in Sikkim and hybridises freely. It is pictured here to the left, while on the extreme right is *Rhododendron wallichii*. In the centre is their progeny, a hybrid perfectly intermediate between both parents.

Rhododendron × *candelabrum* is so far only known from north Sikkim and Bhutan. It forms sizeable populations in the Yumthang valley.

Primula denticulata, the drumstick primula, was an early introduction from the Himalaya, having been raised by the English nursery firm, Messrs Veitch, from seeds received from the British East India Company, collected in the western Himalaya by John Forbes Royle. The Veitch firm first flowered it in March 1842, and through years of breeding created a range of hybrids that look positively obese in comparison to the graceful wild Himalayan plant.

I was also very pleased to find several bushes of *Rhododendron fulgens* in bloom. W. H. Fitch's depiction of this species in *Rhododendrons of the Sikkim Himalaya* is wonderful, giving a good indication of the vividness of its rounded trusses of blood-red

One of the most intriguing plants in this area is a relatively recent discovery, *R.* × *sikkimense*, a bush to about 2 m tall, that at the time of our visit bore trusses of broadly funnel-shaped, blood-red flowers. Known only from this region of Sikkim, its current taxonomic status is somewhat in question. It is most likely a natural hybrid involving *R. thomsonii* and scarlet-flowered variants of *R. arboreum* ssp. *cinnamomeum*, both common in the Lachen valley area.

Higher still, *R. barbatum* appeared, a form with unbearded petioles and trusses of dark crimson-scarlet blossoms held on peeling plum-coloured stems. Further on *R. wallichii* also arrived, in shades of lilac-blue, which were certainly far better than the purples found at lower levels in the valley. *Primula denticulata* became increasingly abundant on the forest verge, its flowers perfectly matching the shade of the surrounding plants of *R. wallichii*.

Rhododendron × *sikkimense* flowering at Shingba. It is most likely a natural hybrid of *R. thomsonii* and scarlet-flowered *R. arboreum* ssp. *cinnamomeum*.

One of my travelling companions, the artist and keen plantswoman Lesley Fennell, pictured in a carpet of *Primula denticulata* at Yumthang.

Primula denticulata, the drumstick primrose. In the Yumthang valley, where it grew in millions, it turned the alpine meadows rose-lilac. It was one of the loveliest scenes we witnessed.

flowers and curved, blue-tinted seed capsules. This species is variable however, as some forms bear only a few trusses hidden by foliage, while others give an exceptional show, and for garden purposes it is worth seeking outstanding clones. In terms of foliage however, it is a lovely species, with a dense fawn indumentum on the undersurface of the leaves. Though the discovery of *R. fulgens* is generally attributed to Joseph Hooker, it had been previously collected in Nepal by Wallich, though Hooker did introduce it to cultivation and it first flowered at Kew in 1862.

At its feet we found a thicket of the dwarf *R. anthopogon* – in this area of the valley all plants were of the white-flowered form. With it grew sheets of the lovely lilac-blue *Primula scapigera* forming carpets of colour. This pretty Himalayan primrose was first described by Hooker from Griffith's Bhutan collection, and from those made by Hooker himself on the summit of Tonglu near Darjeeling.

Higher still, *Euphorbia wallichii* was just awakening and plants were already carrying terminal cymes of lime-green flowers surrounded by showy bracts of a similar hue on stout, short stems. It was strangely reminiscent of the Irish spurge, *Euphorbia hyberna*, which also bears its flowers shortly after emergence, and the Yumthang plants grew in a habitat very similar to that of the Irish spurge, on the Healy Pass, above Glengarriff in the mountains of West Cork.

Rhododendron fulgens, a common species at high altitudes in the Sikkim Himalaya. The undersides of its leaves are covered in a rust- coloured indumentum.

Griffith's cobra lily, *Arisaema griffithii*. Without question the most sinister-looking of its group and certainly deserving of its common name. It is an abundant species around Darjeeling and in north Sikkim. Lovers of the strange and curious will certainly want to grow it.

Wallich's spurge, *Euphorbia wallichii*. Herbaceous plants emerge late in the season in these high valleys and it was just beginning to bloom in early May.

Yumthang — valley of flowers

We drove still further north, deeper into the Yumthang valley, a region aptly compared by Hooker to the European Alps, though many of the peaks and cliff-like valley walls soar far higher than Mont Blanc. Yumthang, 'the valley of flowers', is without doubt, the most spectacular valley in Sikkim. Glaciers and frozen waterfalls descend the cliff-like jagged peaks surrounding the valley and from the upper part of the valley one gets clear views of the Dongkya La, where, in October 1849, Hooker and Archibald Campbell made their unauthorised entry into Tibet, a move that led to their imprisonment by the Rajah of Sikkim.

In the upper Yumthang valley we crossed a gigantic landslide, caused initially by the Sikkim earthquake (which had its epicentre in Chungthang) of September 2011. It hit this part of the valley with such force that it cracked an entire mountain face, and tremors in the spring of 2015 (just a few weeks before our arrival) brought the entire mountain face down across the valley with such ferocity that it created a mini hurricane on the opposite side, blasting the forests away. We felt very humbled (and nervous) at the scene, and as we crossed the Armageddon-like setting we could hear house-sized boulders splitting under the pressure placed on them.

In one place the landslide had blocked the Yumthang River, creating a wonderful turquoise lake, and on its edge, reflected in the waters, were thickets of surviving *Rhododendron campanulatum*, with blue-purple flowers. Nature soon moves in to heal such a devastated landscape however, and in places we spotted *Primula denticulata* pushing its lilac blossoms through shattered rock and rubble.

Beyond the landslide area the valley narrowed and gigantic cliff walls, streaming with glaciers and frozen waterfalls, loomed ominously above us. We soon passed the old British Dak bungalow, used by the late 19th- and early 20th-century explorers. Suddenly the valley widened again, with the Dongkya La and Pauhunri in the distance, and in the foreground, as far as the eye could see, the alpine meadows were painted lilac by the drumstick flowers of tens of millions of the dainty *Primula denticulata*. The true wild plant is certainly daintier, and more graceful, than the highly bred versions found in European and North American gardens. The sight of yaks and their calves grazing among hundreds of acres of plants in those alpine meadows bowled us over that warm, sunny May evening, and there must be few other floral scenes in the Himalaya to match it.

Another plant we were delighted to see was *Viburnum grandiflorum*, a common plant at this altitude, where it grew on the edge of *Abies*

Dramatic scenery in the Yumthang valley. A Dak bungalow with jagged snowy peaks rising sharply overhead. *Abies densa* has almost reached the tree-line and the tiny *Primula denticulata* colours the meadows in the foreground.

The flowers of *Viburnum grandiflorum* are beautifully fragrant. Alongside the Chinese *V. farreri* it is one of the parents of the popular *Viburnum* × *bodnantense*.

densa forest with *Rhododendron campanulatum* and *R. wallichii*. This species is very similar to the Chinese *Viburnum farreri*, and at the time of our visit its leafless stems were densely covered with fragrant, rose-pink tubular blossoms. It is not at all as popular in cultivation as its Chinese ally, since its flowers are apt to be hit by spring frosts in colder gardens, though it is a good plant in milder British and Irish woodland gardens and was introduced to cultivation from Bhutan by Roland Cooper for A. K. Bulley in 1914.

In his description of *R. wallichii* Hooker didn't give an exact location for its habitat in Sikkim, giving instead 'interior of Sikkim' by which he meant the Lachen and Lachung valleys. However, his type specimen for the species in the herbarium at the Royal Botanic Gardens, Kew is dated May 1848. Considering the preferred altitude of this species and the date of collection, the only place he could have found it was on the summit of Tonglu.

Rhododendron wallichii is no longer found on Tonglu, and Hooker reported that during his visit the forests had been cleared to make way for grazing; it is likely, therefore, that Hooker collected from the last remaining plants on Tonglu. In current floras for Darjeeling District it is considered critically endangered on a local scale, with small scattered populations found on the higher peaks, like the Sandakphu-Phalut ridge and on Mount Singalila.

Viburnum grandiflorum flowering its heart out in the Yumthang valley. In the foreground is Rhododendron wallichii.

It is abundant however in the highest valleys of the interior of Sikkim where it generally forms rounded bushes up to 3 m tall. It is rather similar to *R. campanulatum*, from which it can quickly be differentiated in the field by the lack of a continuous, thick brown indumentum on the underside of its leaves. It is a handsome plant in its best forms, with flowers varying in colour from rose-purple, mauve-blue and pale blue fading to white. Hooker named it for Nathaniel Wallich, whom he considered 'Botanicorum Indicorum facile princeps' and sent seeds to Kew in 1850. *Rhododendron wallichii* first blossomed in cultivation at Kew in 1856, though when featured in *Curtis's Botanical Magazine* (t.4928), the history of the plant was simply that it had been raised from Himalayan seed, so perhaps the Kew plant was not one of Hooker's Sikkim collections, but had been grown from seeds sent by another collector based in the Himalaya.

R. hodgsonii re-appeared on the scene, and higher still, in *Abies densa* and *Betula utilis* forest, we met one of the loveliest of all Sikkim's rhododendrons, the incomparably beautiful *R. wightii*, bearing rounded trusses of lemon-yellow, bell-shaped flowers spotted crimson on the upper interior part of the flower (see chapter 9, p. 141).

While exploring nearby later that evening we stopped beneath a forest of the Sikkim fir, *Abies densa*, where we were told there was a population of *R. × decipiens*. As previously mentioned, this is a naturally occurring hybrid that forms where *R. hodgsonii* and *R. falconeri* overlap. It is rare in Sikkim, but was said to occur in the Yumthang valley.

That evening our group had become scattered somehow, leaving just myself, Lesley Fennell and Bruno Nicolai together, alongside our Tibetan guide Thupden Tsering. We entered the woods and there before us, beneath the dark lichen-draped fir trees, was a natural rhododendron garden with flowers in a range of colours. It soon became very apparent to me that we were not in fact looking at *R. × decipiens*, but a new, important natural hybrid, that had remained misnamed and unidentified for a considerable period.

Nature plays on a vast scale in the Yumthang valley. This landslide caused devastation and it will take decades for the vegetation to recolonise the area.

Bruno Nicolai, the author and Thupden Tsering. Before us are three rhododendrons. In the centre is *R. × thupdenii*, with its parents on either side. To the left is *R. hodgsonii*, to the right is the lemon-yellow flowered *R. wightii*.

Rhododendron wightii, the true wild plant in the Yumthang valley. It is far superior to the impostor that long grew in gardens under its name (see chapter 9, p. 140).

Rhododendron wallichii is highly variable in terms of flower colour. In 1852 seedlings from Hooker's Sikkim collections were sent from Kew to the Vice Regal Lodge in Dublin's Phoenix Park.

To begin with, *Rhododendron falconeri,* one of the parents of *R. × decipiens,* was nowhere to be seen, although *R. hodgsonii* abounded, and with it grew the yellow-flowered *R. wightii.* The penny dropped: we had just discovered a new natural hybrid, which needed to be named, described and separated from the allied and rather appropriately named *R. × decipiens,* which is now far rarer than was previously thought.

A study of all populations of the so-called *R. × decipiens* throughout Sikkim and Darjeeling District now needs to be made, to determine which is the *R. falconeri* hybrid and which are hybrids of *R. hodgsonii* with *R. wightii.* Certainly the plants we saw at Yumthang are the latter.

In these woods, this lovely hybrid had formed an extensive population of small bushy trees with characteristics intermediate of both parents. In habit it forms an upright, sparingly branched small tree much like *R. hodgsonii,* and also inherits the same stout branches, bud scale colour and the dense silvery-white tomentum on the underside of the petioles. The flowers too, are midway between the parents, carried in flat-headed, hemispherical trusses similar in shape to *R. wightii.* The soft-pink blossoms have a conspicuous deep-rose blotch at the base of the corolla and rose-coloured spotting on the upper side of the corolla (a trait inherited from *R. wightii*).

This new addition to the rhododendron flora of the Sikkim Himalaya needs to be distinguished from other naturally occurring *Rhododendron hodgsonii* hybrids, and I choose to name it for Thupden Tsering (b. 1960), whose family fled Tibet during the upheaval of the Cultural Revolution, crossing remote mountain passes with their herd of yaks into Yumthang to make a new life and home in Sikkim. Thupden was later adopted by that great mountaineer, Tenzing Norgay (1914–1986), who, alongside the New Zealander, Sir Edmund Hillary, was among the first men to reach the summit of Everest.

Rhododendron wallichii, a particularly fine colour form seen here in the Yumthang valley.

Rhododendron × *thupdenii*, a new addition to the *Rhododendron* flora of the Sikkim Himalaya. It is named in thanks to our Tibetan guide in Sikkim, Thupden Tsering and is also dedicated to Keshab and Sailesh Pradhan.

Rhododendron × *thupdenii* seen here in the centre of the photograph. It is perfectly intermediate between both its parents.

Rhododendron × *thupdenii*. For decades the Yumthang plants were wrongly assumed to be *R.* × *decipiens*, a *R. falconeri* hybrid.

Thupden proved to be a knowledgeable and patient guide throughout our travels together in Sikkim, and I also dedicate this new hybrid to his employer, my good friend Sailesh Pradhan and his father, the great Indian plantsman Keshab Pradhan, the brightest light in the world of plants on the Indian subcontinent.

Rhododendron × *thupdenii* S. O'Brien Hybrida Nova

A sparsely branched tree to 4 m tall, branches erect, stout, tomentose, bark flaking in longitudinal strips to expose flesh-coloured bark beneath. Branchlets with a thin white tomentum. Foliage buds large, conical, green with a white tomentum. Petioles rounded, with a silvery-white tomentum. Leaves leathery, slightly rugulose, held on a horizontal plane, decurrent at flowering time, oblanceolate, 20 × 7.5 cm, midrib grooved, apex apiculate, upper surface glabrous, dark green, with 16 pairs of veins, undersides with a thick silvery-fawn indumentum, midrib prominent. Bud scales light brown, deciduous. Inflorescence a lax, flat-headed racemose umbel of up to 17 tubular-campanulate, flowers, corolla 6-lobed, up to 5cm long, soft pink, fading to almost white, with a deep rose blotch at the base and with rose-coloured spotting on the upper side of the corolla. Stamens 10–12.

Distribution and habitat: India: Sikkim, upper Yumthang valley, in *Abies densa* and *Betula utilis* forest, with *Rhododendron hodgsonii*, *R. wallichii* and *R. wightii*.

Flowering period: Early May.

Conservation status: Forming stable populations, though to date only known from this single location and as such should be considered rare.

With *R.* × *thupdenii* we found great drifts of the stunning *Primula calderiana*, then carrying loose umbels of dark purple blossoms over wonderfully silver farinose foliage. I had previously met this species in the mountains of south-east Tibet. While a beautiful plant, it does have one major drawback – it smells of fish!

Mist moving rapidly north through the Yumthang valley. Trees of *Tsuga dumosa* are silhouetted on a steep mountain ridge to the right. The late evening light in this part of the Himalaya can be magical.

Primula calderiana is relatively common at high altitude in the eastern Himalaya, though never growing in the same great gregarious masses as *Primula denticulata*.

Though described from collections made in Sikkim by William Wright Smith, and from those of his nephew Roland Edgar Cooper in 1910 and 1913 respectively, Hooker made the earliest collection in the upper reaches of the Lachen valley in June and July 1849, while Francis Younghusband found it in the mountains above Thangu in July 1903.

Above there we found Yumthang's high-altitude conifers, *Juniperus recurva* and *Larix griffithii,* growing alongside *Euphorbia sikkimensis*, a fine spurge that in this valley assumes wonderful amber and scarlet autumnal tones.

One evening during an autumn visit we witnessed one of the great Himalayan scenes for which this part of the Yumthang valley is famous. All about us were soaring jagged peaks, glaciers, frozen waterfalls and enormous landslides. Suddenly, as dusk descended, the valleys and dark fir and hemlock-covered ridges beneath us were enveloped in a sea of mist and finally, after a brief wait, the upper snow-clad peaks of the mighty mountains were swallowed in a dense cumulus. That night, we fell into a well-earned sleep in the village of Lachung to the roar of the glacial green torrential river.

William Griffith's larch, *Larix griffithii*, its distinctive crown seen here silhouetted against a cold blue sky in the Yumthang valley in November 2012.

The upper Yumthang Valley and Yume Samdong

The upper Yumthang Valley is a good two-hour drive from the village at Lachung. We first explored this alpine landscape in November 2012 and we began our ascent of the valley at 3,900 m. We could see the Dongkya La, where Hooker and Campbell had made their unauthorised excursion into Tibet in September 1849. The mountain forests there were entirely composed of *Abies densa*, and where landslides had cleared these dark forests, the yellow-flowered *Rhododendron campylocarpum* had recolonised and carpeted vast areas, alongside scattered plants of *R. cinnabarinum*, *R. campanulatum* and dwarf species such as *R. anthopogon*. *Sorbus microphylla* formed small bushy trees still bearing pink-tinged clusters of pearl-like fruits and was one of the few deciduous species in this cold, alpine valley. Gaining height, at 4,150 m, trees disappeared to be replaced by occasional dwarf plants of *R. nivale* and sheets of *Cassiope fastigiata*.

At 4,627 m we reached Yume Samdong, as far as foreign visitors are allowed to venture, and at that point we were to encounter a truly high-alpine flora. Glaciers clung to the highest peak, and the Lachung River (in this area also called the Yumthang River), at that point a small stream, had pushed its way from the Tibetan Plateau, just 12 km away; in the near distance we could clearly see the massive flat-topped, ice-covered bulk of Kangchengyao, where Hooker had explored and botanised in 1849.

Crossing the ice fields, past steaming thermal springs, we made our way towards the screes, passing the stout, fat farinose overwintering buds of *Primula denticulata* on the way. *Primula obliqua* grew alongside the dainty *Lilium nanum* and two of Hooker's finest alpine discoveries – the blue poppy, *Meconopsis simplicifolia* and *Rheum nobile*.

A stupa in the Yumthang valley. A freezing fog fills the air around it.

Late that evening, the sun dipped behind a glacial ridge and within minutes the temperature dropped from a balmy 7°C to –2°C. This was our signal to make a retreat and, having admired a red-billed chough (*Pyrrhocorax pyrrhocorax*), a high-altitude resident of this part of the Himalaya, we began our descent.

CHAPTER 15

A HOSTAGE CRISIS AT TUMLONG

On 3 November 1849, Hooker, Campbell and their team set off for Tumlong, where the Durbar or Sikkim Court was based. Campbell, in the hope of gaining an audience with the Rajah, sent the Chebu Lama in advance with letters announcing his approach. Their route brought them across a steep ridge rising over the east side of Teesta, bristling with enormous trees of *Rhododendron arboreum*, and this led into the Dik (Rycott) River valley, described by Hooker as one of the most lovely and fertile landscapes in Sikkim.

The valley was 16 km long by 5 km broad and was flanked on all sides by lofty mountains, including the snowy Chola range, from which silvery rills of snow and ice descended into the black fir forests beneath. It was a well-populated part of Sikkim, with houses everywhere scattered on the warm valley floor, surrounded by fields of purple buckwheat, wheat, yellow millet, bananas and orange groves. Proceeding eastwards towards the little village of Ramtang (Rangang), they caught sight of the Rajah's palace.

Rhododendron 'Thomas Acton' was raised in 1860 by David Moore at the Royal Botanic Gardens, Glasnevin by crossing a white-flowered plant of *Rhododendron arboreum* ssp. *cinnamomeum* with *Rhododendron campanulatum*, both Hooker seedlings. His son Frederick Moore named it for the owner of the Kilmacurragh Estate in 1880. The original seedling at Kilmacurragh is now a tree.

The royal residence was a long, solidly built stone structure of Tibetan design, with slanting walls, and small windows high up beneath a broad thatched roof, heavily embellished with symbols of Tibetan Buddhism including cylinders of gilded copper. It was surrounded by stupas and enormous prayer flags fluttering from long poles. The building sat on an extensive terrace with spectacular views of the valley and mountains beyond.

The region's political and spiritual importance was made apparent by the many monasteries and convents surrounding the Rajah's palace, including the Lagong nunnery where the lady abbess, the Rajah's daughter, presided. Close by, to the west, the gilded pinnacles and copper canopy of the 18th-century Phodong (Phadong), or royal, monastery gleamed through the surrounding forests.

A large party of armed Lepchas escorted Campbell, Hooker and their men to the village of Tumlong, at that time the capital of the kingdom of Sikkim. Hooker's campsite was on flat ground about 70 m below the palace, close to houses belonging to the Chebu Lama and Meepo. The population on the north side of the Dik River (on which the palace stood), was primarily Sikkim Bhutias and Tibetans, while on the southern side the population was overwhelmingly Lepcha.

A white-flowered tree of *Rhododendron arboreum* ssp. *cinnamomeum* at the National Botanic Gardens, Kilmacurragh. Joseph Hooker made a concentrated effort to collect *Rhododendron* seeds in November 1849, and these were to lay the basis of the future famous *Rhododendron* collection at Kilmacurragh.

The Chebu Lama did his best to arrange an audience with the Rajah, who was surrounded by court officials and councillors, or Amlah, all of whom were loyal to the Dewan, who was in Tibet.

The route to the Cho La pass lay at the head of the Dik valley, crossing the range to the south of Chola peak (6,168 m), and from there leading to a route into the Chumbi valley in Tibet. Hooker wasted no time and set off the following day to visit the pass, soon overtaking Campbell who had started before him. He found the latter being accosted by a court official, asking him to return to Tumlong to meet the court officials. This Campbell refused to do, insisting the officials should meet *him* at his base at Rongpa (Rungpo), on the south side of the valley. The resthouse at Rongpa, where Campbell proposed to meet, lay on a well-maintained road that led to the Rajah's summer palace in the Chumbi valley at an altitude of 1,831 m. Rongpa commanded spectacular views to the north, across the deep river valley, the temples, gompas, hamlets, the royal residence and cultivated fields far below.

The following morning the officials arrived bringing presents and asking Campbell to wait and discuss business with them. Campbell had been put out about the lack of a grand welcome on arrival at Tumlong, and let this be known. He refused to meet the officials, though claimed he was 'ever ready to do so with the Rajah' and promptly set off for the Cho La and Yak La passes.

Seed collecting at Lagyap — the impact of Hooker's work on European gardens

Their route followed a long, steep, narrow ridge leading to Lagyap (Laghep), where a stone-built resthouse sat on a narrow terrace at 3,193 m. By that late season *Rhododendron* seeds were fully ripe and Hooker was kept extremely busy. From Lagyap he sent back to his father, Sir William, at Kew, the seeds of 24 species, which were further distributed by Hooker senior to the major botanical collections of Europe. For example, on 22 April 1850, the Royal (now National) Botanic Gardens, Glasnevin, in Dublin received a consignment of 18 packs of Hooker's Sikkim *Rhododendron* seeds.

From Dublin, seedlings raised from this batch were sent to Thomas Acton at Kilmacurragh in Co. Wicklow, forming the basis for what was to become the largest collection of Himalayan *Rhododendron* species in Europe by the turn of the 20th century. Kilmacurragh was just one of dozens of gardens in Britain and Ireland to be gripped by 'rhododendromania', the Victorian passion for the genus, as a result of Hooker's visit to Sikkim.

During the 1850s Sir William Hooker distributed seedlings across the globe from his son's Sikkim travels, and in the archives at Kew the 'Plants Outward from 1848–1859 Book' records where his seedlings were sent; it makes for fascinating reading. Sir William was determined to raise his son's reputation and placed these young plants in the hands of the best commercial growers and the leading families of the European gentry and aristocracy.

Closer to Kew, many of the foremost London nurseries, including Messrs Knight and Perry, Messrs Rollison of Tooting, Standish and Noble of Bagshot, Hugh Low and Messrs Veitch, all received plants, and English nurseries wasted no time in creating new hardy hybrids from them. For example, Standish and Noble were to use

Mountain peaks at Lagyap as seen from the Nathu La in east Sikkim. Joseph Hooker and Archibald Campbell made a rich seed collection here in November 1849, though it was also in this area that troubles came to a head with the Sikkim court.

Rhododendron thomsonii as a breeding parent, creating the widely popular *Rhododendron* 'Ascot Brilliant', and it is hard to overestimate the effect Hooker's new introductions had on European gardens.

Wardian cases, packed with Sikkim plants, were dispatched to the colonies: Canterbury in New Zealand, Kingston in Jamaica, to Sir William Thomas Denison (1804–1871), the Governor of Van Diemen's Land, James McGibbon (*d.* 1886) of the Cape Botanic Garden and to Charles Moore, the Director of the Royal Botanic Garden, Sydney, to mention a few.

Large commercial nurseries like Messrs Veitch were only too well aware of the profits that could be reaped from the flow of new exotic Himalayan rhododendrons.

Veitch of Exeter, one of the largest and most famous nurserymen of their day. Their catalogue lists several of Joseph Hooker's Sikkim *Rhododendron* species.

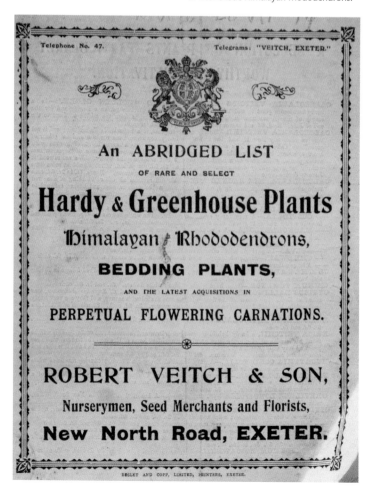

In Ireland the largest collection of 'Hooker rhododendrons' grew at Kilmacurragh, where brother and sister team, Thomas and Janet Acton, kept a large, well-stocked garden. Janet is pictured here peering from the front door of Kilmacurragh House in July 1895.

In England miscellaneous lots of Sikkim plants were sent to various gardens in Cornwall, particularly to William H. Strangways, then enthusiastically developing his garden at Abbotsbury, and Sir Charles Lemon (1784–1868) who received an enormous consignment of rhododendrons for his garden at Carclew. Among these would later be found a natural hybrid of *R. arboreum* ssp. *cinnamomeum* and *R. campanulatum*, a superb plant today known as *Rhododendron* 'Sir Charles Lemon'. Sir Charles was related to the Tremaynes of Heligan, another Cornish property, where several of Hooker's rhododendrons remain to this day. In Devon Sir Thomas Acland (1787–1871) was sent a wide range of rhododendron seedlings for his famous garden at Killerton.

One of the earliest hybridisers of rhododendrons in England was James Robert Gowen (*d.* 1862), a gentleman of independent means who advised on improvements at Highclere, the seat of the Earl of Caernarvon. One of his earliest hybrids, *Rhododendron* 'Altaclarense', was created in 1828 by crossing *R. arboreum* with a *R. ponticum* × *R. catawbiense* hybrid, and is still found in older British and Irish gardens where it makes enormous specimens. He too received a large consignment, and so, by the mid-1850s, thousands of Hooker's Sikkim collections were in the hands of the best British growers. Close friends were similarly looked after. On 2 April 1855, Charles Darwin was sent a tidy batch of a dozen Sikkim rhododendrons alongside a plant of *Berberis darwinii*.

In Ireland, plants of *Larix griffithii* were sent to James Townsend Mackay (1775–1862), Curator of the Trinity College Botanic Gardens in Dublin. At Kew's sister garden, the Royal Botanic Gardens, Glasnevin, David Moore received a plant of *Rhododendron maddenii* in September 1855. He may have requested this particular species since he knew Edward Madden and Madden had visited Glasnevin during home leave from India. Also in Dublin, the Vice Regal Lodge in the Phoenix Park received a number of consignments in 1852. These new introductions were most likely on display in the Turner

curvilinear range of glasshouses there during Queen Victoria's 1853 visit to Dublin. Daniel Ferguson (c. 1801–1864), Curator of the Belfast Botanic Gardens, twice received large miscellaneous lots of Sikkim seedlings.

William Hooker was also well acquainted with the Irish aristocracy and on friendly terms with Vicountess Doneraile who kept a well-stocked garden at Doneraile Court in Co. Cork. In 1854 she received a consignment of orchids, ferns, greenhouse plants and several of Hooker's rhododendrons including *Rhododendron niveum*.

European royal families were similarly treated. The Grand Duchess of Mecklenberg-Strelitz (Princess Augusta of Cambridge) in Germany received eight species of Sikkim rhododendrons in 1853. The Duke of Montpellier brought a larger batch of *Rhododendron* species to Spain. Sir William was aware that benefits could be reaped for Kew by providing gifts like these to the aristocracy.

Joseph Hooker could never have guessed the stir his Himalayan seed collections would make as he explored and harvested seeds around Lagyap. His collections would have a profound effect on gardens of the temperate region. The finest rhododendron garden in England at the time was at Embley Park in Hampshire where William Edward Nightingale (1794–1874) had laid out entire drives through extensive groves of rhododendrons. Joseph Hooker knew Nightingale, and his daughter, Florence (1820–1910). The famous 'Lady with the Lamp' supplied him with a long description of the garden, which he quoted at length in *Rhododendrons of the Sikkim Himalaya*. It is said that Hooker doubled the number of species in cultivation as a result of his visit to Sikkim, and in the following decades, woodland gardens full of choice *Rhododendron* species and hybrids sprang up all over Britain and Ireland as a result of his endeavours.

Doneraile Court in Co. Cork.
Viscountess Doneraile was a friend of
Sir William Hooker and received several
consignments of plants from Kew
including *Rhododendron niveum*.

alpine slopes. He also stressed the need for good sharp drainage for many of the species, though stressing that these plants were exposed to the Indian monsoon, thus needing a moist, humid climate. He also went into detail on the occurrences of natural hybrids and the epiphytic nature of several of his introductions. In the same work he described a new species from Sikkim, one he had not initially identified, the lovely mauve-flowered *Rhododendron niveum*. This must have been a good guide to those pioneering cultivators of his plants, especially those without the means to purchase *Rhododendrons of the Sikkim Himalaya*.

Archibald Campbell was also keen to introduce Sikkim plants to British and Irish gardens. In the archives of the Royal Botanic Garden Edinburgh, there is an entry for 1840, just a year after he had moved to manage the hill station at Darjeeling – 'From the Himalayan mountains, seeds from Mr Campbell – white rhododendron very scarce and handsome; a second species of white rhododendron.' What species it was remains to be guessed.

The Lady with the Lamp, a painting by Henrietta Rae (1891). Florence Nightingale and her father grew many of Joseph Hooker's Sikkim rhododendrons in their Hampshire garden.

Frontispiece of *Illustrations of Himalayan Plants* (1855). Alongside *Rhododendrons of the Sikkim Himalaya*, it more than whetted the appetite of keen plantspeople to grow Hooker's new introductions.

He listed the species found near Lagyap, indicating their altitudinal sequence:

> … in the following order in ascending, commencing at 6000 feet. – 1. *R. dalhousiae*; 2. *R. vaccinioides*; 3. *R. camelliaeflorum* [*sic*]; 4. *R. arboreum*. Above 8000 feet; – 5. *R. argenteum*; 6. *R. falconeri*; 7. *R. barbatum*; 8. *R. campbelliae*; 9. *R. edgeworthii*; 10. *R. niveum*; 11. *R. thomsonii*; 12. *R. cinnabarinum*; 13. *R. glaucum*. Above 10,500 feet: – 14. *R. lanatum*; 15. *R. virgatum*; 16. *R. campylocarpum*; 17. *R. ciliatum*; 18. *R. hodgsonii*; 19. *R. campanulatum*. Above 12,000 feet: – 20. *R. lepidotum*; 21. *R. fulgens*; 22. *R. wightianum* [*sic*]; 23. *R. anthopogon*; 24. *R. setosum*.[1]

This gives a very good idea of the astounding biodiversity of a single valley in Sikkim, taking one genus, *Rhododendron* in this case, as an indicator. *R. argenteum*, as previously stated, is synonymous with *R. grande*. *R. campbelliae* is nowadays *R. arboreum* ssp. *cinnamomeum* Campbelliae Group, *R. glaucum* of Hooker is correctly *R. glaucophyllum* and *R. wightianum* was a simple error for the wonderful golden yellow *R. wightii*.

The Glasnevin gardens were renowned throughout the 19th century, and under David Moore's guidance there was great success raising Hooker's Sikkim *Rhododendron* seedlings. This was not the case elsewhere however, and in *The Journal of the Horticultural Society of London* (1852) Hooker bemoaned the fact that whole batches had been lost through injudicious treatment. As a result, he wrote a long paper detailing each species, their elevation and horticultural potential, highlighting the different cultural needs of species growing in the warmer valleys surrounding Darjeeling from those of the highest

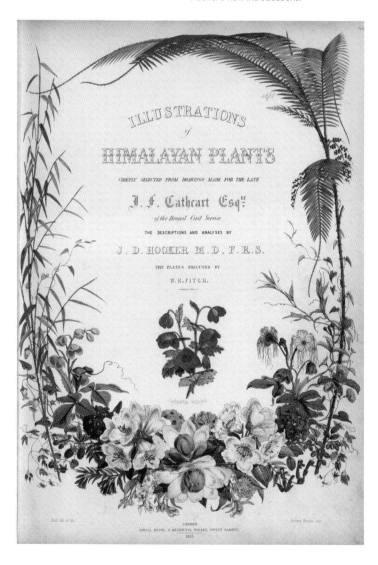

Decaisnea insignis, a plate from *Illustrations of Himalayan Plants*. One of William Griffith's Bhutan discoveries, Hooker collected seeds on the Cho La.

The gardens at Stonefield Castle in Argyll are still full of veteran rhododendrons, as large as those on Himalayan slopes. According to Sir Herbert Maxwell (1845–1937), the Stonefield plants originated from seeds sent from Darjeeling by Archibald Campbell to the then owner, George Campbell (1811–1887); these consignments were later joined by seedlings from Hooker's collections.

In the lower reaches of the valley *Decaisnea insignis* was common on warm, sheltered ridges and covered with long, yellow, broad bean-like fruits, full of a soft, sweet milky pulp, and large black seeds. William Griffith had discovered it a decade earlier in Bhutan, and Hooker collected it three times in Sikkim: in the Lachen and Lachung valleys, and again between Lagyap and the Cho La pass. *Iris clarkei* was abundant around Lagyap, as was another discovery, the lovely *Berberis concinna*, by then covered in edible red berries.

They stayed that night in a stone hut surrounded by a dense forest of *Abies densa* and *Rhododendron hodgsonii*. As darkness fell the sky became brilliant and starlit. A strong north-east wind blew down the valley and a thick hoar frost even powdered the backs of black yaks. The thick, fleshy leaves of *R. hodgsonii* curled into cigar shapes,

as a result of the expansion of frozen fluids in the layer of cells on the upper surface of the leaves.

Along his route towards the pass Hooker collected several new plants, including *Sorbus insignis*, a remarkably large-leaved rowan bearing large corymbs of scarlet fruits. George Forrest sent seeds from the Yunnan–Myanmar border in 1912, though it has not proved hardy in many parts of Britain and Ireland. Passing the tree-line into alpine scrub at 3,353 m he found the stately *Angelica nubigena*, a wonderfully architectural plant rising to 2 m, carrying broad, handsome umbels covered in ripe seeds. The cloud angelica – *nubigena* means 'born among the clouds', an apt description for a plant soaked in summer by the monsoon – is an extreme rarity. It is represented in herbaria by just two or three specimens and is poorly known.

Above 3,962 m the mountains became bleak and bare, the ground was covered in ice, and streams were frozen. Hooker eventually reached the Cho La, where he recorded its altitude as 4,549 m above sea level.

Berberis concinna, a dwarf shrub discovered by Hooker in the Lachen valley. He collected seeds there and near Lagyap. The illustration is from *Curtis's Botanical Magazine* (t.4744) and features the first flowering in 1853.

Another foray into Tibet

Once again Campbell threw caution to the wind. He and Hooker had planned to yet again cross the border (for which they had no permits) into the Chumbi valley in Tibet, and return to Sikkim via the Yak La (where he would find *Saussurea yakla*), a pass that lay a little to the south. It is hard to understand what convinced Campbell in particular, being a political representative of the British East India Company, to break every international border law and create such upset in an independent kingdom like Sikkim, or Tibet, then a protectorate of China. One wonders if there was a touch of European Imperial superiority in both their characters. Their crossing the border on this occasion was to have enormous repercussions.

Descending across the Tibetan side of the pass, a party of Sikkim sepoys tried to send Hooker back forcibly, though he persisted, and about 305 m down found Campbell surrounded by 90 Tibetan border guards armed with matchlocks and bows and arrows. They had finally met their match. The Dingpun, or commandant, marched them politely back to the pass where a frustrated party of Sikkim sepoys was waiting to bring Campbell back to the court at Tumlong. Campbell was stubborn throughout, remonstrating with the men as they approached the pass marking the border.

On the Sikkim side they met an ashen-faced Sentam Subah, who greeted Hooker, though not Campbell. Tempers were frayed and the two men must have known by then that they had finally gone too far. From the windswept Cho La pass they were escorted by a large group to a bleak stone hut at nearby Chumanako, where they were meant to sleep. It was bitterly cold and the little hut soon became overcrowded, so Campbell went out to arrange to have tents pitched instead. He was hardly out of the door before Hooker heard him calling loudly "Hooker! Hooker! The savages are murdering me!"

Hooker had been restrained by several men, but from the door of the hut saw Campbell put up a brave fistfight; he was soon overpowered, however, and thrown to the ground where he was trampled upon. He was then tied hand and foot, the cord doubled and he was shaken violently. The Sentam Subah had Campbell brought bound before him, and asked him through the Chebu Lama if he would write from dictation. Campbell quite rightly answered that if they continued to torture him (this was done by twisting the cords round his wrist using a bamboo-wrench), he might do anything, but his government would recognise nothing thus extorted.

He was then taken to his tent, and all his porters were also tied up, their captors claiming them as Sikkimites, and run-away slaves, and as such, subjects of the Rajah. This was a major blunder. Many of the men were from Darjeeling and therefore under the protection of the British East India Company. All this had been ordered by the Sentam Subah, and while no violence was directed towards Hooker, it was obvious that Campbell, the political figure, was the object of their anger.

Within a few minutes of the incident, the Sentam Subah, pale and trembling like a leaf, and the Chebu Lama, also looking rather frightened, entered the hut. The Subah ordered the men to let go of Hooker and stand guard on either side, and tried to explain that Campbell was now a prisoner by order of the Rajah, who was not happy with his behaviour as a political officer. He was to be taken

Frontier guardsmen; Lepcha Sepoys with Tibetan Sepoys in the background. A wood engraving from *Himalayan Journals*. Hooker and Campbell pushed their luck a little too far by crossing the Cho La pass. The Sikkim Rajah paid dearly for their indiscretion.

to the Sikkim court and confined there until the government of the British East India Company at Calcutta subscribed to a number of articles. The action of capturing the representative of a neighbouring power was common practice along these frontier areas, though the Dewan was ill-advised to take on an organisation so powerful as the British East India Company.

A third person had also entered the tent, and, staring at Hooker, asked if he did not know him? It was not until he stated his name, Dingpun Tinli, that it suddenly dawned on Hooker that this was one of the men sent by the Dewan to guide him and Campbell to the summit of Mount Maenam the previous year. It was also the same person who had been convicted of kidnapping the two Brahmin girls from Nepal and had vowed to take revenge against Campbell who had brought him back within reach of the law. That day had now finally arrived.

Such a crime was of course a violation of the Treaty of Titalia (1817), which, following the Anglo-Nepalese war, guaranteed the security of Sikkim by the British East India Company and returned vast tracts of land annexed by the Nepalese over the centuries to the Sikkim Rajah. In exchange, the latter agreed to abstain from aggression towards the Gorkhas and to allow the British to mediate in any disputes with its neighbours. Further articles agreed that any absconders from British justice, whether criminal or civil, would be arrested in Sikkim.

Therefore, Campbell, as British Political Officer at Darjeeling, had considerable pressure placed on him by the Nepalese court at Kathmandu demanding restitution, which he effected, thus incurring the Dingpun's wrath.

The night spent at Chumanako was extremely cold, with temperatures plummeting to –17°C, and as most of his party had no shelter, Hooker took them into his tent, and managed to convince a few to bring correspondence, by stealth, to Campbell. It emerged that Campbell was to be held prisoner until the Dewan returned from Tibet. The latter had made a grave mistake in engaging the Sentam Subah and the violent Dingpun Tinli, to carry out his plans, which were merely to confine Campbell, not torture and beat him.

On the march back to Lagyap, Hooker kept as close to Campbell as possible, gathering rhododendron seeds along the way. They returned to Tumlong on 10 November, and Hooker camped for the following week close to the Phodong monastery, while Campbell was imprisoned in a cage-like room in a little hut beneath the Rajah's palace. Hooker warned the Subah of the inevitable consequences of such an outrage against the representative of a friendly power, unarmed and without escort.

For the next few weeks any opportunity for extensive exploration was impossible and so he wrote letters to friends at home: George Bentham, Miles Joseph Berkeley and Charles Darwin all received lengthy letters (posted after his release) written during his captivity at Tumlong. He also wrote to his fiancée, Frances Harriet Henslow (1825–1874), daughter of Charles Darwin's mentor, the Reverend Professor John Stevens Henslow (1796–1861). Hooker was desperate to let her know that the reports of his and Campbell's death, then circulating in India, were untrue.

A few days later Hooker was interrogated in the Phodong temple by the Rajah's councillors, a small group consisting of the Dewan's elder brother, a younger brother of the very wealthy headman of the nearby village of Gangtok, and an old Lama. They got very little information from Hooker, but he learned that they hoped to force through a number of articles with the government at Calcutta, which included altering the slavery laws, to draw a new border with Nepal and to institute direct contact between the Sikkim court and the Governor-General of Sikkim. The latter was a contentious subject; due to an oversight at Calcutta many years before, the status of the Agent to the Governor-General, i.e. Archibald Campbell's role, had been reduced considerably, and this was the reason for the Sikkim court's seemingly stubborn silence over the years.

Charles Darwin, a portrait by George Richmond following his safe return from the voyage of the *Beagle*. Hooker wrote to Darwin from Tumlong to assure him he was alive and well.

Joseph Hooker's first wife, Frances Harriet Henslow. This portrait is dated 1867, sixteen years after their marriage. She was among a close circle who received letters from Tumlong.

The scene of the hostage crisis at Tumlong. The Dewan has finally returned, and, wearing a wide-brimmed hat, is carried on a chair, in state, to the Rajah's palace on the brow of a low hill. In the foreground is the small thatched hut in which Archibald Campbell has been imprisoned following his torture. A pair of fine stupas dominate the bottom right-hand corner of Hooker's sketch.

Rajah's residence

Hooker was finally allowed to visit his friend, and actually stay with him, and was relieved to find Campbell in good spirits, though strictly guarded in a small thatched hut constructed from bamboo wattle and clay. Inside, the hut consisted of a dark, black room with a single small window and a fire in the middle on a stone. They slept in a narrow compartment to the rear, which was the cage in which Campbell was originally imprisoned.

In the middle of all this chaos Hooker found time to admire the scenery from this jail house, including some fine stupas close by and a blacksmith's forge at the foot of the hillock below the Rajah's palace.

Despite reassurances, Hooker's men, apart from his Luso-Indian servant and Nepalese porter, were bound and placed in stocks, and were charged with being subjects of Sikkim with no authority to enter service at Darjeeling. They were actually registered as British subjects and had been earlier recognised as such by the Rajah. The real reason for this behaviour was the fear that they would escape to Darjeeling, bringing with them news of Hooker and Campbell's capture.

Campbell, meanwhile, worried that they would carry out a night-time attack on Darjeeling, since threats of sacking the hill station had been made in the past. These rumours were coupled with reports that both Hooker and Campbell had been murdered and that the Rajah had 50,000 Tibetan soldiers marching south to drive the English out of Sikkim. Such reports spread enormous panic at Darjeeling. Guards had been called in from all the surrounding outposts and ladies huddled into a single house for protection, while male inhabitants stood in defence. A major hostage crisis had been initiated, and news eventually reached the senior council of the British East India Company in Calcutta.

On the evening of 15 November 1849, the Dewan arrived at Tumlong in state, carried in a chair given to him by Campbell several years before, and wearing an enormous straw hat with red tassels. Half a dozen Sepoys formed his guard, and approaching Tumlong they bawled out his titles and dignities. He didn't meet the captives on first arrival, feigning sickness. Messengers carrying letters from

Despite the situation at Tumlong, Hooker continued to collect seeds and herbarium specimens for his father at Kew. Pictured above is one of the old Victorian wings of the Kew Herbarium.

Darjeeling were kept in ignorance of Hooker and Campbell's situation on arrival, and were then imprisoned in the same hut, so over time the prison space became overcrowded.

On 3 December 1849 Hooker wrote to his father at Kew, in the hope of having the letter smuggled away to Darjeeling. It was a long, clear letter indicating his situation. He was desperate to know if the report of his and Campbell's death had reached England, a story he claimed, had been deliberately circulated by the Rajah.[2]

It was five days before the Dewan made an appearance, accompanied by the young Gangtok Kaji (headman), and the old monk who had been present at Hooker's interrogation at the Phodong temple. Campbell remonstrated with the Sikkim Prime Minister, pointing out his enormous indiscretions, namely the seizure and imprisonment of the agent of a friendly power, the behaviour of the Sikkim court officials, the disastrous consequences of the possibility of war and finally, the impossibility of the government at Calcutta taking seriously any letters from the Sikkim court while Campbell and Hooker were held prisoner.

The oldest and most venerated Lamas from Pemayangtse travelled to Tumlong requesting their immediate release. Fearing a reprisal from Calcutta, Hooker and Campbell were finally released on 7 December; the Rajah and Ranee sent costly parting gifts, possibly due to news that an English regiment was on its way to Sikkim and that 300 Bhagalpur Rangers had already arrived. The British East India Company had sent another agent to Darjeeling while Campbell was indisposed, thus ruining the Sikkim court's hope of direct communication with the Governor-General, Lord Dalhousie.

On 9 December, Hooker, Campbell and their men left Tumlong, though still under escort as prisoners of the Dewan who led the group. The latter rather coolly led a part of mules and porters loaded with Tibetan merchandise, to trade at Darjeeling and at the Tentulia fair. Hooker was astonished at this behaviour, wondering if this was

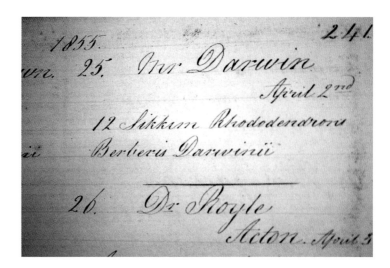

Hooker kept a cool head during the hostage crisis at Tumlong, ensuring his seed collections were not put in jeopardy. These were later raised and distributed from Kew. Charles Darwin received a dozen *Rhododendron* species in April 1855, alongside a plant of the Chilean *Berberis darwinii*.

impudence or stupidity and if the Dewan fully realised the gravity of his situation.

Two days later they reached the tiny hamlet of Gangtok, destined to become a new capital in the following decades, and home to the Gangtok Kaji, a member of one of the oldest and wealthiest families in Sikkim. Just south of here a letter sent by special messenger reached the Dewan, though directed to the Rajah; it was from the Governor-General, then at Bombay. In it, Lord Dalhousie made it absolutely clear that Hooker and his companions were to be released immediately, or his Raj would be forfeited, and if harmed, his life would be the penalty.

The north view of Brian Hodgson's house 'Brianstone.' This image dates to 1847, so the house appears just as Joseph Hooker would have known it. A second floor was added sometime during the late 19th century. Brian Hodgson gave Hooker and Campbell a rapturous welcome following the events at Tumlong.

Dalhousie was also at pains to ensure Hooker's family knew that Joseph was in no immediate danger. He wrote to Sir William at Kew, informing him that he had received a letter from Joseph telling of his capture, giving details of this 'monstrous outrage'. He also let it be known that a force of soldiers had been sent towards Tumlong to compel the captives' liberation, insisting that Joseph was in no immediate danger.

The Dewan, who had deliberately slowed the group's return to Darjeeling, never really understood the anger of the heads of government in India, delaying the group again for no reason. His mood though, changed as they approached Darjeeling, becoming very nervous and depressed; he must have finally realised that his role as Dewan or Prime Minister of Sikkim was fast approaching a bitter end. News also came through that the Rajah's gifts of yaks and ponies had been refused at Darjeeling, and that many of his followers had deserted him following the news of the large bodies of troops building up at the hill station.

The game was over; on Christmas Eve 1849, an exhilarated Joseph Hooker rode across the cane-bridge crossing the Great Rangit River, to Darjeeling, arriving at Brian Hodgson's house, where he was taken for a ghost, and received with shouts of relief, laughter and welcome by his friend, Hodgson, and his guest, Dr Thomas Thomson (1817–1878), who had been waiting for Hooker's return for almost a month. Hodgson had been dreadfully worried, expecting his friend

Rhododendron thomsonii. Hooker named some of the most beautiful Himalayan plants after close friends and colleagues, and for those who smoothed the way for his extensive travels.

Thomas Thomson (1817–1878), Joseph Hooker's childhood friend and travelling companion in India. Hooker named the blood-red flowered *Rhododendron thomsonii* in his honour.

to be carried off in a wooden cage to Lhasa or Peking.[3] Hodgson wrote immediately to inform Sir William Hooker that his son and Campbell had escaped from the hands of the Sikkim barbarians.

Thus ended Joseph Hooker's Sikkim captivity and his travels and exploration of this part of the Himalayan range. His success had been greater than he could ever have dreamed of and his travels had culminated in an unexpected, exciting, though unwanted crisis with the Sikkim court.

Three days after Christmas he wrote to his mother, letting her know that he had safely escaped and that the incident had been exaggerated at Darjeeling and in the Indian newspapers. The news of his confinement had made it to every corner of the subcontinent.

As a result of the hostage crisis at Tumlong, the British East India Company took a swift and harsh revenge on the Sikkim Rajah. The threat of marching an army on Tumlong never materialised, since the dense, forest-clad mountainous landscape of Sikkim was thought to be impracticable for a British army, and because of the fear of Tibetan troops coming to the aid of the Rajah.

Instead, the East India Company modified their threat by seizing the entire southern part of Sikkim lying between the Great Rangit River and the plains of India, including the valuable Terai lands; it was one of the greatest land grabs in the Company's history. The Rajah lost his annual grant of £300 per year for the use of Darjeeling, and the Dewan was disgraced and thrown out of office.

Life at Darjeeling soon returned to normal. As a reward for his support and protection, the Chebu Lama was granted an estate near Darjeeling, where he set up a new residence. Archibald Campbell resumed his duties, and the lands annexed from the Sikkim Rajah were placed under his jurisdiction. The hill station now had a vast hinterland and would make rapid economic strides in the following decades.

Calcutta and the Khasi Hills — an Assam adventure

Joseph Hooker spent much of the first two months of 1850 in Darjeeling, arranging and sending his vast collections to Calcutta, completing manuscripts on his travel and notes for *Rhododendrons of the Sikkim Himalaya*, and also making good his maps and various surveys of Sikkim.

Thomas Thomson had joined him in Darjeeling since they planned to spend the following year travelling and botanising together. Thomson was astounded by Hooker's Sikkim collections, amounting to a hundred porter loads.

They had yet to decide where they should, or could, travel. Bhutan offered the greatest field for exploration, particularly for new botanical discoveries, though it was absolutely inaccessible for Europeans. Nepal was therefore another option, though this plan fell through since the Nepalese Prime Minister, Jung Bahadur Rana (1816–1877) was leaving to travel to England at the time. In any case, one wonders how uneasy the Nepalese court felt about Hooker returning to Nepal, following the Tumlong incident and the British East India Company's seizure of the southern part of the Sikkim Rajah's lands. Many of the Himalayan kingdoms must have looked on nervously.

The final, and seemingly only practical option, was to head for the Khasi Hills (Khasia mountains), then part of Assam, in NE India.

To organise the expedition Hooker travelled to Calcutta, staying at Government House. There he spent several days at the botanic gardens, where Hugh Falconer had returned as Superintendent and was busily restoring and replanting the grounds after William Griffith's rather destructive changes.

The gardens had been laid out originally in the form of an English landscape park, and in a very short period Falconer had carried out enormous improvements and the Calcutta gardens seemed to be on a rapid road to recovery. The lakes had been restored and great sheets of water once again reflected the blue skies. Griffith had swept away the famous avenue of *Cycas circinalis*, once the finest avenue of its kind in any tropical garden, as he also did the experimental groves of teak, mahogany, clove, nutmeg and cinnamon trees.

However, growth was rapid in the Calcutta climate and Hooker could report that the park-like aspect of the grounds was once again reappearing. Broad shady walks had been reinstated and an extensive Palmetum, with a wide range of tall and elegant palms, was in place.

Falconer was also in the process of laying out his 'Thamno-Endogenarium', a novel new scientific garden containing tropical bamboos, tufted palms, rattan palms, bananas, screw-pines and other genera of tropical monocotyledonous plants.

The great banyan tree, *Ficus benghalensis*, still held pride of place. This enormous tree originated as a sapling that sprouted in the crown of a date palm in 1782, and at the time of his visit 68 years later, Hooker estimated that it would almost entirely cover the space occupied by the Great Palm House at Kew.

A view of Garden Reach, Hugh Falconer's residence. This image appeared as a wood engraving in *Himalayan Journals* based on Hooker's original sketch.

Calcutta's famous banyan, *Ficus benghalensis*. In 1849 Hooker reckoned it would cover the space occupied by Kew's Great Palm House. Today it has the largest tree canopy in the world.

Calcutta's famous tree of *Amherstia nobilis* had almost died through ill treatment, but was now recovering. Discovered in Myanmar by Nathaniel Wallich in 1827, this was one of the plants the Duke of Devonshire had sent his gardener, John Gibson, to collect for his conservatory at Chatsworth. The Chatsworth plant died, but was replaced in 1839 by the British East India Company, who obviously obtained their stock from the Royal Botanic Garden, Calcutta. Named for Lady Amherst, the wife of a former Governor-General, it was the botanical sensation of its day.

The only plant to outshine the *Amherstia* was the giant Amazonian waterlily, *Victoria amazonica*, whose giant pads could reach almost 2 m across and for which the botanic gardens at Kew in London and Glasnevin in Dublin had built specially designed aquatic houses in a race to be among the first places in Europe to flower it. Growing it in Calcutta's hot humid climate was child's play, and in *Himalayan Journals* Hooker reported that by the spring of 1853 no fewer than five plants of the splendid *Victoria* waterlily were in bloom in the lakes at the Royal Botanic Garden.

The Calcutta gardens were a great distribution centre of both ornamental and economic plants. During Hooker's stay Hugh Falconer received a box of living plants packed in moss and transported from America on an ice ship. The day after their arrival they burst into leaf and were promptly packed into Wardian cases and despatched to the Himalaya.

Joseph Hooker certainly made his mark on Indian botany. Many of the principal avenues of the Calcutta Botanic Garden are named in honour of the founding fathers of India botany. Hooker's name is among them.

In May 1850 Joseph Hooker set off with Thomas Thomson to explore the mountains and forests of Assam in north-east India. The Brahmaputra River brought them within view of the Khasi Hills. Another adventure had begun.

In mid-April, Hooker returned for the very last time to Darjeeling, to make final arrangements before setting off with Thomas Thomson on a nine-month-long expedition to the Khasi Hills. He departed on 1 May 1850 with a heavy heart; Sikkim had been a marvellous adventure, surpassing even his travels in the Antarctic, and his immense collections were full of exciting new species. He would be for evermore remembered by botanists and geographers for having blazed a trail across the kingdom's great snowy mountains and deep valleys. He would miss the kindly Lepcha people, and he held great hopes for Darjeeling in its development as a hill station, predicting a great future for the place:

> We started on the 1st of May, and I bade adieu to Dorjiling with no light heart; for I was leaving the kindest and most disinterested friends I had ever made in a foreign land, and a country whose mountains, forests, productions, and people had all become endeared to me by many ties and associations. The prospects of Dorjiling itself are neither doubtful nor insignificant. Whether or not Sikkim will fall again under the protection of Britain, the station must prosper, and that very speedily. I have seen both its native population and its European houses double in two years; its salubrious climate, its scenery, and accessibility, ensure it so rapid a further increase that it will become the most populous hill-station in India.[4]

And so it did, just as Hooker had predicted.

My travelling companions, Wyn Hughes and Esther Shickling, seen here strolling along the famous avenue of *Roystonea regia* at the Calcutta Botanic Garden.

CHAPTER 16

TO GANGTOK AND TUMLONG

It was late November 2012 when we left Darjeeling and began our journey towards Gangtok, the state capital of Sikkim. We left the hill station bathed in sunshine, as the great *Dahlia imperialis* put on a riotous display along street-side embankments.

Our route took us in a north-easterly direction, soon following the course of the Great Rangit River, where troops of rhesus macaques, *Macaca mulatta,* had emerged from the surrounding jungle to beg from passing vehicles. Tropical plants abounded, *Bombax ceiba* made enormous forest trees and beneath them grew showy bauhinias, dwarf date palms like *Phoenix acaulis* and squat plants of *Cycas pectinata.*

At Melli Bazaar, a small town straddling the states of West Bengal and Sikkim, we had our passports stamped and from there visited the viewpoint overlooking the confluence of the Teesta and Great Rangit Rivers. Beyond there we continued north, through the tropical Teesta River valley where enormous screw pines, *Pandanus furcatus,* dominated the hot, humid valley floor. Near Temi we met dozens of brightly dressed women on the roadsides with bamboo wicker baskets tied to their foreheads. They were employed to pick tea in the nearby Temi tea gardens, and were on a well-deserved morning break.

Ladies from the nearby Temi tea gardens on a break from picking tea. Tea is big business around Darjeeling, though the garden at Temi, near the Teesta valley, is the only such enterprise in Sikkim.

The following day we set out from Gangtok, a large, modern hill-top town, and drove north along the Mangan road, taking us back once again into the valley of the Teesta River. Thanks to the steepness of this rain-drenched valley, Mangan is a major landslide zone and we met many obstacles during our return visits over the course of the next several years. In November 2012 we encountered an enormous landslide that had torn forests into the valley several hundred metres below.

Cicadas made a deafening racket in the surrounding forests, while giant bamboos covered the higher mountain slopes in places. The most abundant shrub in this warm part of the valley was the lovely *Oxyspora paniculata*, with long pendent panicles of pinkish-purple flowers. One of the prettiest plants on the edge of the forests was the *Loxostigma griffithii*, a large, showy shrubby gesneriad, carrying cymes of yellowish-purple, foxglove-like blossoms with purplish-brown spotting on the interior.

With it grew sheets of ginger lilies and *Miscanthus nepalensis* (covered in silky pendent plumes), and, suspended from the branches of trees overhead, the rope-like vines of *Trichosanthes tricuspidata*, were laden with red, fleshy tomato-like fruits.

The various *Terminalia* species were the most spectacular trees in the warmer part of the valley, particularly *T. myriocarpa*, bearing spectacular panicles of pinkish-red flowers. In the Himalaya, where the tree is employed in traditional herbal medicine, the bark is used as a cardiac stimulant and as a mild diuretic.[1] *Engelhardtia spicata*, another of my favourite Sikkim trees, appeared on the scene, bearing long, pendent spikes of *Carpinus*-like fruits. It is a great shame that this lovely tree cannot withstand the rigours of the Irish or British winter, so we have to content ourselves with seeing the fine 20 m

The walnut relative *Engelhardtia spicata*, a common tree in the warmer valleys of Sikkim. In parts of Asia the bark of wild trees is harvested as a source of tannins. It generally forms medium-sized trees but occasionally reaches up to 40 m.

The Dewan's return to Tumlong. A wood engraving from *Himalayan Journals* based on Hooker's sketch. The figure standing by the gable wall of the Rajah's palace gives scale to the scene.

tall trees, with their rugged, fissured bark, in the upper Teesta valley. *Engelhardtia* is an interesting palaeo-tree that once had a far wider distribution, including in NW Europe. Its pollen is preserved in the fossil record in Denmark, where it grew during the Miocene.

Rising higher into the valley, we were suddenly rewarded with a spectacular view of one of Sikkim's most beautiful snowy peaks, Siniolchu (6,888 m), giving us, during our first visit to Sikkim, an idea of the breathtaking landscape that lay further north-east on the Tibetan frontier.

Tumlong — the ruined capital

Tumlong lies north-west of Gangtok, and, at the time of Campbell and Hooker's confinement there during the famous hostage crisis in November 1849 had been the kingdom's capital since 1793. The settlement centred on the Rajah's palace and was the smallest capital of any of the Himalayan kingdoms.

Tumlong was an important part of our itinerary, and it was there perhaps, more than any of the other locations we visited in Sikkim, that we felt we were indeed travelling 'in the footsteps of Joseph Hooker'. We arrived in the early afternoon, and found the site of the Royal Palace. We were led there by two young Buddhists monks from the nearby Phodong monastery.

Tumlong was abandoned following a major earthquake, when the capital was shifted to Gangtok in 1894, and now lies in ruins, though elements of the buildings from Hooker's sketch in *Himalayan Journals* remained, including the Rajah's crumbling palace and several

Stupas feature prominently in Hooker's sketch of Tumlong, including the structure seen to the far right with a stepped base.

The same stupa in November 2012. Like the rest of Tumlong it lies in ruins and the Mexican marigold, *Tagetes erecta*, has colonised the shattered base. *Alnus nepalensis* now covers the steep slopes on which the Rajah's palace once sat.

Buddhist stupas. We descended to the stupas along the same path that led to the hut in which Campbell was imprisoned 163 years before our visit.

Our route was lined by fine trees of Roxburgh's fig, *Ficus auriculata*, a common tree in the warmer valleys. The steep grassy slope on which the palace once stood is now a dense copse of *Alnus nepalensis*, obscuring the view of the ruins of the palace above. The Buddhist stupas, sketched by Hooker in November 1849, as his colleague lay in confinement in a hut close by, were crumbling, and on their shattered whitewashed remains, the Mexican *Tagetes erecta* had naturalised itself. Tumlong, its palace and stupas are slowly disintegrating, and we were surprised that a site like this, with such an important place in Sikkim's history, should be allowed to fall into further ruin.

The most abundant plant at Tumlong was *Edgeworthia gardneri*, a bushy shrub of about 4 m tall, named for the Irish botanist M. Pakenham Edgeworth, from Edgeworthstown, Co. Longford.

We spent several hours at Tumlong, revelling in the atmosphere of the place, standing on the site of Campbell's makeshift jail and scrambling up and down the slopes in search of buildings and landmarks that appeared in Hooker's sketch. As we climbed our way past the palace, I wondered if Hooker ever regretted the events he and Campbell precipitated at Tumlong and the terrible price the Sikkim Rajah paid for the foolish decisions made by his Dewan.

Gangtok — the Residency, Lagyap and the Nathu la Highway

Gangtok was to be our base on several occasions, and on each of our visits we stayed at the Hidden Forest Retreat, a guest house founded by S. T. Lachungpa, who until his recent retirement was Principal Chief Conservator of Forests and Secretary to the Government of Sikkim. He was extremely knowledgeable on his subject and gave us a good briefing on modern forestry husbandry in Sikkim. His daughter Kesang managed the guest house, and, like her father, had a superb knowledge of plants, particularly orchids.

Gangtok lies at just 1,650m, low for the eastern Himalaya, and in this damp, warm climate, growing orchids is child's play. Near a terrace overlooking the city suburbs and distant mountains was a display of species and hybrids of *Coelogyne, Dendrobium* and *Vanda,* which would have easily scooped a medal at the Chelsea Flower Show. Inside, guests had full access to the family's impressive botanical library, while on the walls were several botanical paintings by Hemlata Pradhan, the Kalimpong-based artist who trained at Kew and has won gold medals from the RHS at their London shows.

The 1.2 ha gardens were equally impressive and proved a perfect base for our group, with a lovely mix of perfectly grown citrus fruits and vegetables, bordered by exotics (to us) including two Sikkim tree ferns, the slim-trunked *Cyathea brunoniana* and *C. chinensis.*

The Sikkim tree fern, *Cyathea brunoniana,* seen here in the garden of the Hidden Forest Retreat in Gangtok. This part of Sikkim possesses a benign climate and, like Darjeeling, it enjoys views of Kangchenjunga.

Gangtok, the modern capital of Sikkim. During Hooker's visit it was a tiny unimportant village which doesn't even feature on his map of Sikkim.

During our visits to Gangtok we were also always greeted with great hospitality by the Indian forester and plantsman Keshab Pradhan and his wife Shanti (whose son Sailesh was guiding our expeditions). An expert on Himalayan rhododendrons and primulas, he was recently awarded the Veitch Memorial Medal by the Royal Horticultural Society. It has always been a humbling experience to meet this legendary figure, and ramble around his 2.4 ha garden and nursery, where he and Sailesh hybridise azalea and day lily cultivars, and grow an eclectic mix of exciting garden plants, with an extensive collection of citrus fruits, orchids and rare exotics including the spectacular *Curculigo orchioides.*

During our 2013 visit to his home we brought two Veitch memorialists beneath one roof (a rare occurrence in India), when Helen Dillon, another medal holder, visited as part of our group. Needless to say, they had a lot to talk about and much more in common than their distinguished awards.

One of the oldest buildings in Gangtok is the Raj Bhavan, currently the official residence of the Governor of Sikkim, though in the heyday of the British Raj it was known as 'The Residency' and provided a home for the British Political Officer in Sikkim.

The first of these officers, Calcutta-born J. Claude White (1853–1918), was appointed in 1889, relieving the Sikkim Rajah of any administrative role and transforming Sikkim's economy in a remarkably short period. Thus Britain finally controlled the affairs of state of Sikkim and treated it as though it was part of the Bengal Presidency. White enhanced the wealth of this Himalayan kingdom by carrying out a major survey of Sikkim's natural resources, through road building and through organised taxation. He also encouraged mass migration of labour from neighbouring Nepal to work the land, taking care not to interfere with the dynamics of ethnic groups in north Sikkim, which to this day is still a stronghold of Bhutia and Lepcha groups.

The original Residency was built by White, and in his account of his life in the princely states of the eastern Himalaya he reminisced about moving to Gangtok (then a tiny village) and building a house there:

> One of the first things to be done on my appointment to Sikhim [*sic*] was to build a house, not an easy task in a wild country where masons and carpenters were conspicuous by their absence, where stone for building had to be quarried from the hill-sides and trees cut down for timber. In my jungle wanderings round Gangtak [*sic*], I came across a charming site in the midst of primeval forest which seemed suitable in every way, so I determined to build on it, felling only the trees which might possibly endanger the safety of the house, a necessary precaution, as many of them were quite 140 feet high... By levelling the uneven ground and throwing it out in front, I managed to get sufficient space for the house, with lawn and flower beds round it. Behind rose a high mountain, thickly wooded, which protected us from the storms sweeping down from the snows to the north-east, and in front the ground fell away with a magnificent view across the valley, where, from behind the opposite hills, Kangchenjunga and its surrounding snows towered up against the clear sky, making one of the most beautiful and magnificent sights to be imagined, and one certainly not to be surpassed, if equalled, anywhere in the world.[2]

White, and his wife Jessie Georgina, were passionate gardeners, and having moved into their new house at Christmas 1890, they set about creating a garden. This they did with great gusto, importing roses from England and France and planting against the walls of the

J. Claude White, Sikkim's first British Political Officer. He built 'The Residency' in 1890 and surrounded it with a beautiful garden.

house old favourites like 'Gloire de Dijon', 'Reine Marie Henriette', 'Cloth of Gold' and 'Devoniensis', while across the lawns they scattered roses in hundreds, again using classics like 'Souvenir de la Malmaison', 'Paul Néron', 'Madame de la Roche-Lambert' and the wonderfully scented 'Marie Van Houtte', an old-fashioned tea rose bred in France in 1871.

In the Sikkim climate these roses soon flourished, flowering with a profusion unmatched in Europe. In the summer months, great drifts of the native *Lilium wallichianum* scented the air, and in a tree fern ravine within the grounds, *Cardiocrinum giganteum* carried its great trumpet-like blossoms on 3.6 m tall stems.

In spring, daffodils, primroses, wallflowers and schizanthus created colour in the borders surrounding the house; in summer mignonette, sweet-pea, sunflowers, montbretias and cannas continued the show. This little piece of colonial Arcadian paradise soon drew the attention of the local people, and the Whites became accustomed to visits by cheerful groups wanting to see how they lived and what European furniture looked like. He was also frequently visited by the Rajah and Ranee, despite getting off to an initial bad start with the royal couple.

The Residency in 1921. A well-tended garden surrounds the house. It was the first European-style residence to be constructed in Sikkim.

Today, The Residency is known as the Raj Bhavan, and is the seat of the Governor of Sikkim. In recent times it has been retrofitted to protect it from earthquake damage, though the original building is still recognisable.

The Whites' interest in plants wasn't just restricted to garden exotics. They loved the many epiphytic orchids that grew in the trees surrounding the Residency, and soon after settling in Mrs White discovered a new species, later named in her honour, *Cymbidium whiteae*, one of Sikkim's rarest endemic orchid species, and to date known only from Gangtok where it grows on *Schima wallichii* and various *Castanopsis* species.

J. Claude White also collected plants in Bhutan, and his most famous discovery, *Primula whitei*, is perhaps best-known by growers of alpine plants. He found this wonderful blue-flowered petiolarid on the Pele La, one of the high passes of Bhutan in May 1905. It is extremely rare in cultivation, though the famous Irish plantsman, David Shackleton, grew over 300 plants in his walled garden at Beech

Park House near Dublin, right up to the time of his death in 1988. By chance his daughter-in-law, Daphne Levinge Shackleton, was travelling as a member of my group when we visited the Raj Bhavan in May 2015, allowing her the opportunity to see where White had lived and gardened.

Gangtok was to become strategically important, when in 1903, Lord Curzon, the Viceroy of India, made White Deputy Commissioner of the Tibet Frontier Commission under Francis Younghusband, which led to the 1903–4 British Expedition to Tibet. The original aim of the expedition was to settle ongoing disputes about the Sikkim–Tibet border, though Younghusband far exceeded his orders, and the visit became a *de facto* invasion of Tibet, with monasteries looted *en route*. The expeditionary force left Gangtok in 1903, travelling along the Nathu La road into the Chumbi valley in Tibet, and as it progressed, White, who was on good terms with the Tibetan leaders, became horrified by Younghusband's arrogance, and wrote to Lord Curzon and Lord Kitchener, to have his orders cancelled.

In a fit of rage Younghusband exacted his revenge by having White sent back to Sikkim, and left him in leech-infested jungles to arrange porter and mule transport to Tibet. In 2014, one of my travelling companions, the Co. Westmeath-based garden consultant, Octavia Tulloch, visited Gangtok, and the hill-top town had special resonance for her, as her great-grandfather had been a civilian member of Younghusband's Expedition.

In the 20th century the Residency became a base for many visiting botanical explorers. Later Political Officers included Frederick Marshman Bailey, the explorer who discovered the lovely Himalayan blue poppy, *Meconopsis baileyi* in south-east Tibet, and the low-growing *Rhododendron baileyi* from the same region. Bailey's rhododendron is also found in the Yumthang valley in Sikkim, though it is rare there, and was missed by Hooker during his visit to the valley. Bailey famously became a double agent for British Intelligence and was one of the last great protagonists of the Great Game.

In 1933 Frank Ludlow and Major George Sherriff stayed as guests at the Residency before botanising along the Nathu La and venturing further into Tibet's Chumbi valley and the mountains of Bhutan. Both men had a long-standing relationship with India and the Himalaya, and their botanical adventures began in Gangtok. In 1942 Sherriff married Betty Graham, the daughter of a well-known Kalimpong-based Scottish vicar and missionary, Dr John Anderson Graham (1861–1942), and for many years the Sherriffs kept a summer house in the hill station at Kalimpong.

In the following years, George and Betty Sherriff botanised the Himalaya together, and her most famous finds must be Betty Sherriff's dream poppy, a form of *Meconopsis gakyidiana*, and the lovely climbing *Rosa brunonii* 'Betty Sherriff'.

Our group at the Raj Bhavan in May 2015. Left to right: Bruno Nicolai, Averil Milligan, Daphne Levinge Shackleton, Gráinne Larkin, the author, Lesley Fennell, Bruce Johnson, Orlaith Murphy, Derek Halpin, Kristin Jameson, Thupden Tsering. In the background, to the left, is *Wisteria sinensis*, an original 1891 planting by J. Claude White.

Frank Ludlow (standing) with Major George Sherriff and the British Political Officer, Frederick Williamson. Ludlow and Sherriff used The Residency as a base for botanical exploration, and their very first collection was made near Changu in east Sikkim. Frank Kingdon Ward and Lord Cawdor also stayed there in March 1924 before setting off on their now-famous expedition to the Tsangpo Gorge in SE Tibet.

In April 1933 George Sherriff and Frank Ludlow arrived in Gangtok in preparation for their first plant-hunting expedition together. Gangtok was then still a small town and one of the world's smallest capitals. They stayed in the Residency with the new Political Officer Frederick Williamson (1891–1935) and his bride, Margaret Dobie Marshall. The Williamsons' wedding was performed by Sherriff's father-in-law at the Residency and guests included the royal families of Bhutan and Sikkim. Following the ceremony, the couple spent part of their honeymoon with Ludlow and Sherriff in Bhutan.

At Gangtok the two plant hunters made time to visit His Highness, Sir Tashe Namgyal (1893–1963), the eleventh ruler of the Namgyal dynasty of Sikkim, who had been crowned by the 13th Dalai Lama. They were also on the hunt for birds, and their stay at the Residency offered a perfect opportunity to boost their collections for the Natural History section of the British Museum. By then it was early May, and the pre-monsoon rains had begun, bringing with them the swarms of leeches for which the warmer valleys of Sikkim are famous:

> ... Ludlow and Sherriff were depressed and disheartened by the leeches – there were great swarms of them infesting the jungles near the Residency. 'We dared not move off the stony path. Immediately we did so we were attacked by myriads of these obnoxious creatures. Sometimes we shot a bird which dropped above or below the pathway. When this happened we made a dash for the bird and after retrieving it fled back to the path and picked off the leeches. If we could not find the bird immediately we had to leave it and beat a hasty retreat. We lost several specimens in this way'.[3]

On 5 May 1933, Ludlow, Sherriff and their party with several Lepcha collectors set off from Gangtok along the Nathu La road, soon reaching the temperate zone where they encountered trees of *Rhododendron arboreum* 15–18 m tall, with several trees of huge white-flowered *Magnolia campbellii*, twice that height, growing among them. Their first plant collection together, from the mountains near Changu,

was the dainty pink-purple *Primula gracilipes* (L. & S. 1), a species first discovered in east Sikkim by Charles Baron Clarke, and later collected in the Yumthang valley and at Changu by George H. Cave, Curator of the Lloyd Botanic Garden in Darjeeling.

In May 2015 we were allowed to visit the private estate surrounding the Raj Bhavan (Royal Palace), and, entering a heavily guarded gate, were ushered through a long avenue to the main building where we were granted an audience with the Governor and were presented with white silk scarves, as is the custom in Sikkim, Tibet and Bhutan.

Octavia Tulloch, pictured here on the trail leading to the Tibetan refugee village of Tshoka, in west Sikkim, in 2014. Her great-grandfather travelled as the only civilian member of Younghusband's expedition to Tibet. She is admiring the black-fruited *Skimmia melanocarpa*, one of Hooker's Sikkim discoveries.

The Governor of Sikkim, Shriniwas Patil, makes a presentation of a traditional white silk scarf and gifts to Kristin Jameson. His aides-de camp stand in the background.

Though enormously extended in recent times, the house retains much of its charm and it was wonderful to see the place in which Bailey, Ludlow, Sherriff and the Whites were once based. Roses still grew on the manicured lawns, though not in the same profusion as when the Whites lived there, and on the house was a thick-stemmed *Wisteria sinensis*, planted there by Claude White in 1891.

The estate surrounding the house is famous today for its biodiversity, and near the front door of the house we were startled when one of Hooker's Sikkim discoveries suddenly appeared on the scene: *Japalura variegata*, the east Himalayan mountain lizard, a native of Darjeeling District, the Jalpaiguri District of West Bengal

Hooker also collected reptiles for the British Museum. While our group were standing by the door to the Raj Bhavan one of his discoveries, the east Himalayan mountain lizard, *Japalura variegata*, made a sudden appearance.

and the extreme east of Nepal. On his return from Sikkim, Hooker sent his reptile collections to the British Museum, and several proved to be new species, being named and described by John Edward Gray (1800–1875). Hooker found it near Darjeeling and it was the last thing we expected to see at Gangtok that day.

Behind the house lay dense forest dominated by evergreen trees including *Beilschmiedia roxburghiana*, *Cinnamomum obtusifolium*, *Elaeocarpus lanceifolius*, *Exbucklandia populnea* and *Magnolia doltsopa*. *Rhododendron arboreum* (the blood-red flowered variant) also grew on the hillside, and, in the ravines, were tree ferns, *Cyathea spinulosa*, forming enormously tall specimens alongside *Himalayacalamus hookerianus*, with its lovely blue culms.

Several climbers scaled the trees to great heights including the rampant yellow-flowered *Mucuna macrocarpa*, though the most impressive liana by far was *Rhaphidiphora decurvisa*, a robust climbing aroid, with leaves like that of a Swiss cheese plant.

Alnus nepalensis and *Quercus lamellosa*, two superb host trees for orchids, also inhabited the forests and a wide range of epiphytic orchids clothed the canopy overhead, including *Cymbidium hookerianum*, one of Joseph Hooker's Darjeeling discoveries named for him in *The Gardeners' Chronicle* in 1866, on the first New Year's Day of his Directorship at Kew. It has become extremely rare in the wild due to over-collecting.

Gyalmo Hope Namgyal. Born in San Francisco in 1940, Hope Cooke married the last Rajah of Sikkim. She is seen here among rhododendrons in north Sikkim.

Several branches of the Indian army wear highly decorative and colourful headgear. Seen here is an army officer at his security post at the Raj Bhavan.

The last king's uncle, the scholarly, strong-willed, Oxford-educated Sidkeong Tulku Namgyal (1879–1914), enjoyed a very brief reign, dying of heart failure at just 34 under exceptionally suspicious circumstances. It seems he was too clever for British India. In December 1914, while feeling ill, he was treated by a British physician from Bengal who administered a heavy transfusion of brandy, placed him beneath a heavy layer of blankets, and kept a strong fire near his bed. Death came within an hour, thus vanquishing a promising career and a time of positive change for the little Himalayan kingdom. As a contemporary put it, 'His spirit was too strong you see. The British liked their rulers to be compliant.'[5]

Sikkim's geopolitical importance was to ultimately lead to its downfall. Events in Tibet during the 1950s and the Sino-Indian war of 1962 convinced Independent India to complete the work begun over a century previously by the British East India Company: the 1975 annexation of Sikkim as a state of India and the abolition of the monarchy. Three years later, Crown Prince Tenzing, heir to the Namgyal dynasty, was killed when his car left a steep road in Gangtok and plunged into the hills below. It was a sad end to over three centuries of rule. Sikkim entered a time of much change and modernisation.

Sikkim is now a state of India and a statue at the Raj Bhavan honours the father of the Indian nation, Mahatma Gandhi (1869–1948).

Another dominant feature of the Gangtok cityscape is the Royal Palace, formerly the home of the Sikkim Royal Family. Its last occupant was Palden Thondup Namgyal (1923–1982) and his Irish-American wife Hope Cooke (*b.* 1940), the final gyalmo or Queen Consort of Sikkim. They first met during the summer of 1959 at the Windamere Hotel in Darjeeling and married in a spectacular fairytale-like ceremony covered by the American press four years later, capturing global attention. Their reign, however, was during a time of tumultuous change in Sikkim. Due to Claude White's policy of mass migration of Nepalese workers into Sikkim, the indigenous peoples, the Lepchas and the Bhutias, became a minority in their own country.

This ultimately led to the Sikkim Nepalese demanding change, one-voter parity and union with India. In her autobiography, Hope Cooke describes a smear campaign orchestrated from government offices in New Delhi and how former friends to whom they had shown great kindness became adversaries. Publications in her native America described Hope Cooke as a Himalayan Marie Antoinette and Sikkim's relationship with India became more and more complex.

The Sikkim Royal Family had seen much humiliation throughout the 20th century, particularly when the British Political Officer J. Claude White supplanted the then Rajah Thutob Namgyal (1860–1914), placing the royal couple more or less under house arrest and relieving them of their administrative powers.[4] Thus began the slow decline of the ruling house of Sikkim.

Botanising around Gangtok

Though a relatively new city, Gangtok has grown rapidly in recent years and is an interesting place for a brief stop; its hinterlands make for excellent botanising. For us it proved to be a perfect base for exploring SE Sikkim. In the autumn of 2013, we drove east along the Nathu La Highway to visit Lagyap (Laghep), from where, in just two days in November 1849, Hooker collected the seeds of 24 *Rhododendron* species, these collections being the parent material of many of the surviving 'Hooker rhododendrons' at Kilmacurragh and elsewhere.

The roadside cliffs beyond Gangtok were home to hundreds of plants of *Curculigo crassifolia*, a fantastic foliage plant with bold, pleated 1.5 m long sword-shaped leaves wonderfully silver-felted beneath. The pretty yellow flowers are held in small rounded racemes at the base of the plant, though are often hidden by foliage, which is the showiest feature of the plant in any case. It is perfectly hardy at Kilmacurragh, where it is planted close to the walled garden, at the base of a large old tree of the Chilean *Luma apiculata*. Apparently the secret of growing it in British and Irish gardens is to keep its roots relatively dry in winter, and this is achieved by planting it on the sunny side of an evergreen tree.

Lagyap is close to the Cho La, the type locality for Hooker's lovely yellow-flowered poppy, *Cathcartia villosa,* and we were pleased to find sheets of this within the confines of the nearby Kyongnosla Alpine Sanctuary at 3,193 m. Hooker found it on mountain clefts at about 3,048 m, on 6 November 1849. He also collected seeds, sending them to Kew where it first flowered in June 1851.

He chose this pretty little plant to commemorate his friend, the retired Bengal judge, John Ferguson Cathcart, whose Indian artists had painted it at Darjeeling. Following his retirement from the Indian civil service, Cathcart was returning to Europe when he died

Curculigo crassifolia. In gardens this fantastic foliage plant needs sharp drainage. Seeing its cliff-side native habitat explains why.

suddenly at Lausanne, Switzerland in 1851, aged just 49. Following his death, Cathcart's family donated his botanical art collection of almost 700 folio coloured plates of Himalayan plants to Kew, where, on his return, Joseph Hooker had them re-worked by Kew's botanical artist, Walter Hood Fitch. They were published by Hooker (who provided text for the plants featured) in *Illustrations of Himalayan Plants* (1855), one of the most beautiful florilegia ever published, a rare work that today fetches astounding sums at auction.

We found Cathcart's lovely poppy growing beneath a wide range of *Rhododendron* species with *Satyrium nepalense* (then in full bloom), *Phlomoides macrophylla* and the tiny creeping *Gaultheria nummularioides*. *Satyrium nepalense* is a common orchid across the Himalaya. Sir Henry Elwes collected tubers during his travels in Sikkim, presenting some to Kew in 1881, where they flowered for the first time in Europe in the following year. Elwes was always keen to introduce new, exciting garden-worthy plants, sending the roots and tubers of several other Sikkim plants from the same trip, and it was by this means that he re-introduced the spectacular *Hedychium gracile* to gardens.

Elwes also presented large consignments of seeds to Kew, many of which featured in the pages of *Curtis's Botanical Magazine*. In 1883 Joseph Hooker featured *Aster diplostephioides*, a stunning little blue aster I was once fortunate enough to see in the mountains of west Bhutan. Always keen on economic botany, Hooker stressed

Our vehicles on the Nathu La Highway in November 2013. This route afforded us the opportunity to botanise in the mountains to the east of Gangtok.

that in Kashmir the roots of this species were extensively used in washing clothes. This was soon followed by a plate of the spectacular *Dracocephalum speciosum*, another of Elwes's introductions to flower at Kew during the early 1880s.

Interestingly, we also found the tomato-relative *Anisodus luridus* (syn. *Scopolia stramonifolia*) growing alongside *Iris clarkei*; the latter plants are often dried for winter fodder, though the *Anisodus* is poisonous when fresh.

The Nathu La is an offshoot of the ancient Silk Road, and the highway we travelled along led us along a parallel route south of Lagyap. The Lagyap mountains, clearly visible to the north, were steep and densely forested. Looking on, it was easy to visualise Hooker and Campbell exploring the valley in 1849. It was beyond Lagyap where Campbell met his misfortune, being beaten and tortured; we met no such mistreatment during our visit to the region!

Cathcartia villosa, formerly *Meconopsis villosa*. This perennial Himalayan poppy is still common in the mountains of east Sikkim.

Joseph Hooker discovered Cathcart's Himalayan poppy on 6 November 1849, just before the famous hostage crisis and his forced return to Tumlong. He sent seeds to Kew, where it first flowered in June 1851, when it was promptly illustrated by Walter Hood Fitch for *Curtis's Botanical Magazine*.

During our travels in India we visited several places that never made it on to Hooker's itinerary, including, in November 2013, a trip to New Delhi to see the work of the Edwardian architect, Sir Edwin Lutyens (1869–1944), and in the old city we made time to take in the Red Fort and many of the city's Mughal mausoleums. The most beautiful of these, the 16th-century Humayun's Tomb, brought Persian style to Delhi and was later to become the blueprint for the Taj Mahal. It was at Humayun's Tomb, following the Indian Rebellion of 1857, that the last Mughal Emperor surrendered to the British East India Company, thus ending over three centuries of Muslim rule.

The Lahore Gate to Delhi's famous Red Fort, residence of the Mughal Emperors for almost two centuries until 1857. Today it is a UNESCO World Heritage Site.

Humayun's Tomb in Delhi, the inspiration for the Taj Mahal at Agra. Built by Humayun's son, Akbar, it is surrounded by wonderful Mughal gardens.

Our first view of the Taj Mahal was at dawn as we entered through the East Gate. We stood in awe, just as millions before us have, at a building that has been described as 'a teardrop on the cheek of eternity'. Marianne North was also suitably impressed.

From Delhi we travelled south-east, through the flat, productive farmlands of Uttar Pradesh, reaching Agra as the sun was beginning to set on the city suburbs. The following day we rose early to visit the majestic Taj Mahal, its vast marble surface bathed in glowing, soft, pearly-cream sunlight.

The Taj was built by a heartbroken Shah Jahan (1592–1666), the fifth Mughal Emperor, as a memorial to his third wife, Mumtaz Mahal, who died in 1631. Construction began the following year and was completed 21 years later. It was a project on a colossal scale, employing over 20,000 people from across India and Central Asia, and the white marble was quarried in Rajasthan, and carted by elephants for 473 km (294 miles).

The detail of the building is incredibly intricate; favourite Persian flowers like tulips and the crown imperial, for example, are delicately carved into the marble, while elsewhere flowers are similarly replicated using inlaid semiprecious gemstones. The Taj Mahal and its extensive gardens are a perfect fusion of Indian and Persian art.

Finally we reached the Persian-style Taj gardens, laid out by Ali Mardan Khan, a noble at the Emperor's court, and, strolling through its Mughal gardens, we saw fine trees of *Swietenia macrophylla* (big-leaf mahogany) and *Mimusops elengi*, whose flowers are popular in India for making festive garlands and for stuffing pillows, since they retain their fragrance long after being dried.

And so, our travels in India came to a close in the gardens of the Taj Mahal, without a doubt the most beautiful building in the world. What a pity Joseph Hooker never made it there. He too would have marvelled, though another of my great heroes, the English botanical artist Marianne North, did travel there in 1878 and she too was captivated:

Favourite Persian flowers carved in marble. Here the crown imperial, *Fritillaria imperialis*, is immediately recognisable with the blossoms of narcissi beneath.

The finer details of the Taj Mahal are worthy of study. Here a tulip is exquisitely carved into white marble that was carted on the backs of elephants from distant Rajasthan.

The Taj Mahal in late morning light. By then its marble gleams a pure white and the building reflects in the shallow waters of the canal, which represents one of the four rivers of Paradise.

I started at night for Agra, which I reached on the morning of the 14th of March. The ground all round the city was pure dust, – one ate it, breathed it, drank it, slept in it, – but the place was so glorious that one forgot the dust entirely. I went that same afternoon to the Taj, and found it bigger and grander even than I had imagined; its marble so pure and polished that no amount of dust could defile it, the building is so cleverly raised on its high terrace, half-hidden by the gardens on one side, and washed on the other by the great river Jamuna. The garden was a dream of beauty; the bougainvillea there was even finer than I ever saw it in its native Brazil... Sugar-palms and cocoa-nuts added their graceful feathers and fans, relieving the general roundness of the other trees. The Taj itself was too solid and square a mass of dazzling white to please me (as a picture), except when half hidden in this wonderful garden, though on the river side it was relieved by wings and foundations of red sandstone. The gates, which are chiefly of that beautiful material, would in themselves be worth a journey to see, so graceful and exquisitely finished are they. It was some days before I mounted the terrace and went inside. Like a great snow-mountain, I felt I wanted to know it well before I dared approach nearer; but the more I studied it, the more

I appreciated its marvellous detail and general breadth of design. The interior was most elaborately inlaid with jasper, serpentine, amethyst, and other half-precious stones, many which have been ruthlessly picked out by the barbarians – of different tints.[6]

A few months later I was at the Royal Botanic Gardens, Kew, to deliver a lecture about our travels in Sikkim, and one of my companions in India, Kristin Jameson, had travelled with me from Ireland. We went for a short walk opposite the front gates of Kew, where, for many years, Sir Joseph Dalton Hooker had been a very capable Director. We strolled into St Anne's Church, where on the walls was mounted the famous marble and Wedgwood memorial tablet to Joseph Hooker, and on it, modelled by botanical artist Matilda Smith (1854–1926), were several plants representing his global botanical travels. On the bottom right we immediately identified one of his most important Himalayan discoveries, the blood red-flowered *Rhododendron thomsonii*. Visions of the Yumthang valley and the great thickets there of Thomas Thomson's rhododendron came to mind.

Joseph Hooker's tomb at St Anne's Church on Kew Green. He lies buried with his parents close to the main gates of the Royal Botanic Gardens, Kew.

Joseph Hooker pictured at home in 'The Camp'
with Lady Hooker in 1904. His memories of Sikkim
remained vivid to the end. He died in 1911.

Outside, in the old graveyard, by a low wall dividing the church grounds from the busy road crossing Kew Bridge and the Thames, we paid our respects at the Hooker family tomb, where he lies alongside his father. It was a simple burial place, far removed from the grandeur of the magnificent mausoleum at the Taj Mahal, though the lawns surrounding it were full of scented *Viola odorata*, a charming scene and a fitting resting place for the greatest botanical explorer of the 19th century.

When he died on 10 December 1911, the Dean and Chapter of Westminster Abbey offered a grave near Darwin's in the nave. His widow declined, and perhaps rightly so. Instead he rests just a short distance from the gates of the world's greatest botanic garden, a place he and his father brought to prominence. At Westminster Abbey, a simple white marble tablet forms a suitable memorial in the north choir aisle to Kew's greatest ever Director.

We owe much to Joseph Hooker; without him our gardens would be lesser places. As I write these closing words in April 2016, the gardens at Kilmacurragh are a riotous display of colour, in many cases provided by veteran Hooker rhododendrons. My own favourite, *Rhododendron grande*, now a 12 m mound of creamy-white flowers, certainly transports me back to the wild forests of Sikkim, and I remind myself that when it was planted at Kilmacurragh in 1862, Charles Darwin and Florence Nightingale were growing sister seedlings. Long may it continue to thrive.

EPILOGUE

Looking back, our Sikkim expeditions have been among the greatest adventures in which I have ever participated, only matched perhaps, by past expeditions to south-east Tibet. Highlights were innumerable, though if I were to choose a few magical moments, those that stand clearly in mind include our very first view of the great mass of Kangchenjunga from our base in Darjeeling, our reception at St Paul's School, in the house formerly owned by Brian Hodgson, the sun setting in a fiery amber sky across east Nepal as we descended the western slopes of Tonglu, the heady excitement of locating the ruins of the Rajah's Palace at Tumlong, where Joseph Hooker and Archibald Campbell had been held as prisoners and finally, after a great struggle, reaching the Goecha La, where Kangchenjunga, that great deity of Sikkim, raised her colossal eastern flank just a short distance from the pass.

It has been a wonderful learning experience. Unlike botanical and horticultural explorers of the past, we were not there to collect, merely observe and learn, which we did, and it was fascinating to witness so many of Darjeeling District and Sikkim's complicated ecosystems. We left with just photographs and notebooks crammed full of the names of places, plants and people, and the results are to be seen in this book.

Prayer flags draped from *Rhododendron arboreum* var. *roseum* at the National Botanic Gardens, Kilmacurragh. In the background is the Highclere rhododendron, *Rhododendron* 'Altaclarense'. A Himalayan scene transplanted to an Irish setting. We owe much to Joseph Dalton Hooker and his Sikkim travels.

Sikkim's is a remarkable flora, though like many other places on the planet it is feeling the effects of a changing climate and a burgeoning population. The eastern Himalaya is a fragile region; its glaciers are retreating at an alarming rate, and the ecology of this upland area is delicately balanced. Hydroelectrical projects on the Teesta River, though bringing wealth to the inhabitants of the State, have brought with them much environmental damage, and it is worrying to see another new massive dam project at Chungthang, within view of the original type population of *Rhododendron maddenii*, one of the rarest species in Sikkim.

Of course nature can be extremely destructive too, particularly in the form of the mammoth landslides for which Sikkim is justly famous. Historically, natural calamities like these have obliterated not one but two entire populations of *Rhododendron niveum* in north Sikkim, placing this vulnerable species under further threat.

The population explosion of southern Sikkim and Darjeeling District has had a major impact on the flora of those regions, though to their credit, the last king of Sikkim, and later, the government of India, have established vast National Parks and Sanctuaries that provide some measure of protection to the region's valuable natural resources.

Ecotourism has become a major industry in recent times, though at a price. In places, formerly pristine alpine landscapes have seen much degradation. One hopes that the appropriate authorities can bring this damage to a halt, so that future generations may enjoy the dramatic and spectacular beauty of this stretch of the Himalaya, as Hooker did, and as our group encountered many years later.

Joseph Hooker is a well-known figure in modern-day Darjeeling District and Sikkim. Despite his unauthorised forays into Tibet, which ultimately caused the annexation of the lower half of the kingdom of Sikkim into British India, he is remembered in a positive light. It was Hooker, through his publications, particularly *Rhododendrons of the Sikkim Himalaya, Himalayan Journals* and the lavishly illustrated *Illustrations of Himalayan Plants*, who brought attention to this little Himalayan kingdom's spectacular landscape and its remarkable biodiversity. It is as a result of Hooker's legacy that Sikkim's ecotourism industry has grown in recent times at such a remarkable rate, and tour operators make constant reference to Joseph Hooker's historic journey through the region's deep valleys and its snowcapped mountains.

I remember some years ago, sitting in the gardens of our hotel in Darjeeling. A red-flowered *Rhododendron arboreum* loomed large nearby, though dominating the far distance was Kangchenjunga, an enormous wall of rock and ice, with great glaciers streaming down its face. We had reached the end of a tiring expedition and I was glad of a few hours of solitude and rest. Beneath me, the hill station spilled across undulating ridges, and in the distance I spied the ridge on which Archibald Campbell's house, Beechwood, sat, while in the woods directly above my position, on the Jalapahar Ridge, lay St Paul's School and the bungalow once occupied by Brian Hodgson.

It was late evening and the low November sunlight bathed the face of Kangchenjunga in a warm glowing light, forever changing in intensity as the day progressed. Thus surrounded by a landscape that would have been extremely familiar to Joseph Hooker, I wondered, if, when back at his base at Kew, he ever regretted his unauthorised incursion into Tibet, and the terrible price the Sikkim Rajah had to pay, because of the foolish treatment of his captives at Tumlong by 'the mad Prime Minister' (the Dewan). We will never know, but I doubt it. Shortly afterwards I promptly fell into a deep sleep, badly needed after several weeks of trekking through some of the highest, most remote corners of the Himalaya. Our journeys were inspired by the pioneering adventures of a young Joseph Hooker, and I can wholeheartedly recommend this spectacularly beautiful landscape to future travellers, now 170 years on!

HOOKER'S RHODODENDRONS

Since his travels in east Nepal and Sikkim, enthusiasts have spoken of 'Hooker's rhododendrons'. What species are his rhododendrons? What were his discoveries? Which had been found by earlier collectors? Which species did he introduce? The columns below list the taxa Hooker collected and highlight their history of discovery and introduction.

Though it has been impossible to list all the organisations and individuals who received seedlings from Kew during the 1850s, I have listed a representative selection, which indicates just how far Hooker's seedlings travelled at this time.

Rhododendrons discovered and introduced by Joseph Hooker

Rhododendron camelliiflorum
Rhododendron campanulatum subsp. *aeruginosum*
Rhododendron campylocarpum
Rhododendron ciliatum
Rhododendron cinnabarinum Blandfordiiflorum Group
Rhododendron cinnabarinum Roylei Group
Rhododendron hodgsonii
Rhododendron lanatum
Rhododendron × lancifolium
Rhododendron maddenii
Rhododendron pendulum
Rhododendron pumilum

Rhododendron thomsonii
Rhododendron vaccinioides
Rhododendron wallichii
Rhododendron wightii
Rhododendron xanthocodon
Rhododendron 'Sir Charles Lemon'

Rhododendrons discovered by Hooker

Rhododendron × candelabrum
Rhododendron lindleyi
Rhododendron nivale

Rhododendrons introduced by Hooker

Rhododendron cinnabarinum
Rhododendron dalhousieae
Rhododendron edgeworthii
Rhododendron fulgens
Rhododendron glaucophyllum
Rhododendron grande
Rhododendron griffithianum
Rhododendron lepidotum
Rhododendron niveum
Rhododendron triflorum
Rhododendron virgatum

RHODODENDRONS COLLECTED BY JOSEPH HOOKER — IN DETAIL

Rhododendron anthopogon

Described in 1821 from collections made on Gosainkund in Nepal by Nathaniel Wallich. Introduced to cultivation as early as 1820, probably by Wallich. Re-introduced by Hooker from east Sikkim, where he made seed collections in November 1849.

Rhododendron arboreum

Rhododendron arboreum is a widespread species across the Himalaya, represented by various different geographical subspecies and varieties, and a number of horticultural 'Groups'. Several were collected by Hooker in Nepal, Darjeeling District and Sikkim.

The taxonomy of these variants continually changes, with flower colour ranging from white, pink, rose, scarlet to blood-red, and, with the undersides of leaves plastered with silver or fawn-coloured indumentum. The different variants collected by Hooker are outlined in the following pages.

Rhododendron arboreum
(crimson and blood-red flowered)

Commonly found in the warmer valleys around Darjeeling. Two different forms, one of bushy habit, the other of columnar shape, and both Hooker seedlings, grow at Kilmacurragh. The crimson-flowered plant had previously been collected by Nathaniel Wallich and by Francis Buchanan-Hamilton, and was introduced by the latter. It first flowered in Hampshire in 1825. In May 1917, while visiting the gardens at Kilmacurragh, the Marquess of Headfort measured the best of Hooker's original seedlings, stating that they were then between 7.6 and 9 m tall. Today, the tallest tree is a staggering 28 m tall, though this tree may pre-date Hooker's introduction and has a very upright, tree-like habit with blood-red flowers. He also singled out for mention exceptional plants of *R. lanatum, R. campylocarpum, R. triflorum, R. niveum, R. grande* and *R. hodgsonii*.

Rhododendron arboreum, a contemporary watercolour of an old Kilmacurragh plant by Lynn Stringer.

Rhododendron arboreum var. *roseum*

Discovered by Nathaniel Wallich in Nepal in 1820, and introduced by him in the spring of 1821. This variety first flowered in 1828 at Knight's Nursery, Chelsea. Joseph Hooker re-introduced it in 1850 and recorded that it grew near the summit of Tonglu, west of Darjeeling. One of his original seedlings, *Rhododendron arboreum* var. *roseum* 'Fernhill Silver', belongs here and was raised at Glasnevin in 1850 from Hooker's Sikkim seed collection.

On 6 June 1851, Kew sent two Wardian cases of Hooker's Sikkim seedlings to Canterbury, New Zealand. The consignment was mostly rhododendrons and included *Rhododendron arboreum* var. *roseum*. Another case of plants, sent from Kew to Valencia, Spain on 22 April 1853, also included this variety, and a further plant was also sent to the botanic gardens in Dijon, France on 6 June 1853. Later that year, on 18 August 1853, a Wardian case, packed with plants, including this variety, was dispatched from Kew to Saint Helena for trial and acclimatisation.

Rhododendron arboreum forma *album*

Discovered by Dr Francis Buchanan-Hamilton in Nepal in 1803. Introduced by Nathaniel Wallich in 1814 when he sent seeds to Mr Shepherd of Liverpool. It first flowered in cultivation at Chester

in 1831. Joseph Hooker re-introduced this white-flowered variant in 1850, and an original tree, raised from Hooker's Sikkim seed collection, grows at Stonefield in Argyllshire, Scotland.

In his 1919 article on Plants and Rhododendrons at Kilmacurragh in *The Rhododendron Society Notes*, the Marquess of Headfort mentions a 12 m tall tree of the white-flowered *Rhododendron arboreum*, of the upright form, growing beneath the house on the Pond Vista, where it had been planted in 1864.

Rhododendron arboreum subsp. *cinnamomeum* (scarlet flowered)

Another of Nathaniel Wallich's Nepal discoveries and introductions, dating to 1822. Sometimes 'lumped' with *Rhododendron arboreum* var. *roseum*, but distinct. Common in the upper Lachung valley, where it makes a spectacular display. Hooker re-introduced this very fine plant from Sikkim in 1850.

Rhododendron arboreum subsp. *cinnamomeum* Campbelliae Group

This particular Group, sometimes relegated into synonymy with *Rhododendron arboreum* subsp. *cinnamomeum*, was first found by Nathaniel Wallich in 1820, in Nepal, and was introduced by him the same year. Hooker re-introduced it to cultivation in 1850, and original trees from his collections still grow at Kilmacurragh. The Marquess of Headfort measured one of Hooker's original seedlings at Kilmacurragh in May 1917, recording a height of 6 m. It is still common near the summit of Tonglu, where Hooker made his collections.

Kew distributed several seedlings of this plant during the early 1850s. On 3 February 1852, James Robert Gowen, the *Rhododendron* enthusiast and garden advisor to Highclere in Hampshire, received a very large batch of *Rhododendron* seedlings, including Hooker's *Rhododendron campbelliae*.

Rhododendron arboreum subsp. *cinnamomeum* Campbelliae Group, seen here in bloom at Kilmacurragh.

During the 1870s, Thomas Acton was enthusiastically hybridising several of Joseph Hooker's rhododendrons. In 1919 the Marquess of Headfort recorded an interesting hybrid between *R. barbatum* and *R. arboreum* subsp. *cinnamomeum* Campbelliae Group. It grew on the present-day Bleach Green at Kilmacurragh and bore deep-pink blossoms with frilled edges and was very fragrant. The cross was made at Kilmacurragh in 1874.

Rhododendron arboreum subsp. *cinnamomeum* var. *roseum*

Thought to have been introduced to cultivation by Wallich from Nepal in 1822, Hooker re-introduced it to cultivation in 1850. Old veterans, thought to be Hooker originals, exist at Kilmacurragh.

Rose-flowered variants of *Rhododendron arboreum* are relatively common in the Himalaya, with indumentum transitioning from brilliant silver to rusty-brown. For this reason, this plant is sometimes 'lumped' with *Rhododendron arboreum* var. *roseum*.

Rhododendron arboreum subsp. *cinnamomeum* var. *roseum* forma *album*

Another of Nathaniel Wallich's Nepal discoveries dating to 1822. This white-flowered variant first flowered under glass at Messrs Rollison's nursery, Tooting, London in 1836. Joseph Hooker re-introduced it to Western gardens in 1850. There is an old plant at Kilmacurragh that is reputedly a Hooker original. David Moore, Curator of the Royal Botanic Gardens, Glasnevin, crossed this white-flowered variant (said to be a Hooker original) with *Rhododendron campanulatum* (another Hooker original) to create *Rhododendron* 'Thomas Acton'. The original seedling still grows at Kilmacurragh.

Rhododendron barbatum

Discovered about 1829 at Gosainkund, Nepal, by Nathaniel Wallich, and introduced the same year. Hooker re-discovered it on the summit of Tonglu and in the interior valleys of north Sikkim, and sent seeds from Sikkim in November 1849. In May 1917, the Marquess of Headfort measured one of Hooker's original seedlings at Kilmacurragh. It was then 4.5 m tall.

Rhododendron × *lancifolium*

Rhododendron lancifolium Hook. f. *Rhododendrons Sikkim Himalaya* t. 4. (1849). Type: India, Sikkim, without locality, Hooker no. 4. May 1848 (K.)

Also, Sikkim: Khechopari Lake, in a marsh. Hooker no. 17. January 1849.

In 1849, in *Rhododendrons of the Sikkim Himalaya*, Hooker described a new species, *Rhododendron lancifolium*. At present this 'species' is placed in synonymy with *Rhododendron barbatum*. However, Hooker's plant is clearly a natural hybrid between *Rhododendron arboreum* subsp. *cinnamomeum* Campbelliae Group and *Rhododendron barbatum,* and is best treated as a nothospecies, *Rhododendron × lancifolium* Hook. f.

His earliest collection of this hybrid, in the herbarium at Kew, is a flowering specimen dated May 1848. No locality is given, but at this time he was collecting on Tonglu, where both parents of the

Rhododendron 'Thomas Acton'. The original plant depicted by the Irish botanical artist, Lynn Stringer.

Rhododendron × lancifolium from *Rhododendrons of the Sikkim Himalaya.*

hybrid grow side by side, and it is highly likely that he gathered his material there.

In *Himalayan Journals*, Hooker mentions finding *Rhododendron barbatum* in the marsh by Khechopari Lake. His dried specimens from this region also belong to this hybrid. The Khechopari plants were very definitely planted by monks from the nearby monastery, having been transplanted from higher altitudes. The lake is at low altitude, and on our last visit there in November 2014 I found young saplings of *Rhododendron griffithianum* planted in the same marsh, obviously by Buddhist monks from the same monastery.

Hooker collected seeds of *R. × lancifolium*, and seedlings of this hybrid were raised and distributed from Kew. For example, on 27 March 1852, *R. × lancifolium* was among a batch of Sikkim *Rhododendron* seedlings sent to East India House, the London headquarters of the East India Company.

This natural hybrid was deliberately recreated in 1874 at Kilmacurragh by Thomas Acton.

Rhododendron camelliiflorum

Discovered by Hooker in east Nepal in 1848. He introduced this curious species from the Lagyap valley in east Sikkim, where he made a rich collection of seeds of several *Rhododendron* species in November 1849.

Rhododendron campanulatum

One of Nathaniel Wallich's finest discoveries. Introduced by him in 1825. Abundant in north Sikkim, from where Hooker collected his seeds in 1849.

Rhododendron campanulatum subsp. *aeruginosum*

Discovered by Hooker in the upper Lachen valley on 13 June 1849. He introduced it to cultivation in the spring of 1850, and one of his seedlings bloomed, for the first time in cultivation, in April 1862.

Rhododendron campylocarpum

Discovered by Hooker in east Nepal during the autumn of 1848, and introduced by him to cultivation from seeds collected in Sikkim in November 1849. A handsome species that often hybridises in the wild with *Rhododendron thomsonii*.

Several Scottish gardens received seedlings raised at Kew from Hooker's Sikkim collections. For example, on 13 May 1852, Lord Rutherford, who gardened at Lauriston Castle, Edinburgh, received a very large batch of plants including *R. campylocarpum*. On 10 May 1853, Kew also dispatched a Wardian case packed with Sikkim plants to Mr McHenn, Natal, South Africa. Two rhododendrons are listed, *R. campylocarpum* and *R. × lancifolium*.

Rhododendron × candelabrum

A hybrid discovered by Joseph Hooker near Lachen and named by him as *Rhododendron candelabrum* in 1851. He had suspected it was a mere variant of *Rhododendron thomsonii*. In fact it is a natural hybrid between that species and *Rhododendron campylocarpum*. Introduced to cultivation by Ludlow and Sherriff from Bhutan in 1936.

Rhododendron cinnabarinum Blandfordiiflorum Group from *Curtis's Botanical Magazine* t. 4930 (1856)

Rhododendron ciliatum

One of the loveliest of the low-growing Sikkim rhododendrons. Discovered by Joseph Hooker in the Lachen valley in June 1849, he sent seeds to Kew, gathered in November of the same year. From that seed consignment, several plants flowered there in March 1852.

This species was included in a batch of seedlings sent from Kew to Sir Thomas Acland at Killerton in Devon on 13 May 1852. Two years later, on 13 November 1854, a Wardian case was dispatched from Kew to Charles Moore (1820–1905), Director of the Sydney Botanic Gardens. This included *R. ciliatum* and *R. × lancifolium*. Charles Moore was a brother of David Moore, Curator of the Glasnevin Botanic Gardens, Dublin.

On 4 October 1855, Kew also supplied several of Hooker's Himalayan plants to the Grand Duchess of Mecklenburg-Strelitz in Germany. These included *R. ciliatum* and *R. arboreum* subsp. *cinnamomeum* Campbelliae Group.

Rhododendron cinnabarinum

Discovered by William Griffith in Bhutan in 1838. Described as a new species in 1849 by Joseph Hooker and introduced to cultivation by him in 1850.

Rhododendron cinnabarinum Blandfordiiflorum Group

Discovered by Hooker in 1849, introduced by him in 1850.

Rhododendron cinnabarinum Roylei Group

Discovered by Hooker in 1849, introduced by him in 1850.

Rhododendron dalhousieae

Discovered by William Griffith in Bhutan in 1838. Described as a new species in 1849 by Joseph Hooker and introduced to cultivation by him in 1850. During the early 1850s, when Kew was distributing thousands of Joseph Hooker's seedlings across the globe, Daniel Ferguson (*c*. 1801–1864), Curator of Belfast Botanic Gardens, received a large consignment, including *Rhododendron grande, R. dalhousieae, R. thomsonii, R. cinnabarinum* Roylei Group, *R. glaucophyllum, R. hodgsonii* and *R. maddenii*.

Rhododendron edgeworthii

Discovered by William Griffith in Bhutan in 1838. Introduced to cultivation by Joseph Hooker in 1850. During the 1850s, Kew sent large quantities of Hooker's Sikkim seedlings, primarily rhododendrons, to leading nursery firms in England and Scotland. Messrs Standish and Noble received seedlings on 15 March 1852, including *R. edgeworthii*.

Rhododendron falconeri

Introduced from the Himalaya by William Henry Sykes (1790–1872) in 1830. Described as a new species by Hooker in 1849 and re-introduced by him in 1850. In May 1917, the Marquess of Headfort, a keen *Rhododendron* enthusiast and personal friend and sponsor of George Forrest, visited Kilmacurragh, highlighting Hooker's rhododendrons and measuring some of the best specimens. He recorded *Rhododendron falconeri* as having reached 7.6 × 6 m.

Rhododendron falconeri in bloom at Kilmacurragh.

Rhododendron grande, a watercolour of Hooker's original plant at Kilmacurragh by Lynn Stringer.

Rhododendron fulgens

Though the discovery of *Rhododendron fulgens* is generally attributed to Joseph Hooker, it had been previously collected in Nepal by Nathaniel Wallich during the 1820s, though Hooker would introduce it to cultivation in 1850. It first flowered at Kew in 1862. On 21 July 1851, Kew sent a Wardian case of plants to Sir William Denison (1804–1871), Governor of Van Diemen's Land (Tasmania). Among the plants sent to Hobart was *Rhododendron fulgens*.

Rhododendron glaucophyllum

Discovered by William Griffith in Bhutan in 1838. Introduced to cultivation by Joseph Hooker in 1850. Three years later, on 14 September 1853, Kew dispatched a Wardian case full of Sikkim *Rhododendron* seedlings to Sir Henry Barkly (1815–1898), Governor of Jamaica. These included *R. ciliatum, R. dalhousieae, R. glaucophyllum, R. campylocarpum, R. × lanciflolium* and *R. cinnabarinum*.

Rhododendron grande

Described in 1849 by Hooker as a new species, *Rhododendron argenteum*, a name that persisted in gardens for decades afterwards. It was first described in 1847 under its present binomial from the original collections made in Bhutan in 1838 by William Griffith. Introduced to cultivation by Joseph Hooker in 1850, plants first flowered under glass at Kew in 1858.

On 9 November 1851 Kew sent a consignment of Hooker's Sikkim seedlings, including *Rhododendron grande*, to Messrs Veitch of Exeter. The Marquess of Headfort measured Hooker's original seedling at Kilmacurragh in 1919 when it was 7.6 m tall. It still grows on the Bleach Green at Kilmacurragh where it makes a magical floral display every February.

Rhododendron griffithianum

Discovered by William Griffith in Bhutan in 1838. Described as a new species, *Rhododendron aucklandii*, by Joseph Hooker in 1851, before introduced to cultivation by him in 1850.

Rhododendron hodgsonii

Discovered by Hooker in east Nepal in the autumn of 1848 and introduced by him in the spring of 1850, from seeds collected in November 1849 at Lagyap in east Sikkim.

On 6 October 1852, Kew sent a plant of *Rhododendron hodgsonii* to James McNab, Curator of the Royal Botanic Garden Edinburgh. The Edinburgh gardens had, just two years previously, received a seed consignment of ten of Hooker's Sikkim rhododendrons, including *R. falconeri*, *R. grande* and *R. fulgens*, thus laying the foundation for the establishment of the world-famous *Rhododendron* collection now based at Edinburgh and her satellite gardens, Benmore, Dawyck and Logan.

In 1919, the Marquess of Headfort measured several of Hooker's rhododendrons at Kilmacurragh, stating that the best specimen of *Rhododendron hodgsonii* had reached 8 m tall by 6 m.

Rhododendron lanatum

Discovered by Hooker at Dzongri in west Sikkim in January1849. He collected seeds near the Cho La in November of the same year, and it was by this means that the species entered cultivation.

Rhododendron lepidotum

Described in 1835 from Nathaniel Wallich's Nepal collections, this charming dwarf species was introduced by Joseph Hooker from his November 1849 seed collections from east Sikkim. Hooker described a number of new species in his seminal work, *Rhododendrons of the Sikkim Himalaya*, a number of which have been placed into synonymy with *R. lepidotum*. These include *R. elaeagnoides*, *R. obovatum* and *R. salignum*. *R. lepidotum* is a highly variable species, and Hooker, generally conservative in establishing new species, may be forgiven for assuming he had found several new taxa.

Rhododendron lindleyi

Discovered by Hooker near Darjeeling in 1848, though he initially confused it with *R. dalhousieae*. The species wasn't described till 1864, when the English gardener and botanist Thomas Moore (1821–1887) studied a plant in Standish's nursery at Ascot, which had been raised from seeds collected in Bhutan by Thomas Jonas Booth (1829–1879). Booth was based in Bhutan between 1849 and 1860.

Rhododendron griffithianum, Hooker's plant seen here in bloom at Kilmacurragh.

Rhododendron nivale, from *Rhododendrons of the Sikkim Himalaya*.

Rhododendron pendulum, Fitch's re-working of Joseph's Hooker's Sikkim field sketch.

Rhododendron pendulum

Discovered by Hooker in the Lachen valley in June 1849, introduced by him through Kew, where seedlings were raised in the spring of 1850.

Rhododendron pumilum

Discovered by Hooker in the Zemu valley, north Sikkim, on 12 June 1849. He sent seeds to Kew where seedlings were raised in the spring of 1850. A charming, dwarf species.

Rhododendron setosum

This species was described in 1821 from a collection made in Nepal by Nathaniel Wallich. It was first introduced to cultivation in 1825, perhaps from seeds sent to Europe by Wallich. Reintroduced by Joseph Hooker from east Sikkim from seeds collected there in November 1849. Abundant on the higher mountain slopes of Sikkim and Bhutan.

Rhododendron thomsonii

Discovered by Joseph Hooker in east Nepal in the autumn of 1848, and he introduced it to cultivation in 1850, from seeds collected in November 1849 near Lagyap in east Sikkim. A superb species.

Rhododendron triflorum

Discovered by William Griffith in Bhutan in 1838. Introduced to cultivation by Joseph Hooker in 1850.

Rhododendron maddenii

Discovered by Hooker in 1849, and introduced by him from seeds collected in November of that year. Very rare in Sikkim, though widely distributed in the eastern Himalaya. Taxonomically, a very complex and variable species.

On 11 September 1855, a plant of *Rhododendron maddenii* was sent from Kew to David Moore, Curator of the then Royal Botanic Gardens, Glasnevin, Dublin. In October of that year a plant was also sent to James Veitch of Exeter, from where it was propagated and further distributed.

Rhododendron nivale

Discovered by Joseph Hooker in 1849, introduced by Roland Edgar Cooper in 1915, and first flowered at the Royal Botanic Garden Edinburgh in March 1920.

Rhododendron niveum

Discovered by William Griffith in Bhutan in 1838. Introduced to cultivation by Joseph Hooker in 1850. During the distribution of Hooker's seedlings in the early 1850s, Kew sent a large batch to Mr Bitzean, Copenhagen, Denmark, on 3 May 1853. Several rhododendrons are listed including *R. niveum*. In the following year Kew sent seedlings of *R. niveum* to Vicountess Doneraile in County Cork, Ireland.

Rhododendron virgatum, from *Rhododendrons of the Sikkim Himalaya*.

Rhododendron vaccinioides

Discovered by Joseph Hooker in the Lachen valley in 1849. He sent seeds to Kew in the spring of 1850. They germinated though they did not persist.

Rhododendron virgatum

Discovered by William Griffith in Bhutan in 1838. Introduced to cultivation by Joseph Hooker in 1850.

Rhododendron wallichii

Discovered by Joseph Hooker in May 1848 (perhaps from the summit of Tonglu), and introduced by him from seeds collected in Sikkim in November 1849.

During the distribution of Hooker's Sikkim seedlings from Kew, several *Rhododendron* species were sent to the Vice Regal Lodge (now Áras an Uachtaráin) in Dublin's Phoenix Park on 6 October 1852. These included *R. wallichii*, *R. ciliatum*, *R. cinnabarinum*, *R. glaucophyllum*, *R. campylocarpum*, *R. thomsonii*, *R. fulgens*, *R. edgeworthii* and *R. dalhousieae*. Queen Victoria stayed there during her visits to Dublin and it is highly likely she saw Hooker's rhododendrons on display in the grounds there.

Rhododendron wightii

This superb species was first found by Joseph Hooker in east Nepal in 1848, and was later introduced by him from seeds collected in east Sikkim in November 1849. A tricky species in cultivation.

Rhododendron xanthocodon

Hooker discovered this species in 1849 and introduced it to cultivation in 1850. His father, Sir William Hooker, originally named it *Rhododendron cinnabarinum* var. *pallidum*. It is now considered to be synonymous with *Rhododendron xanthocodon*, a species based on one of Frank Kingdon Ward's 1924 Tibetan collections. From Joseph Hooker's seed consignment plants flowered at Kew under glass in May 1854.

Rhododendron 'Sir Charles Lemon'

A natural hybrid raised at Kew from Hooker's Sikkim seed collections and sent to Carclew, the garden of Sir Charles Lemon in Cornwall. Its parentage is *Rhododendron arboreum* subsp. *cinnamomeum* and *R. campanulatum*, both of which are abundant in the higher valleys of north Sikkim.

Sir Charles Lemon received his first batch of Sikkim seedlings from Kew on 5 December 1851. These included *R. falconeri*, *R. hodgsonii*, *R. griffithianum*, *R. maddenii*, *R. dalhousieae*, *R. cinnabarinum* Roylei Group, *R. glaucophyllum*, *R. niveum*, *R. thomsonii* and four unnamed *Rhododendron* species. It is likely that one of these was this hybrid.

Rhododendron wallichii commemorates Nathaniel Wallich, one of the founding fathers of Indian botany.

Appendix 2

Joseph Dalton Hooker's Nepal and Sikkim plants

Plants named for Joseph Dalton Hooker that are accepted in current botanical nomenclature

Aconitum hookeri
Acronema hookeri
Aeschynanthus hookeri
Agapetes hookeri
Agrostis hookeriana
Amischotolype hookeri
Anaphalis hookeri
Androsace hookeriana
Anthoxanthum hookeri
Argyreia hookeri
Begonia josephi
Berberis hookeri
Carex daltonii
Ceropegia hookeri
Chionocharis hookeri
Christisonia hookeri
Coprinus hookeri (fungus)
Cortiellia hookeri
Corydalis hookeri
Crepidium josephianum
Cyananthus hookeri
Cymbidium hookerianum
Dendrobium hookerianum
Dendrocalamus hookeri

Deutzia hookeriana
Elatostema hookerianum
Eriobotrya hookeriana
Ficus hookeriana
Fragaria daltoniana
Glochidion daltonii
Habenaria diphylla var. josephi
Herminium josephi
Himalayacalamus hookeriana
Ilex hookeri
Inula hookeri
Josephia spp.
Ligularia hookeri
Lindenbergia hookeri
Nannoglottis hookeri
Nervilia hookeriana
Onosma hookeri
Panus hookerianus (fungus)
Pedicularis daltonii
Phalaenopsis deliciosa subsp. hookeriana
Photinia arguta var. hookeri
Pilea hookeriana
Pleione hookeriana
Pleurospermum hookeri

Polygonatum hookeri
Polygonum hookeri
Primula hookeri
Pterocephalus hookeri
Remusatia hookeriana
Rhaphidophora hookeri
Rynchospora hookeri
Salix daltoniana
Sarcococca hookeriana
Saussurea hookeri
Saxifraga hookeri
Styrax hookeri
Swertia hookeri
Tetrastigma hookeri
Themeda hookeri
Tipularia josephi
Toxicodendron hookeri
Tournefortia hookeri
Trikeraia hookeri
Vandellia hookeri
Xantolis hookeri
Zehneria hookeriana

Hooker plants named for friends and colleagues

Thomas Anderson, Royal Botanic Garden, Calcutta
Didymocarpus andersonii
Dittoceras andersonii
Globba andersonii
Saussurea andersonii

George Bentham, (Royal) Horticultural Society; Royal Botanic Gardens, Kew.
Codonopsis benthamii

Archibald Campbell, Superintendent of Darjeeling
Acer campbellii
Artemisia campbellii
Magnolia campbellii

Mrs Archibald Campbell, Beechwood, Darjeeling
Rhododendron campbelliae =
    Rhododendron arboreum var. cinnamomeum
    Campbelliae Group

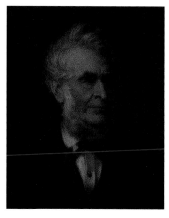
George Bentham

**John Ferguson Cathcart,
Bengal Civil Service**

Asarum cathcartii
Begonia cathcartii
Cathcartia villosa
Cautleya cathcartii
Esmeralda cathcartii
Impatiens cathcartii
Liparis cathcartii
Magnolia cathcartii
Polygonatum cathcartii

**Charles Baron Clarke,
Royal Botanic Garden, Calcutta**

Blumea clarkei
Chirita clarkei
Globba clarkei
Iris clarkei
Isachne clarkei
Ophiopogon clarkei
Pedicularis clarkei
Persea clarkeana
Phyllanthus clarkei

**Sir James Colvile,
Supreme Court of Bengal**

Buddleja colvilei

Susan, Marchioness of Dalhousie

Rhododendron dalhousieae

**M. Pakenham Edgeworth,
Bengal Civil Service**

Rhododendron edgeworthii

**Sir Henry Elwes,
naturalist and plant collector**

Aconitum elwesii
Gentiana elwesii
Pedicularis elwesii

**Hugh Falconer,
Royal Botanic Garden, Calcutta**

Himalayacalamus falconeri (syn.
   Thamnocalamus falconeri)
Rhododendron falconeri

**James Sykes Gamble, British
Imperial Forest School, Dehradun**

Olea gamblei

**James Alexander Gammie,
Cinchona plantations, Darjeeling**

Machilus gammieana
Pedicularis gammieana

**William Griffith,
Royal Botanic Garden, Calcutta**

Larix griffithii
Peliosanthes griffithii
Potentilla griffithii
Rhododendron griffithianum
Ribes griffithii
Sorbus griffithii

**Brian Hodgson,
Brianstone, Darjeeling**

Hodgsonia heteroclita
Magnolia hodgsonii
Rhododendron hodgsonii

**Andrew T. Jaffray,
Cinchona plantations, Darjeeling**

Herminium jaffreyanum

**Sir George King,
Royal Botanic Garden, Calcutta**

Ilex kingiana
Litsea kingii
Saussurea kingii

**John Lindley,
(Royal) Horticultural Society**

Galeola lindleyana
Rhododendron lindleyi

**Edward Madden,
Royal Bengal Artillery**

Maddenia himalaica
Rhododendron maddenii

**Robert Pantling, Royal Botanic
Garden, Calcutta**

Habenaria pantlingiana
Pedicularis pantlingii

John Lindley

**Thomas Thomson,
Royal Botanic Garden, Calcutta**

Berberis thomsoniana
Cremanthodium thomsonii
Impatiens thomsonii
Oenanthe thomsonii
Ophiorrhiza thomsonii
Rhododendron thomsonii
Rubus thomsonii
Sorbus thomsonii
Strobilanthes thomsonii

**William John Treutler,
Darjeeling tea plantations,
plant collector in Sikkim**

Aniselytron treutleri

**Nathaniel Wallich,
Royal Botanic Garden, Calcutta**

Colpodium wallichii
Pimpinella wallichii
Rhododendron wallichii

**Robert Wight,
Madras Botanic Garden**

Rhododendron wightii

Hooker plants named for regions of Sikkim and Darjeeling District

Sikkim

Acer sikkimense
Actinodaphne sikkimensis
Agapetes sikkimensis
Agrostis sikkimensis
Allium sikkimense
Anchusa sikkimensis
Anthoxanthum sikkimense
Arcuatopterus sikkimensis
Aster sikkimensis
Ampelocissus sikkimensis
Astragalus sikkimensis
Begonia sikkimensis
Beilschmiedia sikkimensis
Berberis sikkimensis
Blumea sikkimensis
Commelina sikkimensis
Corydalis sikkimensis
Cyathopus sikkimensis
Cyperus sikkimensis
Draba sikkimensis
Elatostema sikkimense

Ephedra gerardiana subsp. *sikkimensis*
Epilobium sikkimense
Euphorbia sikkimensis
Gentiana sikkimensis
Hedysarum sikkimensis
Juncus sikkimensis
Litsea sikkimensis
Malus sikkimensis
Musa sikkimensis
Pimpinella sikkimensis
Pitardella sikkimensis
Pleurospermopsis sikkimensis
Poa sikkimensis
Primula sikkimensis
Ranunculus sikkimensis
Rubia sikkimensis
Rubus sikkimensis
Saccharum sikkimense
Salix sikkimensis
Stellaria sikkimensis
Vaccinium sikkimense

Darjeeling

Edgaria darjeelingensis
Peziza darjeleensis (fungus)

Dongkya La (Donkiah La), Sikkim–Tibet frontier

Saussurea donkiah

Kongra La, Sikkim–Tibet frontier

Astragalus kongrensis

Lachen

Polystichum lachenense
Strobilanthes lachenensis

Tonglu

Clematis tongluensis

Yak La, Sikkim–Tibet frontier

Saussurea yakla

Tamur River, Nepal

Strobilanthes tamburensis

APPENDIX 3

PLACE NAMES

Bhutan mountain ranges

Jomolhari (Chumulari)

West Bengal and Darjeeling District

Darjeeling (Dorjiling)
Kurseong (Kursiong)
Pankhabari (Punkabaree)
Siliguri (Siligoree)
Sonada (Pacheem)

Darjeeling mountains and mountain ranges

Ghoom (Goong) ridge
Senchal (Sinchul)
Singalila (Singalelah)
Tonglu (Tonglo)
Tukvar (Tukvor) spur

Darjeeling District and Terai rivers

Great Rangit (Rungeet) River
Mahananda (Mahanuddy) River
Rimbi (Rungbee) stream

Sikkim — towns, villages and places

Bakhim (Buckeem)
Brom (Broom)
Chonpung (Tchonpong)
Chungthang (Choongtam)
Dzongri (Jongri)
Gar (Gorh)
Lachen (Lamteng)
Lachung (Lachoong)
Lagyap (Laghep)
Lingdam (Lingcham)
Lingdong (Lingo)
Lingkiang (Lingtuim or Lathiang)
Lingmo (Lingo)
Mikola (Mikk)
Namchi (Namtchi)
Namphak (Nampok)

Phalung (Palung)
Rabdentse (Phieungoong)
Ramtang (Rangang)
Rongpa (Rungpo)
Samdong (Bhomsong Samdong)
Sentam (Singtam)
Talam (Tallum Samdong)
Thangu (Tungu)
Thingling (Tengling)
Tshoka (Choka)
Tumlong (Tumloong)
Yuksam (Yoksun)
Yume Samdong (Momay Samdong),
Yumthang (Yeumtong)
Zema (Zemu Samdong),

Sikkim monasteries and religious sites

Dubdi (Doobdi)
Norbugang (Nirbogong)
Pemayangtse (Pemiongchi)
Phodong (Phadong)
Sanga Choeling (Changachelling)
Tashiding (Tassiding)
Yangang (Neongong)

Sikkim rivers and lakes

Cholamu or Tso Lhamo (Cholamoo) Lake
Dik (Rycott) River
Gurudongmar Lake (Yeumtso Lake)
Kalek (Kulhait) River
Khechopari (Catsuperri) Lake
Lachung (Lachoong) River
Lhonak (Thlonok) River
Rathong (Ratong) River
Samiti (Sungmoteng) Lake
Tarum (Taktoon) River
Teesta (Tista) River
Teesta River (Lachen-Lachoong)

Mount Pandim seen here from Mans
Lepche. Hooker camped nearby at
Dzongri and was greatly impressed by
Pandim's massive cliff face.

Sikkim mountains and mountain passes

Chameringu (Chungthang Peak)
Changmekang (Chango-khang)
Chiwa Bhanjyang Pass (Islumbo Pass)
Chomoyummo (Chomiomo)
Dongkya La (Donkia pass)
Kabru (Kubra)
Kabur (Gubroo)
Kangchengyao (Kinchinjhow)
Kangchenjunga (Kinchinjunga)
Lamo Angdang (Tukcham)
Mount Maenam (Mainom)
Pandim (Pundim)
Pauhunri (Donkia mountain)
Sebu La (Sebolah)
Singalila (Singalelah) Range
Tangkra (Tunkra)

India — general

Bardhaman (Burdwan)
Bhagalpur (Bhagulpore)
Bijalgarh (Bidjegur)
Chennai (Madras)
Ghazipur (Ghazepore)
Hooghly (Hoogly)
Kishanganj (Kishengunj)
Madhuban (Maddaobund)
Mirzapur (Mirzapore)
Parasnath (Shikharji)
Purnia (Purnea)
Sunderbans (Sunderbunds)

West Bengal

Baikunthapur (Bai-kant-pore)
Jalpaiguri (Jeelpigoree)

Bangladesh

Tentulia (Titalya)

Indian rivers

Mahananda (Mahanuddee) River

Indian mountain ranges

Kaimur (Kymore) Range

Nepal

Kambachen (Kanglachem)
Hunza or Khunza La (Kambachen or Nango La)
Walungchung (Wallanchoon)
Yalung (Yalloong) valley

Nepalese mountains and passes

Gosainkund (Gossain-Than)
Jannu (Junnoo)
Kanglachen Pass (the Kang La or Khangla Deoral)
Khang La (Kanglanamo pass)
Mirgin La (Hooker's Choonjerma)
Mount Everest (Tsungau)

Nepalese rivers

Kabeli (Khabili) River
Koshi (Cosi) River
Tamur (Tambur)

Tibet

Bamcho (Bhomtso)
Kampa Dzong (Kambajong)
Pagri (Phari)
Yum Cho (Yeumtso)

Tibetan mountains

Qang La (Kiang-La)

Tibetan rivers

Yarlung-Tsangpo (Yaru)

Indian/Tibetan terms

Amlah (court officials)
Chu (river)
Dewan (Prime Minister)
Dingpun (Captain)
Gompa (monastery/temple)
La (mountain pass)
Phipun (chief officer)
Sepoys (soldiers)
Soupun (Lieutenant)
Stupa (shrine)
Subah (chief officer)
Terai (lowlands)

Appendix 4

Select glossary

Achlorophyllous, without chlorophyll, and as a result, unable to photosynthesise.

Aroid, plants belonging to the Arum family, Araceae.

Bract, a modified protective leaf associated with an inflorescence.

Bullate, when the surface of an organ, most commonly a leaf, is covered in rounded, blister-like swellings.

Calyx, a collective term for the sepals, which form the outer whorl of the perianth or floral envelope.

Campanulate, bell-shaped, in terms of a corolla.

Cauliflorus, the production of flowers directly from older wood, branches and trunks of trees.

Cordate, heart-shaped, e.g., the shape of a leaf.

Corolla, the petals of a flower, typically forming a whorl within the sepals and enclosing the reproductive organs.

Corymb, a flat-topped flowerhead with the outer blossoms opening first; *see also* cyme and umbel.

Cryptogamist, a person proficient in the study of cryptogamic botany, i.e. the study of ferns, mosses, algae, fungi and lichens.

Culm, the stems of the grass family, Poaceae.

Cyme, a flat-topped inflorescence with the inner blossoms opening first; *see also* corymb.

Decurrent, where the base of a leaf blade extends down and is adnate to the petiole and the stem.

Dichotomous, branching regularly by forking repeatedly in two.

Elliptic, ellipse-shaped, with equally rounded or narrowed ends.

Epilithic, growing on rocks.

Epiphyte, a plant which grows on another plant, but not in a parasitic manner and not depending on it for nourishment.

Ericaceous, may mean belonging to the Ericaceae, or needing a growing medium with a low pH.

Fall, one of the drooping petals belonging to an Iris perianth.

Farinose, having a mealy, granular texture or appearance.

Fascicle, a cluster or bundle of flowers or leaves arising from a common point.

Glabrous, smooth, hairless.

Hemicryptophytic, a plant whose buds rest on the surface, and are protected in various ways.

Herbarium (*pl.* **herbaria**), a systematic collection of dried plants used for scientific study.

Holoparasite, a plant that is completely parasitic on other plants, and lacking chlorophyll, cannot exist without a host.

Indumentum, a dense covering of hair, scurf or scales.

Labellum, a lip, especially the enlarged distinctive third petal of an orchid.

Lanceolate, lance-shaped, widening above the base and long, tapering towards the apex.

Lepidote, alludes to *Rhododendron* species with scales on their leaves or with scales on their shoots, leaves and floral parts.

Liana, a woody climbing vine.

Lignotuber, a swollen woody base found on some shrubs adapted to withstand drought and fire.

Lithophytic, a plant evolved and adapted to growing on rocks, deriving its nourishment from rainwater and nearby decaying plants.

Mesophytic, plants that are adapted to neither a particularly dry nor particularly wet habitat.

Monocarpic, dying after bearing fruit, flowering just once.

Mycology, the study of fungi.

Nothospecies, a hybrid which is formed by direct hybridisation of two species, not other hybrids.

Orchidologist, a person specialising in the study of orchids.

Ovate, broadest below the middle (like a hen's egg).

Paleobotany, the study of fossil plants.

Palmate, a leaf with five or more lobes whose midribs all radiate from one point.

Panicle, a branching raceme.

Pantropic, belonging to the tropical regions of the Old and New Worlds.

Pedicel, the stalk of an individual flower in an inflorescence.

Peduncle, the stalk of an inflorescence.

Peltate, a leaf that is more or less circular and flat with the stalk inserted in the middle.

Petiole, the leaf stalk.

Pinnate, with leaflets arranged either side of a central stalk.

Placodiod, applied to a lichen thallus which is roughly circular and crustose, with a determinate margin and radiating peripheral lobes.

Pseudostem, an erect aerial stem apparently furnished with leaves but composed of compressed or overlapping sheaths and stalks of basal leaves, e.g., as seen in *Arisaema*.

Pteridologist, a specialist dealing with the branch of botany covering ferns, horsetails and club mosses.

Raceme, a simple elongated inflorescence with stalked flowers.

Rugulose, finely wrinkled with irregular lines and veins.

Saprophytic, a plant deriving its nutrients from dissolved or decayed organic matter, i.e., not producing its own food.

Spadix, the fleshy axis of a spike, often subtended by a spathe. Commonly seen in the aroids.

Spathe, a large bract often ensheathing an inflorescence; applied only in monocotyledons.

Spike, a simple, elongated inflorescence with sessile flowers.

Taxonomy, the branch of science concerned with the classification of organisms.

Thyrsoid, thyrse-like, a compact, narrow panicle.

Tomentum, a dense covering of matted hairs.

Umbel, a flat-topped inflorescence in which the pedicels all arise from a common point.

Ventricose (corolla), inflated or swollen on one side.

Appendix 5

Notes

The endnote numbering below refers to corresponding citations in the main text; see also Appendix 6, Select bibliography.

Chapter 1

1. Hooker, 1848: 2: 250
2. Hooker, 1847: Personal communication
3. Hooker, 1848: 2: 263
4. Turrill, 1963: 45
5. Hooker, 1848: 2: 315
6. Hooker, 1848: 2: 318
7. Anon, 1847: 14(1): 190
8. Anon, 1847: 14(1): 192
9. Hooker, 1849: 1: 10
10. Hooker, 1849: 1: 223

Chapter 2

1. Gamble, 1922: 66
2. Singh, 2011: 487
3. Hooker, 1850: 2: 59
4. Hooker, 1854: 1: 128
5. Hooker, 1854: 1: 130
6. Hooker, 1854: 1: 136
7. Hooker, 1854: 1: 115
8. Hooker, 1854: 1: 83
9. Hunter, 1896: 328
10. North, 1892: 2: 28
11. Hooker, 1854: 1: 125
12. Hooker & Thomson (1855): 77
13. Andrews, 2005: 10
14. Hooker, 1850: 2: 12
15. Hooker, 1850: 2: 14
16. Hooker, 1850: 2: 215
17. Huxley, 1918: 1: 250
18. Elwes, 1930: 8
19. Elwes, 1930: 73
20. North, 1892: 2: 32

Chapter 3

1. Moore, 1893: 558
2. Blunt & Stearn, 1994: 264
3. Burbidge, 1893: 13: ser. 3. 710
4. Lane Fox, 2012: 1
5. O'Brien, 2008: 162
6. RBG, Glasnevin, 1850
7. RBG, Glasnevin, 1858–59

8. Nelson & Walsh, 2008: 166
9. Nelson & Walsh, 1984: 52
10. Jasanoff, 2005: 50–1, 121
11. Madden, 1847: 16: 245
12. Morley, 1972: 97: 204
13. Nelson & Walsh, 1984b: 68
14. Adair, 1878: 65
15. Rosse, 1979: 141
16. Moore, 1940: 12: 180
17. Madden: 1849: 1: 60.
18. Madden, 1846: 15: 124
19. Colvin & Nelson, 1983: 16: 64
20. Edgeworth, 1846: 20: 23
21. Blatter, 1914: 7(1): 6.
22. Clarke, 1896: 583
23. Moore, 1893b: 559

Chapter 4

1. Pradhan, 2008: 99
2. Anon, 1922: 2
3. Anon, 1911: 4
4. Anon, 1932: 9.
5. Pearce & Cribb, 2002: 264
6. Moore, 1893: 559
7. Hooker, 1854: 1: 162
8. Headfort, 1917: 158
9. Hooker, 1894: 6: 275
10. Bean, 1989: 4: 419
11. Andrews, 2005: 8
12. Hooker, 1849–51: t. 6
13. Gibbons & Spanner, 1998: 42(1): 24–9
14. Gamble, 1922: 418

Chapter 5

1. Hooker, 1854: 1: 179
2. Hooker, 1854: 1: 185
3. Hooker, 1854: 1: 187
4. Hooker, 1854: 1: 217
5. Hooker, 1854: 1: 231
6. Hooker, 1854: 1: 240
7. Huxley, 1919: 2: 343
8. Hooker, 1854: 1: 266

Chapter 6

1. Hooker, 1854: 1: 290
2. Hooker, 1854: 1: 301
3. Hooker, 1854: 1: 323
4. Hooker, 1854: 1: 328
5. Hooker, 1854: 1: 332
6. Hooker, 1854: 1: 369

Chapter 7

1. Hooker, 1854: 1: 280
2. Hooker, 1854: 1: 368
3. Hooker, 1854: 1: 336
4. Bean, 1989: 4: 419

Chapter 8

1. Hooker, 1854: 1: 384
2. Ghosh & Mallick, 2014: 225
3. Hooker, 1854: 1: 381
4. Hooker, 1854: 1: 390

Chapter 9

1. Freshfield, 1903: 20
2. Maity & Maiti, 2007: 18
3. Ghosh & Mallick, 2014: 422
4. White, 1909: 54
5. Storrs, 1998: 119
6. Hooker, 1890b: 5: 606
7. Gamble, 1922: 678
8. Bean, 1992: 3: 488
9. Schilling & Watson, 2014: 13(2): pp. 94–100
10. Stapf, 1929: 153: t. 9176
11. Millais, J. G. 1921: 74
12. Hooker, 1854: 1: 349
13. Freshfield, 1903: 220
14. White, 1909: 55
15. White, 1909: 56
16. Tambe & Rawat, 2009: 8(1): pp. 75–80
17. White, 1909: 55
18. Pradhan, 2010: 100

Chapter 10

1. Huxley, 1918: 1: 286
2. Hooker, 1854: 2: 4
3. Gamble, 1922: 344
4. Hooker, 1854: 2: 17
5. Hooker, 1849–51: t. 18
6. Hooker, 1854: 2: 32
7. Elwes, 1880: t. 11

Chapter 11

1. Freshfield, 1903: 99
2. Hooker, 1854: 2: 50
3. Huxley, 1918: 1: 295
4. Hooker, 1849–51: t. 14
5. Hooker, 1854: 2: 52
6. Hooker, 1854: 2: 58
7. Hooker, 1849–51: t. 25
8. Pradhan, 2012: 220
9. Dykes, 1924: 178
10. Hooker, 1854: 2: 75
11. Hooker, 1854: 2: 75
12. Hooker, 1854: 2: 78
13. Humboldt, 1849: 337
14. Hooker, 1854: 2: 82
15. Hooker, 1849–51: t. 26(B)
16. Davidian, 1982: 1: 197
17. Huxley, 1918: 1: 326

Chapter 12

1. Hooker, 1890: t. 7116

Chapter 13

1. Hooker, 1854: 2: 100
2. Hooker, 1854: 2: 122
3. Huxley, 1918: 2: 183
4. Huxley, 1918: 1: 275
5. Huxley, 1918: 1: 304
6. Hooker, 1854: 2: 143
7. Hooker, 1854: 2: 154
8. Huxley, 1918: 1: 309
9. Hooker, 1854: 2: 157
10. Hooker, 1854: 2: 160
11. Hooker, 1854: 2: 172
12. Hooker, 1854: 2: 184

Chapter 14

1. Long, 1979: 356
2. Pradhan, 2010: 76
3. Bawa & Kadur, 2013: 144
4. Hooker, 1849–51: t. 15

Chapter 15

1. Hooker, 1854: 2: 197
2. Hooker, 1849b:
   Personal communication
3. Hooker, 1849c:
   Personal communication
4. Hooker, 1854: 2: 248

Chapter 16

1. Ghosh & Mallick, 2014: 165
2. White, 1909: 33
3. Fletcher, 1975: 2
4. Cooke, 1980: 146
5. French, 1995: 171
6. North, 1892: 1: 343

SELECT BIBLIOGRAPHY

Adair, J. (1878). *Hints on the culture of ornamental plants in Ireland.* E. Ponsonby, Dublin.

Andrews, S. (2005). Tree of the year: Magnolia campbellii. *International Dendrology Society Yearbook* (7–28). Dendrology Charitable Company, Herefordshire.

Anon. (1847). Biographical Memoir of the late William Griffith. *Madras Journal of Literature and Science,* 14:1 (187–197). J. K. Bantleman, Madras.

Anon. (1911). *Annual report of the Royal Botanic Garden and the Gardens in Calcutta for 1910–11.* The Bengal Secretariat Book Depot, Calcutta.

Anon. (1922). *Annual report of the Royal Botanic Garden and the Gardens in Calcutta for 1921–22.* The Bengal Secretariat Book Depot, Calcutta.

Anon. (1932). *Annual report of the Royal Botanic Garden and the Gardens in Calcutta for 1931–32.* The Bengal Secretariat Book Depot, Calcutta.

Bawa, K. & Kadur, S. (2013). *Himalaya – Mountains of Life.* Ashoka Trust, Bangalore, India.

Bean, W. J. (1989). *Trees and Shrubs Hardy in the British Isles,* vol.4. John Murray, London.

Bean, W. J. (1992). *Trees and Shrubs Hardy in the British Isles,* vol. 3. John Murray, London.

Blatter, E. (1914). Flora of Aden. *Records of the Botanical Survey of India.* Superintendent Government Printing, Calcutta.

Blunt, W. & Stearn, W. T. (1994). *The Art of Botanical Illustration.* Antique Collectors' Club in association with Royal Botanic Gardens, Kew.

Burbidge, F. W. (1893). In an Irish Garden. *The Gardeners' Chronicle* 13: ser. 3. (710), London.

Clarke, C. B. (1896). H. C. Levinge. *Nature.* Macmillan & Co., Ltd., London.

Colvin, C. & Nelson, C. (1988). Building Castles of Flowers; Maria Edgeworth as Gardener. *Garden History: The Journal of the Garden History Society,* 16:1 (58–70). Maney Publishing, Leeds.

Cooke, H. (1980). *Time Change.* Simon and Schuster, New York.

Davidian, H. H. (1982). *The Rhododendron Species,* vol. 1. Timber Press, Portland, Oregon.

Desmond, R. (2006). *Sir Joseph Dalton Hooker – Traveller and Plant Collector.* Antique Collectors' Club Ltd., Suffolk.

Dykes, W. R. (1924). *A Handbook of Garden Irises.* Martin Hopkinson & Co., Ltd, Covent Garden, London.

Edgeworth, M. P. (1846). Descriptions of some unpublished species from north-western India. *The Transactions of the Linnean Society of London,* 20 (23–91). Richard and John E. Taylor, London.

Elwes, H. J. (1880). *A Monograph of the Genus Lilium.* Taylor and Francis, London.

Elwes, H. J. (1882). On a collection of butterflies from Sikkim. *Proceedings of the Scientific Committee of the Zoological Society of London,* 50 (398–407). Messrs. Longmans, Green, Reader and Dyer, London.

Elwes, H. J. & Möller, O. (1888). A catalogue of the lepidoptera of Sikkim. *Transactions of the Entomological Society of London,* 36: 3 (269–465). Messrs. Longmans, Green, Reader and Dyer, London.

Elwes, H. J. (1930). *Memoirs of Travel, Sport and Natural History.* Ernest Benn Ltd., London.

Fletcher, H. R. (1975). *A Quest of Flowers – The Plant Explorations of Frank Ludlow and George Sherriff.* Edinburgh University Press, Edinburgh.

French, P. (1995). *Younghusband: The Last Great Imperial Adventurer.* Penguin Books, London.

Freshfield, D. W. (1903). *Round Khangchenjunga – A Narrative of Mountain Travel and Exploration.* Edward Arnold, London.

Gamble, J. S. (1922). *A Manual of Indian Timbers.* Sampson Low, Marston & Company Ltd., London.

Ghosh, D. & Mallick, J. (2014). *Flora of Darjeeling Himalayas and Foothills.* Bishen Singh Mahendra Pal Singh, Dehra Dun.

Gibbons, M. & Spanner, T. (1998). *Trachycarpus latisectus*: The Windamere Palm. *Principes,* 42:1 (24–29). The International Palm Society.

Headfort, Marquess of (1917). Rhododendrons at Kilmacurragh. *The Rhododendron Society Notes,* 1 (158). W. S. Cowell Ltd., Ipswich.

Hooker, J. D. (1847). Personal communication to Sir William Hooker, 17 December 1847. Director's Correspondence, Royal Botanic Gardens, Kew.

Hooker, J. D. (1848). Dr. Hooker's mission to India. *London Journal of Botany,* 2: (237–321). Reeve, Bentham and Reeve, London.

Hooker, J. D. (1849). Dr. Hooker's Mission to India. *Hooker's Journal of Botany and Kew Garden Miscellany,* 1(1–370). Reeve, Bentham and Reeve, London.

Hooker, J. D. (1849b). Personal communication to William Hooker, 3 December 1849. Director's Correspondence, Royal Botanic Gardens, Kew.

Hooker, J. D. (1849c). Personal communications to Lady Hooker, 28 December 1849. Director's Correspondence, Royal Botanic Gardens, Kew.

Hooker, J. D. (1849–51). *Rhododendron of the Sikkim Himalaya*. Reeve and Co., London.

Hooker, J. D. (1850). Dr. Hooker's Mission to India. *Hooker's Journal of Botany and Kew Garden Miscellany*, 2 (11–249). Reeve & Bentham, London.

Hooker, Joseph (1854). *Himalayan Journals; or Notes of a naturalist in Bengal, the Sikkim and Nepal Himalayas, the Khasia mountains, &c.* John Murray, London.

Hooker, J. D. & Thomson, T. (1855). *Flora Indica*. W. Pamplin, London.

Hooker, J. D. (1890). *Berberis virescens. Curtis's Botanical Magazine* 116: t. 7116. L. Reeve & Co., London.

Hooker, J. D. (1890b). *Flora of British India* 5: 606. L. Reeve & Co., London.

Hooker, J. D. (1894). *Flora of British India* 6: 275. L. Reeve & Co., London.

Humboldt, von, A. (1849). Elevation of the great Table Land of Thibet. In *Hooker's Journal of Botany and Kew Garden Miscellany*, 1 (336–344). Reeve, Bentham and Reeve, London.

Hunter, W. W. (1896). *Life of Brian Haughton Hodgson*. John Murray, London.

Huxley, L. (1918). *Life and Letters of Sir Joseph Dalton Hooker*. John Murray, London.

Jasanoff, M. (2005). *Edges of Empire; Lives, Culture, and Conquest in the East, 1750–1850*. Alfred Knopf, New York.

Lane Fox, R J. (2012). *Financial Times* Weekend Supplement, 12 May, p 1. Pearson, London.

Long, D. G. (1979). The Bhutanese Itineraries of William Griffith and R. E. Cooper. *Notes from the Royal Botanic Garden Edinburgh*. HMSO, Edinburgh.

Madden, E. (1846). Diary of an excursion to Shatool and Boorun Passes over the Himalaya, in September 1845. *Journal of the Asiatic Society of Bengal*, 15: (79–135). J. Thomas, Bishop's College Press, Calcutta.

Madden, E. (1847). Notes on an Excursion to the Pindree Glacier. *Journal of the Asiatic Society of Bengal*, 16: (226–266). J. Thomas, Bishop's College Press, Calcutta.

Madden, E. (1849). On the botany, &c., of the Turaee and outer mountains of Kumaon. *Hooker's Journal of Botany and Kew Garden Miscellany*, 1 (57–62). Reeve, Bentham and Reeve, London.

Maity, D. & Maiti, G. G. (2007). *The Wild Flowers of Kachenjunga Biosphere Reserve, Sikkim*. Noya Udyog, Kolkata.

McQuire, J. F. J. & Robinson, M. L. A. (2009). *Pocket Guide to Rhododendron Species*, p. 360. Royal Botanic Gardens, Kew.

Millais, J. G. (1921). *Rhododendron campylocarpum* and *Rhododendron wightii*, the true forms of. *The Rhododendron Society Notes*, 2: 2 (73–75). W. S. Cowell Ltd., Ipswich.

Moore, F. W. (1893). Himalayan Rhododendrons in Ireland. *The Garden: An Illustrated Weekly Journal of Horticulture in all its branches*, 44 (558–560). William Robinson, London.

Moore, F. W. (1940). Some Reminiscences. *The New Flora and Silva*, 12: 180. Dulau & Co. Ltd., London.

Morley, B. (1972). Edward Madden (1805–1856). *Journal of the Royal Horticultural Society*, 95: 7 (203–6). London.

Nelson, E. C. & Walsh, W. (1984). *An Irish Flower Garden*. Boethius Press, Kilkenny.

Nelson, E. C. & Walsh, W. (2008). *An Irish Florilegium – Wild and Garden Plants of Ireland*. Thames and Hudson Ltd., London.

North, M. (1892). *Recollections of a Happy Life*. Macmillan & Co., London.

O'Brien, S. (2008). Thomas Acton – A centennial celebration at Kilmacurragh. *International Dendrology Society Yearbook*. Dendrology Charitable Company, Herefordshire.

O'Brien, S. (2011). *In the Footsteps of Augustine Henry*. Garden Art Press, Suffolk.

Pearce, N. R. & Cribb, P. J. (2002). *The Orchids of Bhutan*. Royal Botanic Garden Edinburgh and Royal Government of Bhutan, Edinburgh.

Pradhan, K. (2008). *The Life and Times of a Plantsman in the Sikkim Himalayas*. Wayside Gardens, Gangtok.

Pradhan, K. (2010). *The Rhododendrons of Sikkim*. Sikkim Adventure, Gangtok.

Pradhan, K. C. (2012). *Sikkim and the World of Primulas*. Wayside Gardens, Gangtok.

Rosse, Earl of (1979). *Irish Gardening and Horticulture*. Royal Horticultural Society of Ireland, Dublin.

Royal Botanic Gardens, Glasnevin. *Donations to Gardens* (1850). Unpublished Manuscripts.

Royal Botanic Gardens, Glasnevin. *Donations to Gardens* (1858–59). Unpublished Manuscripts.

Schilling, T. & Watson, M. (2014). Clarifying the identities of two Nepalese *Mahonia. The Plantsman*, 13: 2 (94–100). RHS Media, Peterborough.

Singh, S. *et al.* (2011). *India*. Lonely Planet Publications Pty Ltd., London.

Stapf, Otto. (1929). *Syphonosmanthus suavis. Curtis's Botanical Magazine* 153: t. 9176. L. Reeve & Co., London.

Storrs, A. & J. (1998). *Trees and Shrubs of Nepal and the Himalayas*. Book Faith India, Delhi.

Tambe, S. & Rawat, G. S. (2009). Traditional livelihood based on sheep grazing in Khangchendzonga National Park, Sikkim. *Indian Journal of Traditional Knowledge*, 8: 1 (75–80).

Turrill, W. B. (1963). *Joseph Dalton Hooker – Botanist, Explorer and Administrator*. The Scientific Book Club, London.

White, J. C. (1909). *Sikhim & Bhutan – Twenty-one years on the north-east Frontier 1887–1908*. Edward Arnold, London.

Index to people and places

Bold text or page number denotes an illustration. See also Appendices.

Wilson-Wright, Robert ix, xv, **107**, **148**, 200

Windamere Hotel, Darjeeling 63, 78, 79, 279

Winterbottom, James Edward 188

Wushan, Sichuan 169

Y

Yakchey 239, 240, 241, **242**, **243**

Yakchey reserve 239

Yak La pass 37, 232, 258, 263, 297

Yale School of Forestry 59

Yalung (Yalloong) valley 86, 91, 92, 100, 117, 238, 300

Yangang (Neongong) monastery 96, 97, 299

Yangma Gola xix, **88**, 89

    Yangma River 89, 90

    Yangma River valley xix, **89**, 90

Yarlung-Tsangpo River 190, 230, 300

Yemen 5

Yonpu La, Bhutan 235

Younghusband, Lt-Col. Francis 222, 255, 276, 277

Yuksam (Yoksun) xiii, 93, 100, 101, **107**, 108, 109, 117, 119, 121, 125, 126, 129, 130, 131, 137, 141, 147, 148, 152, 299

Yum Cho (Yeumtso) 228, 229, 230, 300

Yume Samdong (Momay Samdong) 35, 189, **219**, **220**, 223, 224, 225, 231, 256, 299

Yumthang 74, 225, 226, **232**, **249**, 250, 299

Yumthang (Yeumtong) valley 35, 57, 74, 175, **211**, **218**, 219, **224**, 225, 226, **232**, **241**, 243, 246, 247, **248**, **249**, 250, **251**, **252**, 253, 254, **255**, **256**, 276, 277, 283, 299

    Yumthang River 250, 256

    yak camp 231

Z

Zema (Zemu Samdong) 169, 170, 173, 180, 226, 299

Zemu River **173**

Zemu River valley 128, 168, 169, 170, 172, **173**, **174**, **175**, 177, **178**, 179–182, 206, 293

    Zemu Glacier 126, 128, 176

Index to plant and animal names

Bold text or page number denotes an illustration. See also Appendices 1 and 2.

Berberis virescens in the Lachen valley